Variation in Mammalian Populations

SYMPOSIA OF THE ZOOLOGICAL SOCIETY OF LONDON
NUMBER 26

Variation in Mammalian Populations

(The Proceedings of a Symposium held at The Zoological Society of London on 14 and 15 November, 1969)

Edited by

R. J. BERRY

Department of Biology,
Royal Free Hospital School of Medicine,
London, England

H. N. SOUTHERN

Department of Zoology,
Animal Ecology Research Group,
Botanic Garden,
Oxford, England

Published for

THE ZOOLOGICAL SOCIETY OF LONDON

BY

ACADEMIC PRESS

1970

ACADEMIC PRESS INC. (LONDON) LTD.
Berkeley Square House
Berkeley Square,
London, W1X 6BA

U.S. Edition published by
ACADEMIC PRESS INC.
111 Fifth Avenue,
New York, New York 10003

Library of Congress Catalog Card Number: 75-129792
ISBN: 0-12-613326-3

PRINTED IN GREAT BRITAIN BY
J. W. ARROWSMITH LTD., BRISTOL

CONTRIBUTORS

ANDERSON, P. K., *Department of Biology, The University of Calgary, Calgary, Alberta, Canada* (p. 299)

BERRY, R. J., *Department of Biology, Royal Free Hospital School of Medicine, London, England* (p. 3)

CHITTY, D., *Department of Zoology, University of British Columbia, Vancouver, Canada* (p. 327)

CORBET, G. B., *Mammal Section, British Museum (Natural History), London, England* (p. 105)

CORYNDON, S. C., *Department of Palaeontology, British Museum (Natural History), London, England* (p. 135)

DADD, M. N., *The Zoological Society of London, London, England* (p. 117)

DELANY, M. J., *Department of Zoology, University of Southampton, Southampton, England* (p. 283)

DEOL, M. S., *Department of Animal Genetics, University College, London, England* (p. 239)

DRUMMOND, D. C., *Infestation Control Laboratory, Ministry of Agriculture, Fisheries and Food, Tolworth, Surrey, England* (p. 351)

FORD, C. E., *Medical Research Council, Radiobiology Unit, Harwell, Didcot, Berkshire, England* (p. 223)

GROVES, C. P., *Duckworth Laboratory of Physical Anthropology, Cambridge University, Cambridge, England* (p. 127)

HAMERTON, J. L., *Department of Genetics, The Children's Hospital of Winnipeg, Manitoba, Canada* (p. 223)

KOWALSKI, K., *Institute of Systematic and Experimental Zoology, Polish Academy of Sciences, Kraków, Poland* (p. 149)

LEWIS, W. H. P., *Department of Pathology, St. Helier Hospital, Carshalton, Surrey, England* (p. 93)

LLOYD, H. G., *Ministry of Agriculture, Fisheries and Food, Infestation Control Laboratory, Field Research Station, Worplesdon, Guildford, Surrey, England* (p. 165)

LUSH, I. E., *Department of Biology, Royal Free Hospital School of Medicine, London, England* (p. 43)

MATHER, K., *University of Southampton, Southampton, England* (p. 27)

MAYNARD SMITH, J., *School of Biological Sciences, University of Sussex, Brighton, England* (p. 371)

MEYLAN, A., *Vertebrate Section, Federal Agricultural Research Station, CH-1260 Nyon, Switzerland* (p. 211)

PUCEK, Z., *Mammals Research Institute, Polish Academy of Sciences, Białowieża, Poland* (p. 189)

RASMUSSEN, D. I., *Department of Zoology, Arizona State University, Tempe, Arizona, U.S.A.* (p. 335)
ROBINSON, R., *St. Stephens Road Nursery, London, England* (p. 251)
SELANDER, R. K., *Department of Zoology, University of Texas, Austin, Texas, U.S.A.* (p. 73)
STODDART, D. M., *University of Oxford, Department of Zoology, Animal Ecology Research Group, Oxford, England** (p. 271)

* Present address: Department of Zoology, University of London King's College, Strand, London, England.

ORGANIZERS AND CHAIRMEN

ORGANIZERS

R. J. Berry and H. N. Southern *on behalf of The Zoological Society of London*

CHAIRMEN OF SESSIONS

A. J. Cain, *Department of Zoology, The University, Liverpool, England*
H. R. Hewer, *Department of Zoology, Imperial College of Science and Technology, London, England*
J. Maynard Smith, *School of Biological Sciences, University of Sussex, Brighton, England*
H. N. Southern, *Department of Zoology, Animal Ecology Research Group, Botanic Garden, Oxford, England*

FOREWORD

This volume contains the proceedings of a two-day symposium (14th and 15th November 1969) sponsored by The Zoological Society of London and the Mammal Society. The aim of the organizers was to bring together geneticists, ecologists, systematists and palaeontologists working on mammalian populations. We believe that the meeting achieved a useful exchange of ideas and put on record much new information; and we hope that this published record will stimulate profitable research based on a wider understanding of evolutionary problems at this level.

We should like to express our thanks to The Zoological Society of London for their support and organization, and in particular to Mrs. Linda René-Martin, and also to Miss C. M. King for preparing the Subject and Systematic Indices.

July, 1970

R. J. Berry
H. N. Southern

CONTENTS

GENERAL

Covert and Overt Variation, as exemplified by British Mouse Populations

R. J. BERRY

The Nature and Significance of Variation in Wild Populations

KENNETH MATHER

CRYPTOMORPHISMS

The Extent of Biochemical Variation in Mammalian Populations

I. E. LUSH

Biochemical Polymorphism in Populations of the House Mouse and Old-field Mouse

ROBERT K. SELANDER

Human Molecular Variation

W. H. P. LEWIS

TAXONOMY AND PHYLOGENY

Patterns of Subspecific Variation

G. B. CORBET

Overlap of Variation in British and European Mammal Populations

MICHAEL N. DADD

Taxonomic and Individual Variation in Gibbons

COLIN P. GROVES

The Extent of Variation in Fossil Hippopotamus from Africa

S. C. CORYNDON

Variation and Speciation in Fossil Voles

KAZIMIERZ KOWALSKI

REPRODUCTIVE AND SEASONAL VARIATION

Variation and Adaptation in Reproductive Performance

H. G. LLOYD

Seasonal and Age Change in Shrews as an Adaptive Process

ZDZISŁAW PUCEK

CHROMOSOMAL VARIATION

Chromosomal Polymorphism in Mammals

ANDRÉ MEYLAN

Chromosome Polymorphism in the Common Shrew, *Sorex araneus*

C. E. FORD and J. L. HAMERTON

VISIBLE VARIATION

The Determination and Distribution of Coat Colour Variation in the House Mouse

M. S. DEOL

Homologous Mutants in Mammalian Coat Colour Variation

ROY ROBINSON

Tail Tip and Other Albinisms in Voles of the Genus *Arvicola* Lacépède 1799

D. MICHAEL STODDART

Biochemical Polymorphisms and Genetic Structure in Populations of *Peromyscus*

DAVID I. RASMUSSEN

Variation in Rodent Populations in Response to Control Measures

D. C. DRUMMOND

CONCLUSION

The Causes of Polymorphism

J. MAYNARD SMITH

GENERAL

Symp. zool. Soc. Lond. (1970) No. 26, 3–26

COVERT AND OVERT VARIATION, AS EXEMPLIFIED BY BRITISH MOUSE POPULATIONS

R. J. BERRY

Department of Biology, Royal Free Hospital School of Medicine, London, England

SYNOPSIS

Taxonomists recognize many sub-species of British small mammals. All but one of these sub-species are limited to a single island. As far as mice, at least, are concerned, this differentiation has resulted from colonization during historical times following accidental introductions by man. For example, the sixteen sub-species of the Long-tailed field mouse (*Apodemus sylvaticus*) recognized on the North Atlantic islands are most closely related to Norwegian mice, and almost certainly arose after the voyages of the Vikings. Since all the islands represent broadly similar environments for the mice, adaptation would be expected to be along similar lines on all of them. Hence the considerable differences that exist between sub-species (even on neighbouring islands) must be at least partly due to the chance genetical characteristics of the small founding populations. This implies that the apparently uniform population from which the different island races originated possessed considerable hidden variability.

In fact if one looks at any one mouse population it is possible to recognize much inherited variation in weight, coat colour, skeletal characteristics, biochemical and physiological traits, reproductive performance and cytological behaviour. In all these systems, there seems to be sufficient variation available for selection to act upon, and, taking into account the probable method of genetical determination of most of the characters, considerably more potentially available.

An example of the maintenance and importance of variation in a population is given by the House mice on the small Pembrokeshire island of Skokholm, which seems to have as much variation as mainland populations, despite being founded by a small number of animals only 70 years ago. It can be shown from four apparently independent systems (metrical and non-metrical skeletal variation, protein polymorphisms, and respiratory physiology) that different genotypes have different survival values at different times of the year. This means that the population can maintain higher numbers throughout the year than would be possible if it was genetically uniform. In other words, the variation acts as a stabilizing influence in the normally varying environment of the mice.

INTRODUCTION

The Bible describes a good shepherd as one who can recognize and respond to individuals in his own flock (e.g. Gospel according to St. John, 10: 2–5). Similarly any worker engaged upon a long-term investigation of a mammalian population soon comes to recognize differences between members of his study group, even among such apparently anonymous animals as House mice (Crowcroft, 1966). The existence of such differences implies the possibility of genetical distinctiveness between populations within a species. It was such differences—particularly with reference to domesticated forms—that Darwin used as one line of evidence for "descent with modification" in the "Origin of Species". This paper is a description of the sort of variation found in British mouse populations. The reason for choosing British mice is that they have been fairly extensively sampled; they have been studied in the laboratory as well as in the field and museum; and the obvious variation they manifest is not so spectacular as to suggest an unusual situation.

TAXONOMIC VARIATION: THE DISTRIBUTION AND ORIGIN OF BRITISH SUB-SPECIES

In the past, the clearest indications of intra-specific variation have been provided by taxonomists. On this evidence, the most polytypic British mammal is the Long-tailed field mouse (*Apodemus sylvaticus* (L.)), in which 16 sub-species have been described (Corbet, 1964*a*). These infra-specific categories are based on minor pelage traits, and on size, particularly as affecting the skull (e.g. Hinton, 1914). There are several reasons for criticizing the grounds on which they were raised, in particular the too few specimens used in originally describing them, and the lack of knowledge of the normal range of variation in any of them (cf. Matthews, 1952). However there can be no doubt that a situation of considerable inter-population differentiation exists in *Apodemus sylvaticus* and that it is to some extent described by these named sub-species. The same conclusion can be drawn for the three British species of small voles: *Clethrionomys glareolus* (Schr.), *Microtus agrestis* (L.), and *M. arvalis* (Pall.), all of which have five sub-species.

Now the interesting fact about the sub-species of British small mammals is that, with the single exception of the Scottish Highland form of the Water vole (*Arvicola terrestris reta*), they are all found on off-shore islands (Fig. 1). This means that the variation must have arisen and been maintained in some other way than by the spread of the variant—presumably mutant—individuals from a centre of origin so as to form a cline (Haldane, 1948; Kettlewell and Berry, 1969). The island races are usually assumed to be distinct from the mainland forms, either because they represent the relicts of a form which has become extinct on the mainland (as a result of competition with a race which only reached the mainland after the islands were separated: this view has been stated most explicitly by Beirne, 1952), or because numbers in the populations on the islands are so small that random genetical changes ("drift") have brought about changes in gene

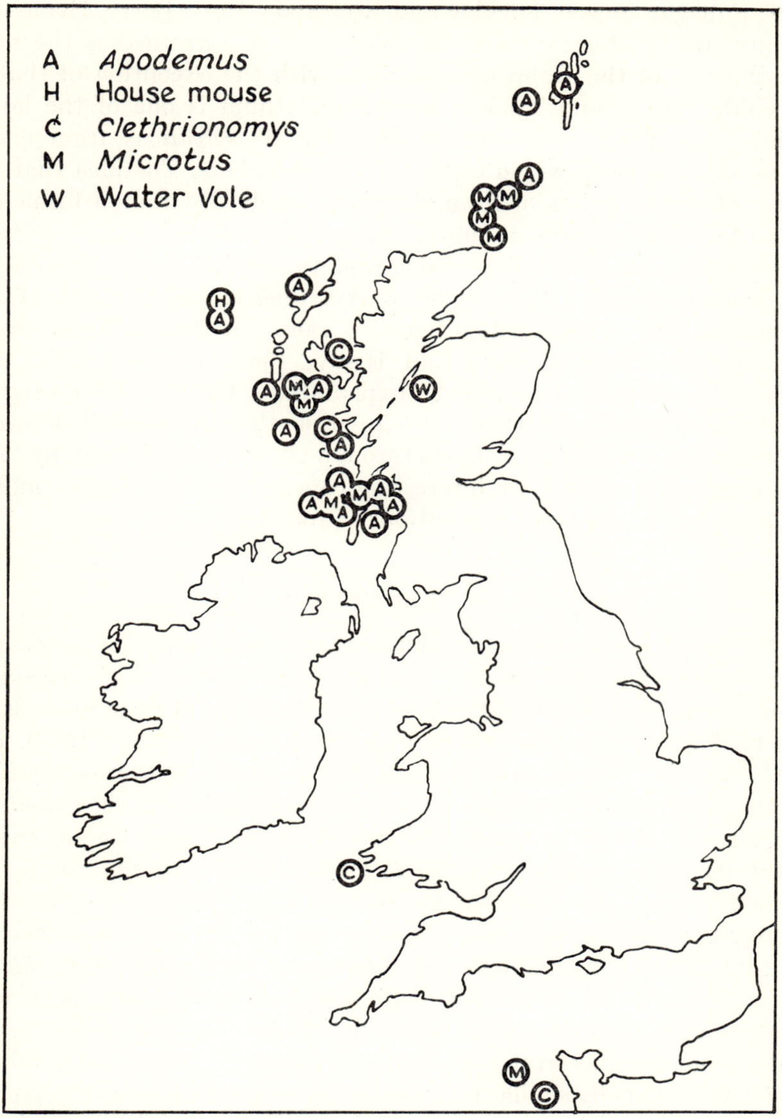

FIG. 1. Distribution of named races of small mammals in the British Isles.

frequencies. Although it is impossible completely to exclude the second possibility (since occasional extreme conditions may reduce a local population to small numbers), nevertheless the *A. sylvaticus* population on most islands is usually of the order of several thousands (Boyd,

1959; Fullagar, Jewell, Lockley and Rowlands, 1963; Berry, Evans and Sennitt, 1967). Moreover the Rhum race of *A. sylvaticus* is the most extreme one of the Hebridean Islands (with the exception of that of St. Kilda: Berry and Tricker, 1969), but Rhum is one of the larger islands, and there is no competition for *A. sylvaticus* from either House mice or voles which could limit its numbers. The idea that the island races are relicts of formerly more widely distributed forms can be confounded on two grounds: the unlikelihood of *A. sylvaticus*—and probably also the voles—being able to survive the Pleistocene on the Scottish islands (West, 1968; Berry, 1969*a*); and the lack of any simple pattern of similarities between races on islands in the same area (Delany, 1964; Delany and Healy, 1964; Berry, 1969*a*). The inadequacy of the "traditional" explanations for the differentiation of the island races led Corbet (1961) to argue that the most likely origin for most of the races of rodents is through chance introduction by man. This suggestion leads to an interesting conclusion about the amount of variation hidden in large populations of rodents.

FOUNDER EFFECTS

If Corbet is correct, most of the island races must be descended from only a few individuals—perhaps from only a single pregnant female—carried onto the islands with the bedding of farm animals, or in seed for crops, or possibly at ship-wreck (e.g. Lockley, 1947). If a large amount of genetical heterozygosity exists in the populations whence these animals came, it is virtually impossible that the colonizing animals will carry the same allelomorphs in the same frequencies as in the parental population (Fig. 2) (Berry, 1967). In the absence of any subsequent genetical change due to adaptation or genetic drift (immigration will almost certainly be on too small a scale to change the gene frequencies in established populations), populations with similar genetical constitutions will be more closely related than those with different ones, but even first generation colonizers may differ significantly from the ancestral population (Livingstone, 1969). Now, adaptation on ecologically similar islands would be expected to be convergent. This means that if the populations are large (Berry, 1964), present inter-population differences may be smaller than those initially established. On these grounds, it is possible to argue that the similarities and differences between *A. sylvaticus* races from islands on the north-west Atlantic sea-board can be largely accounted for by separate introductions from Norway into Iceland, Shetland and the Hebrides (Fig. 3) (Berry, 1969*a*). The implication from this is that the different races have been isolated

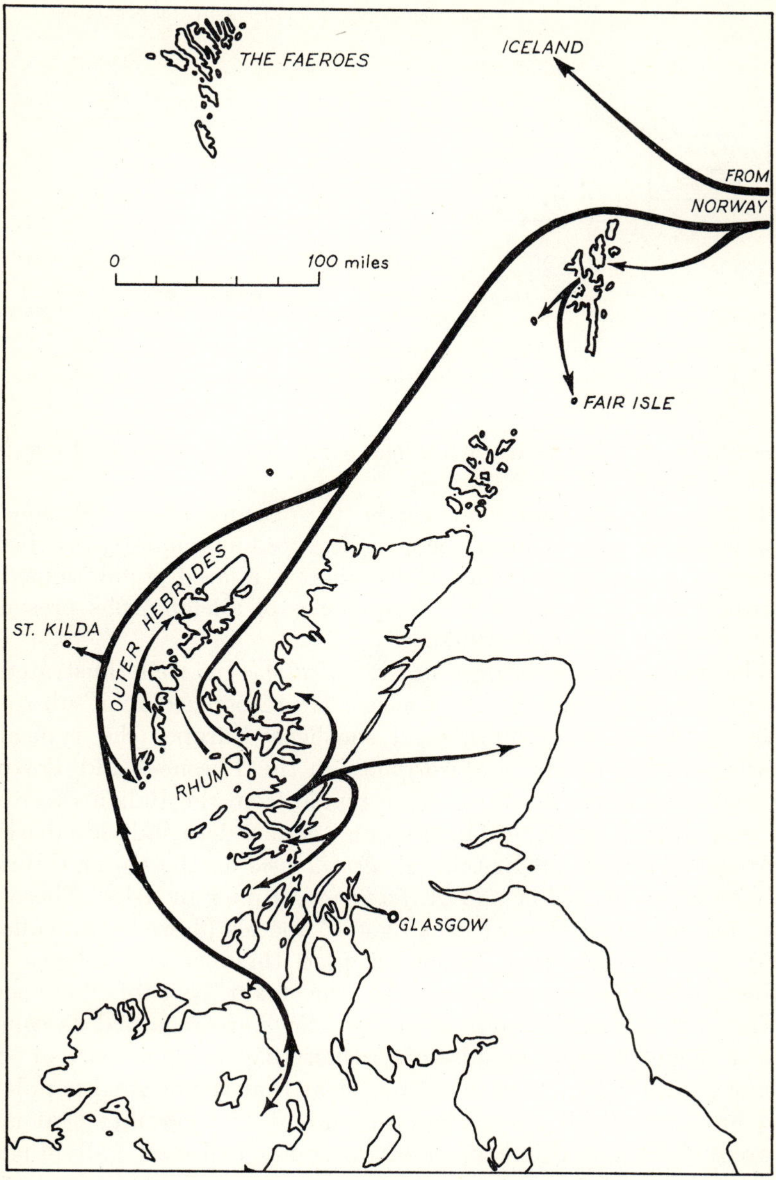

Fig. 2. Relationship between races of *Apodemus sylvaticus*, suggesting that the island forms were the results of introductions by Norwegian voyagers (after Berry, 1969*a*).

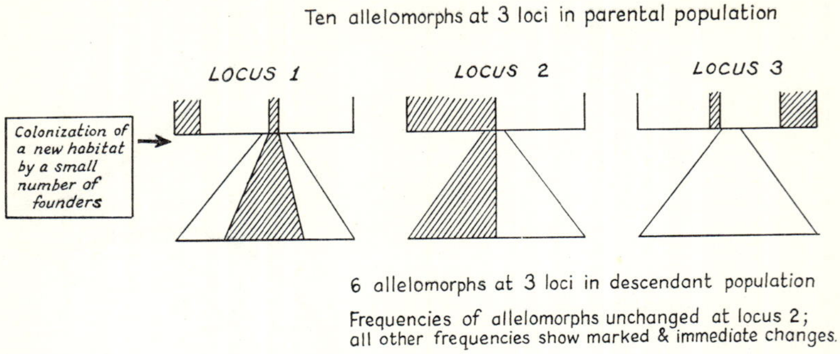

FIG. 3. Instant genetical change: allelomorphic frequencies at three loci in a large population, and after a small number of individuals establish a new population.

a maximum of 1200 years, when the early Viking voyages took place (Jones, 1968).

Taxonomic formalization tends to obscure close relationships between races, since a form becomes accepted as qualitatively distinct as soon as a name is attached to it, whereas related populations differ quantitatively in terms of the frequencies of allelomorphs present in them (Berry, R. J. *et al.*, 1967).

The taxonomic situation in *A. sylvaticus* is well illustrated by Miller's (1912) revision of the species. He recognized four sub-species in the whole of continental and southern Europe (the typical *A. sylvaticus* in the north, *A.s. callipides* in the Pyrenees and Portugal, *A.s. dichrurus* in the Mediterranean region from the Balkans to Spain, and *A.s. creticus* in Crete) but 3 *species* confined to individual islands or groups of islands off Britain (*A. hebridensis* on the Outer Hebrides, *A. hirtensis* on St. Kilda and *A. fridariensis* on Fair Isle). These last forms are no longer regarded as "good" species (Jewell and Fullagar, 1965), but the fact that Miller accepted them as such shows their distinctiveness in comparison with the main part of the species. However, if the island races of *A. sylvaticus* are regarded as samples from a single variable population, rather than as some sort of "sub-specific swarm" (Fig. 4), it is at once clear that the original population must have contained a considerable amount of variability that is not apparent from an examination of randomly collected individuals of the parental race.

The same conclusion can be made on similar grounds about *Clethrionomys glareolus* (Steven, 1953; Corbet, 1963, 1964*b*), *Microtus arvalis* in Orkney (Turner, 1965) and *Mus musculus* L. For example, *Mus musculus* from St. Kilda (*M.m. muralis*) and from Myggenaes

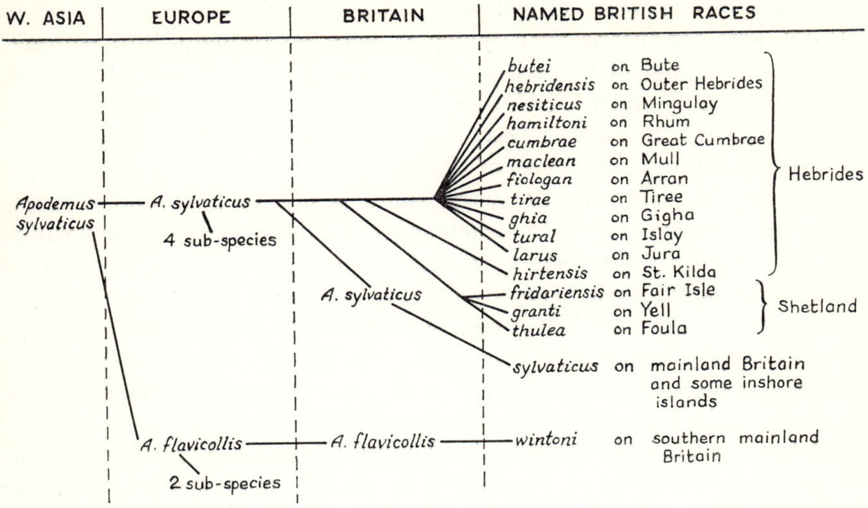

FIG. 4. *Apodemus sylvaticus* at one edge of its range (after Matthews, 1952).

and Nolsø in the Faroes (*M.m. mykinessiensis* and *M.m. faeroensis*) are extremely large and most of them have a narrowing of the mesopterygoid fossa on the ventral aspect of the cranium (Degerbøl, 1942), a character not previously known in any other European mouse. It has been suggested that these similarities could be due to convergent adaptation, to allometric growth, or that both the St. Kildan and Faeroese populations were relicts from a single continuous population. However, the simplest explanation here, as with *A. sylvaticus*, is that small numbers of House mice were introduced to both St. Kilda and the Faeroes by men. Since the Vikings were the earliest people to have sizeable and sturdy ships, and to be known to colonize both island groups, it is not surprising to find that narrowing of the mesopterygoid fossa occurs with different frequencies on different Shetland islands, *via* which came the colonizing Vikings of the Western Isles and North Atlantic (R. J. Berry, unpublished).

It is worth reiterating the point that much of the variability that becomes manifest in the island races is hidden in the parental race and would otherwise not be detected unless extensive breeding tests were carried out (Batten and Berry, 1967; Berry, 1970). It is particularly explicit in the statement of Barrett-Hamilton (1900) with regard to *A. sylvaticus*: "*Apodemus sylvaticus* appears to be a form which, in its longstanding and successful struggle for existence has attained to a height of specialization from which it has either very little power of variation, or else which is such as to fulfil all the needs of the species in almost any conditions with which it may be brought into contact. The possible range of its variations, whether individual or geographical, would seem to be narrow. Unlike some of our common mammals, such as the Squirrel, Rabbit, or Rat, it is not subject to either melanism or albinism. In the whole series of the 'Zoologist' and the volumes of the 'Field' for the last twenty years, there is not to be found a single recorded instance of a well-marked sport of this species—a result which would have been very widely different had the object of the search been *Sciurus vulgaris*, *Talpa europaea*, or *Mus musculus*. This, of course, does not prove that conspicuous sports do not occur, but it undoubtedly emphasises their rarity . . . *Apodemus sylvaticus* is, it seems, a good instance in support of Mr. A. Sedgwick's remarks on the loss of variability in old species."

I have quoted the comments at length because Barrett-Hamilton was a careful and cautious worker. Taken with the preceding discussion about the variation that must exist in north European *A. sylvaticus* in order to have produced so many and so distinct island forms, it well illustrates the paradox that exists between little overt and considerable covert variation. This paradox has been studied by a number of geneticists, but almost entirely in fruit-flies (e.g. Mather, 1941, 1953, 1964; Lerner, 1954; Thoday, 1955; Grant, 1964). *A priori* it seems possible that mammals may have a complex genetical architecture similar in principle to that in *Drosophila* (Cook, 1961; Selander, Hunt and Yang, 1969). The great advantage of mammalian over insect populations is that they offer better opportunities for studying the interplay of ecological opportunity and failure with the whole complex of processes and products that constitutes the phenotype. Hence it is relevant to consider some of the systems where variability has been analysed. Many of the examples are drawn from studies on House mice, because of the vast amount of laboratory information that is available to assist in the analysis of data collected in the field (e.g. Grüneberg, 1952; Green, 1966), although ironically there is less ecological knowledge about wild-living House mice than for most other British small mammal species (but see Southern, 1954; Crowcroft, 1966; Berry, 1968*a*, 1970).

VARIATION AVAILABLE FOR SELECTION

Size

The most consistent feature of island small mammal races is increased body size (Cook, 1961; Foster, 1964): *A. sylvaticus* from St.

Kilda average twice as heavy as specimens from the mainland (Berry and Tricker, 1969); *Mus musculus* from Nolsø in the Faeroes were described by Clarke (1904) as "veritable giants, being considerably larger than the type and also than any of the numerous geographical races"; in the Skokholm House mouse population, body size has increased by about 15% in around seventy generations (Berry, 1964). These size differences from the mainland forms are maintained when animals are bred in the laboratory (Steven, 1953; Berry, 1964). Rensch (1959) has discussed the mechanisms which could confer an advantage on being large: intra-uterine competition puts a premium on speed of development and large size which continues after birth (although see Deol and Truslove, 1957); large animals may live longer, thereby having more time to accumulate antibodies, gain experience, etc.; they may have an advantage in intraspecific competition; there may be an organizational advantage in having an increased number of nerve cells; larger animals are more efficient metabolically than small ones. Corbet (1961) suggested that intraspecific selection may favour large size (for some or all the reasons discussed by Rensch, 1959), while ground predators lead to selection for small size, i.e. the ability to escape into holes inaccessible to the predator. The absence of ground predators on many of the islands round the coasts means that size may increase to what is presumably a more efficient level.

The availability of inherited weight variation for selection was shown by Falconer (1953). He selected for large and small body size in laboratory mice originating from a random-bred population and in one experiment achieved a mean adult weight in the positive selection line of almost twice that in the negative line, after eleven generations. Crowcroft and Rowe (1961), as part of a study of the behaviour, growth and reproduction of House mice living under much more natural conditions than are usually obtained in the laboratory (*q.v.* Crowcroft, 1966), found considerable weight variation in comparable mice raised from the same wild-caught progenitors. Some males were only half as heavy as others. They found that the larger males tended to achieve social dominance, but that the dominance was a consequence and not a cause of their large size. Furthermore they found that crowding caused a fall in the mean size of later born males in three of their four colonies, but in the fourth colony, there was a significant increase in weight of such males. In other words, there seems to be a number of genetical factors involved in the determination of size and these may lead to a considerable inherent variability in weight. (Falconer (1960) calculated that there are 100 genes which affect body weight in laboratory mice.

This is probably an upper estimate—D. S. Falconer, personal communication.) This means, in turn, that overall size can be presumed to be able to respond rapidly to directed natural selection.

Coat colour

Size and pelage characteristics are the obvious features to classify in any small mammal. However, strikingly polymorphic features of the fur are uncommon in British populations. This means most of the existing variation concerns minor features such as precise hue of the dorsal coat or extent of the ventral counter-shading. Delany (1964, 1970) assessed the "darkness" and "richness" of the dorsal fur of *A. sylvaticus* by comparison with an arbitrary scale, and measured the length of the pectoral stripe in the same individuals. Such sophistication is unusual, and some described colour variability can almost certainly be ascribed to differences in age and moult condition of the specimens scored, as in Barrett-Hamilton's (1899) description of the coat of *M. musculus muralis* of St. Kilda, (R. J. Berry and G. M. Truslove, unpublished). For example, Berry, R. J. *et al.* (1967) concluded that the apparently dark coloration of the pelage of *A. sylvaticus* from the Inner Hebridean island of Muck was due to the high proportion of young (although adult) skins in their collection. Indeed Scottish Highland *A. sylvaticus* shows considerable seasonal and secular variation in coat colour, and the only coat colour difference that Berry, R. J. *et al.* (1967) found between mainland *A. sylvaticus* and animals from a group of four Hebridean islands was an increased length of pectoral stripe in one of them. Petras, Reimer, Biddle, Martin and Linton (1969) have summarized the frequency of coat variant allelomorphs in a number of House mouse populations, but most of the reported variation (such as the oft-repeated statement that wild-living House mice are more sandy or russet than commensal ones: e.g. Barrett-Hamilton and Hinton, 1910–21) is almost certainly due to multiple allelomorphs at a few gene loci. Sumner (1930) attempted a quantitative explanation for the sandy-coloured form of *Peromyscus polionotus* in Florida and Alabama, but the genetics of sandy-coloured British House mice—such as those of the Dublin Bull Island (Jameson, 1898, O'Gorman and Von Rizzori, 1969) or of the Spurn Head sand-dunes in Yorkshire (Clegg, 1963)—do not seem to have been analysed (Deol, 1970).

Although it is often claimed that natural selection may change the coat colour of a mammalian population so as to render it less conspicuous on the background on which it lives (Jameson, 1898; Ford, 1964), the only two instances in which selection in action seem to have been shown are one experimental (Dice, 1947), and one commensal situation

(Brown, 1965). Nevertheless there is little doubt that considerable coat colour variation exists in most small mammal populations.

Skeletal

The traditional method of accurately demonstrating variation in a population is from the variance—or coefficient of variance—of skeletal measurements. Classical studies can be criticized on a number of grounds, chiefly the environmental determinants of growth which have to be eliminated with a consequent loss of information (Bader, 1956; Bader and Lehmann, 1965), and the choice of parameters to measure, these usually being highly correlated (Fisher, 1936; Berry, Berry and Ucko, 1967). There are theoretical and empirical reasons for believing that minor non-metrical skeletal traits genetically characterize populations better than metrical ones (R. J. Berry and C. A. B. Smith, unpublished). Nevertheless, the variance of measurements of the skeleton can be used to measure the importance of mutagenic agents in the environment (Grüneberg, Bains, Berry, Riles, Smith and Weiss, 1966), and, in the same closed population in successive years, the influence of ecological factors on variability. This latter use has some general biological interest.

The House mouse population of the small Pembrokeshire island of Skokholm increases about ten-fold in number during the breeding season, although the expected increase on the basis of litter size and breeding interval is twenty-fold or more. Between 50% and 90% of the population die during the winter, the severity of the mortality depending on the coldness of the early months of the year (Berry, 1968*a*). According to theory, the variance of a population increases when selection is relaxed (e.g. Ford and Ford, 1930; Berry and Crothers, 1968). However the skeletal variance of the Skokholm population (measured by the mean coefficients of variation of four dental and four skeletal measurements corrected for age and size differences by the regression on weight) was consistently *higher* in the spring after the period of winter mortality, than it was in the autumn at the end of the breeding season (Fig. 5). Generally speaking, the greater the winter mortality, the higher the skeletal variance. In other words, the more variable animals survive the winter, the less variable are more successful at breeding. However, the mean variance of the population as calculated from non-metrical variants is reduced during the winter (Berry, 1968*b*), reinforcing the conclusion of Berry and Smith (*v.s.*) that skeletal size and non-metrical skeletal variants are controlled by different genes. Taken together the two sets of results appear to show

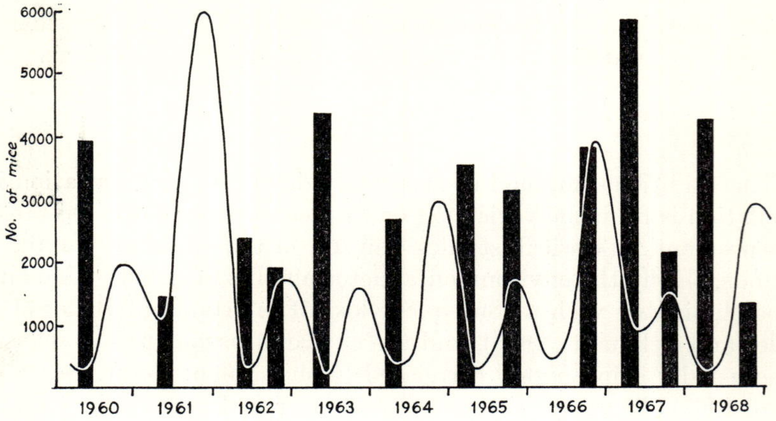

Fig. 5. Osteometric variation and population size on Skokholm: mean coefficients of variance of 4 dental and 4 skeletal measurements (adjusted for body size) of samples of the Skokholm mouse population (solid bars), and population size (continuous line).

that selection is favouring one genotype during the winter and another during the summer. If true, this would be a potent factor in maintaining intra-population variation. This conclusion is considered further below.

A third metrical study involving the Skokholm mouse population was made by Van Valen (1965–1966). He used a single dental measurement, and determined its variance in young and old individuals of the same sample. There was no difference between the age-classes. However the variance of the same measurement in mice collected from corn-ricks in different parts of England and Wales (Berry, 1963, 1964) showed an *increase* in the older group. The extent of this indicated an intensity of what Van Valen calls "de-stabilizing" selection of about 25%. The most likely source of this is the emigration of young adults from the ricks (Southwick, 1958): a closed population like that of Skokholm is atypical in that the whole population can be sampled; most populations will undergo immigration and emigration.

Biochemical

The difficulties about gross anatomical variation lie in isolating the consequences of variation at particular gene loci and of different environmental components. The same difficulties apply only in slight measure to biochemical variants, where the allelomorphs carried by an individual can often be unequivocally determined by electrophoretic treatment of small amounts of fluid. The techniques concerned have only recently begun to be applied to small mammal populations

(Arnason and Pantelouris, 1966; Pantelouris and Arnason, 1967; Selander, 1970; for a review of earlier work, see Lush, 1966). The most often found result is a deficiency of heterozygotes to an expectation based on the Hardy-Weinberg theorem (Rasmussen, 1964, on *Peromyscus maniculatus*: Petras, 1967, on *Mus musculus*), and is taken to indicate inbreeding resulting from division of the population into a number of small units. It is interesting, therefore, to compare gene and genotype frequencies on Skokholm before and at the end of the breeding season (Berry and Murphy, 1970) (Table I).

The Skokholm population is polymorphic at all 6 loci that have been tested—4 of them scored from red cells, 2 of them in the serum. In 1968, the first year in which the population was studied biochemically, there was no change in the frequencies of the allelomorphs at the transferrin locus during the breeding season. There was a significant decrease in the frequencies of the commonest allelomorph at 3 peptidase loci, this having the effect of increasing the proportion of heterozygotes at the 3 loci from 4·3% in the spring to 21·8% in the autumn. However the most interesting results were provided by the remaining 2 loci. At the diffuse/single haemoglobin locus the gene frequency did not change significantly between spring and autumn, but there was a 40% *excess* of heterozygotes over expectation in the autumn, as opposed to an 11% excess (which is statistically non-significant at the 5% level) in the spring. The pre-albumen esterase-2 locus showed the opposite trend: there was an 18% excess of heterozygotes in the spring which had disappeared by the autumn. In other words there is heterosis at the diffuse/single haemoglobin locus during summer breeding but not during the period of winter mortality, and the opposite for the pre-albumen esterase-2 locus. Similar changes were found before and after the 1969 breeding season. The evidence of natural selection acting in different ways at different times of the year (*endocyclically*) which was suggested by the skeletal studies (*v.s.*) is here confirmed from entirely different evidence.

Another interesting point to emerge from the study of biochemical polymorphisms in the Skokholm population is the number of allelomorphs present. The population is descended from a few mice carried over to the island in empty sacks in the mid-1890s (Berry, 1964, 1968*c*) and subsequent introductions have almost certainly been rare, yet it lacked only one of the 14 allelomorphs commonly present in British populations at the 6 loci studied. Far from having the property of genetical variability expected in a population descended from a small founding population, the Skokholm animals are as genetically diverse as most mainland populations.

TABLE I

Biochemical variation at three loci

			Haemoglobin		Pre-albumen esterase-2		Peptidase-C	
		N	% frequency diffuse allele	% excess of heterozygotes over expected*	% frequency *a* allele	% excess of heterozygotes over expected*	% frequency 2 allele	% proportion of heterozygotes
Skokholm	Spring 1968 (after winter mortality)	87	65·7	11	72·4	18	92·4	10
	Autumn 1968 (after summer breeding)	82	57·3	40	80·9	11	80·5	29
	Spring 1969 (after winter mortality)	73	62·3	19	66·9	33	82·3	17
	Autumn 1969 (after summer breeding)	115	53·9	61	69·6	26	67·1	42
Corn rick (Pembrokeshire)		75	15·0	2	90·0	0	78·1	28

*Expected on Hardy-Weinberg distribution.

Physiological

Physiological variation would be expected to be correlated with biochemical variability, although suffering from the same problems of analysis as anatomical variation. However, the immunological and electrophoretic variants studied do not have any obvious connection with adaptive traits, and any association with selective factors (*v.s.*) is likely to be coincidental. Furthermore most biochemical variants are controlled by single genes, and hence a biochemical study is unlikely to survey more than a dozen or so loci, whereas many anatomical and physiological variations may each be affected by 10 to 50 loci. Hence there are often advantages in studies of physiological variation, since traits of apparent adaptive importance can be directly investigated.

Since winter mortality in the Skokholm House mouse population is temperature-dependent, and since there is a differential survival of some genotypes during the winter, a study of the cold tolerance of animals on the island has been undertaken (Berry, Jakobson and Moore, 1969). After testing, the animals were released at the site of capture so that their future fate could be determined. It was found that the variance of the metabolic rate measured at 5°C was reduced by up to 70% during the winter, and that mice with higher basal metabolic rates seemed to have a greater chance of survival to spring. Preliminary data suggest that heterozygotes at the pre-albumen esterase-2 locus (which also have a high over-wintering survival rate) tend to have a higher basal metabolic rate than homozygotes.

Reproduction

The number of young to which a polytocous mammal gives birth is determined by interactions between a limited number of genetical, environmental and maternal (i.e. quasi-genetical) factors. It is affected, for example, by the size and age of the mother, the genotype of the father, by social stress, etc. (Roberts, 1965; Batten and Berry, 1967). Under natural conditions, litter size in animals which produce several litters in a season varies according to the stage of the breeding season, increasing at first and then decreasing. (For example, mean litter size in the Skokholm mouse population rises from 5·0 in March to over 10 in July, falling to 6·0 in October when virtually all breeding stops.) The length of the breeding season itself varies in different years (Smyth, 1966; Lloyd, 1970). The environmental control of breeding is shown by the fact that when Skokholm mice are bred in the laboratory, seasonal variation in both litter size and breeding disappears immediately (Berry, 1968*a*).

Lack (1948) has pointed out that, since litter size is ultimately dependent on genetical factors, it is subject to the action of natural selection, and selection will favour genotypes which give rise to the maximum number of new parents. He argued that the optimum litter size will be less than the maximum, since survival is reduced in large litters. Hence it was interesting to find that the correlation of 60% or so between maternal size and litter size in wild-caught House mice does not mean that the litter size of the large island races is proportionately greater. Indeed there is no correlation between maternal and litter sizes for mice caught on Skokholm, the Isle of May or Foula (Batten and Berry, 1967).

The remarkable polymorphism at the *t* locus in the House mouse can be included under the heading of reproductive variation. Dunn and his co-workers (e.g. Dunn, Beasley and Tinker, 1960) found all populations but one in the United States which they sampled adequately had a high frequency (around 40%) of one of the allelomorphs at the *t* locus (Lyon and Meredith, 1964), despite the fact that *t* homozygotes are usually inviable or cause male sterility. The main reason that these allelomorphs are found at high frequencies is an abnormally high rate of fertilization by *t*-bearing sperm as opposed to $+^t$-bearing sperm (Dunn, 1960). Some of the most illuminating studies of the genetical structure of small mammal populations have used *t*-allelomorphs (e.g. Anderson, 1964, 1970; Anderson, Dunn and Beasley, 1964; Petras, 1967). No search for *t*-allelomorphs seems to have been made in British mice. The variation in number of young born to the females in any population is probably determined by a combination of genetical and environmental factors similar to the osteological and physiological characters considered above; the difference is that the different factors have been more precisely identified (cf. Spickett and Thoday, 1966).

Cytogenetical

Little work on cytologically detectable genetical variation has been undertaken on British mice. Nevertheless there is no doubt that such variation exists. For example, Koller (1941) reported small differences between the karyotype of *A. sylvaticus* from Barra in the Outer Hebrides and mainland mice (R. J. Berry and J. M. Parrington, unpublished, have been unable to detect any differences in karyotypes from mitoses of *A. sylvaticus* from Edinburgh, Rhum, St. Kilda and Fair Isle); Searle, Berry and Beechey (1970) found the number of chiasmata in Skokholm House mice spermatogenesis to be considerably less than that found in laboratory mice, and that the sensitivity of the island mice to the induction of chromosomal re-arrangements (thought to be correlated with chiasma frequency) was approximately one third that in laboratory mice. Paradoxically, M. E. Wallace (personal communication) has found that some recombination fractions are much higher in Skokholm than in laboratory mice. There seems little doubt that the chromosomal complement of mouse populations will show as much variation as do other systems (cf. Gropp, Tettenborn and Lehmann, 1969; Ford and Hamerton, 1970; Meylan, 1970).

Genetically-determined variation exists in the House mouse even in the characters used by ecologists to age wild-caught animals: time of third molar eruption (Grüneberg, 1951); age at fusion of skull sutures (Deol, Grüneberg, Searle and Truslove, 1957); rate of fusion of the epiphyses of the long bones (Silberberg and Silberberg, 1941); and weight of the eye lens (Berry and Truslove, 1968). Many genes affecting behaviour are known in laboratory mice (e.g. Parsons, 1967), and allelomorphs of such loci have been found in wild-caught American *Rattus norvegicus* (King, 1939; Berry, 1969*b*). Chitty (1967, 1970) has discussed some of the implications of this, but studies of genetically determined behavioural differences in wild populations have not yet been intensively undertaken (as opposed to the significance of behavioural differences between age-groups: e.g. Sadleir, 1965). The differences between individuals in feeding rhythm noted by Hellwald (1931) and Southern (1954) may be genetical but no information is available on this point. The discovery of Berry *et al.* (1969) that mice with a high basic metabolic rate have a greater chance of living through the winter on Skokholm than ones with a low rate, presumably reflects, partially at least, the value of activity in survival, but the physiological finding has not yet been followed up with a behavioural study. Crowcroft and Jeffers (1961) concluded that individual reactions towards traps ("trap-proneness" and "trap-shyness") in House mice might involve "inherent differences" since similar reactions were found in both parents and offspring, and were apparently independent of experience in traps.

THE SIGNIFICANCE OF VARIATION IN BRITISH MICE

Theoreticians insist on the importance of "flexibility" for the continued survival of a population in a fluctuating environment. Dobzhansky's studies of *Drosophila pseudoobscura* illustrate this well. However, an example from British mice is provided by the field-mouse of St. Kilda (*A. sylvaticus hirtensis*) during the past 60 years or so. In 1930 the human population of St. Kilda was evacuated, and within 2 years the St. Kilda House mouse had become extinct, almost certainly as a result of competition with *A. sylvaticus* (Berry and Tricker, 1969). Museum collections exist of the St. Kilda field-mouse made between 1910 and 1919, in 1948, 1964 and 1967. The three more recent samples are indistinguishable in the frequencies of 20 non-metrical skull variants; the earlier, pre-human evacuation sample, differs "significantly" from them (at the 5% probability level) in 3 of these frequencies and also in a multivariate statistic (the "measure of divergence") derived from a comparison of the frequencies of all 20 variants. Samples of *A. sylvaticus* trapped on Lewis and Harris 45 years apart, and on Barra 25 years apart (Berry, 1969*a*) did not differ from each other in the frequencies of the same characters.

It is impossible to prove that the genetical change that occurred in the St. Kilda field-mouse between 1910 and 1948 was a consequence of adaptation to a new ecological situation, although this is the sort of assumption necessarily made in palaeontological material. A more clear-cut example of the "use" of variation is from the endocyclic selection experienced by the Skokholm mouse population, in which different genotypes have different survival values at different times of the year—a conclusion derived from four independent lines of evidence (metrical and non-metrical skeletal characters, physiological and biochemical traits (Fig. 6)).

The large variation in the Skokholm population permits higher numbers to be maintained throughout the year than would otherwise

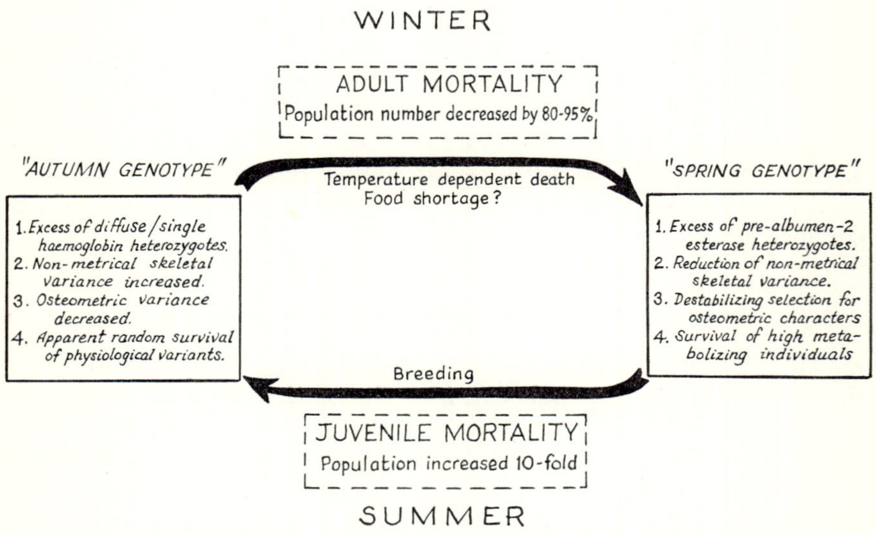

FIG. 6. Endocyclic selection operating on Skokholm mice.

be possible, and thus acts as a stabilizing influence. Moreover it allows fine adjustment to any changes in the environment of the mice. The implication from these considerations is that there will be an advantage in increasing the variation in a Mendelian population if it is below a certain—presumably optimum—level. Clearly the Skokholm mice must have changed the frequencies of existing and assimilated new allelomorphs since their original establishment (*v.s.*, and cf. Mayr, 1954). The average mutation rate in laboratory mice is of the order of 1×10^{-5} per locus (Schlager and Dickie, 1966). This means that a new allelomorph at a particular locus in the population is only likely to appear once every 5 years or so, so that selection must be invoked to explain the frequency of any but the rarest allelomorphs.

CONCLUSION

Selander *et al.* (1969) have estimated that any individual House mouse is heterozygous at about 8·5% of his loci. Similar figures for man have been derived by Harris (1966) and for *Drosophila pseudoobscura* by Lewontin and Hubby (1966). Such a large proportion of variability renders probable the formation of unfit genotypes by segregation. Some workers (e.g. Muller, 1950; Haldane, 1957) have argued that there is an upper limit to the amount of heterozygosity that can be tolerated by a population, beyond which the number of "genetic" deaths proves too much of a "load" for the population. This conclusion has been challenged on theoretical grounds (Sved, 1968; Smith, 1968; Brues, 1969).

The data presented in this paper show the large amount of genetical variation present in mouse populations and we have seen some of the value of this variation in the Skokholm population at least. The great value of such studies is that they are leading us to understand the factors involved in the maintenance of variation in mammalian populations; other papers in this Symposium expand particular themes which I have only touched on because of more or less confining myself to British mice. It is to be hoped that the whole will contribute to a fuller understanding of the biological problem of variability.

Acknowledgements

My thanks are due to Mr. A. J. Lee for drawing the figures.

References

Anderson, P. K. (1964). Lethal alleles in *Mus musculus*: local distribution and evidence for isolation of demes. *Science, N.Y.* **145**, 177–178.

Anderson, P. K. (1970). Ecological structure and gene flow in small mammals. *Symp. zool. Soc. Lond.* No. 26, 299–325.

Anderson, P. K., Dunn, L. C. and Beasley, A. B. (1964). Introduction of a lethal allele into a feral house mouse population. *Am. Nat.* **98**, 57–64.

Arnason, A. and Pantelouris, E. M. (1966). Serum esterases of *Apodemus sylvaticus* and *Mus musculus*. *Comp. Biochem. Physiol.* **19**, 53–61.

Bader, R. S. (1956). Variability in wild and inbred mammalian populations. *Q. Jl Fla Acad. Sci.* **19**, 14–34.

Bader, R. S. and Lehmann, W. H. (1965). Phenotypic and genotypic variation in odontometric traits of the house mouse. *Am. Midl. Nat.* **74**, 28–38.

Barrett-Hamilton, G. E. H. (1899). On the species *Mus* inhabiting St. Kilda. *Proc. zool. Soc. Lond.* **1899**, 77–78.

Barrett-Hamilton, G. E. H. (1900). On geographical and individual variation in *Mus sylvaticus* and its allies. *Proc. zool. Soc. Lond.* **1900**, 397–428.

Barrett-Hamilton, G. E. H. and Hinton, M. A. C. (1910–1921). *A history of British mammals*. London: Gurney & Jackson.

Batten, C. A. and Berry, R. J. (1967). Prenatal mortality in wild-caught house mice. *J. Anim. Ecol.* **36**, 453–463.

Beirne, B. P. (1952). *The origin and history of the British fauna*. London: Methuen.

Berry, A. C., Berry, R. J. and Ucko, P. J. (1967). Genetical change in ancient Egypt. *Man* **2**, 551–568.

Berry, R. J. (1963). Epigenetic polymorphism in wild populations of *Mus musculus*. *Genet. Res.* **4**, 193–220.

Berry, R. J. (1964). The evolution of an island population of the house mouse. *Evolution, Lancaster, Pa.* **18**, 468–483.

Berry, R. J. (1967). Genetical changes in mice and men. *Eug. Rev.* **59**, 78–96.

Berry, R. J. (1968*a*). The ecology of an island population of the house mouse. *J. Anim. Ecol.* **37**, 445–470.

Berry, R. J. (1968*b*). The biology of non-metrical variation in mice and men. In *The skeletal biology of earlier human populations*. 103–133, Brothwell, D. R. (ed.). London: Pergamon.

Berry, R. J. (1968*c*). The Skokholm mouse: chance and change. *Rep. Skokholm Bird Obs.* **1967**, 30–35.

Berry, R. J. (1969*a*). History in the evolution of *Apodemus sylvaticus* (Mammalia) at one edge of its range. *J. Zool., Lond.* **159**, 311–328.

Berry, R. J. (1969*b*). The genetical implications of domestication in animals. In *The domestication and exploitation of plants and animals*. 207–217, Ucko, P. J. and Dimbleby, G. W. (eds) London: Duckworth.

Berry, R. J. (1970). The natural history of the house mouse. *Fld Stud.* (in press).

Berry, R. J. and Crothers, J. H. (1968). Stabilizing selection in the Dog whelk (*Nucella lapillus*). *J. Zool., Lond.* **155**, 5–17.

Berry, R. J., Evans, I. M. and Sennitt, B. F. C. (1967). The relationships and ecology of *Apodemus sylvaticus* from the Small Isles of the Inner Hebrides, Scotland. *J. Zool., Lond.* **152**, 333–346.

Berry, R. J., Jakobson, M. E. and Moore, R. E. (1969). Metabolic measurements on an island population of the house mouse during the period of winter mortality. *J. Physiol.* **201**, 101–102P.

Berry, R. J. and Murphy, H. M. (1970). The biochemical genetics of an island population of the house mouse. *Proc. R. Soc.* (B) (in press).

Berry, R. J. and Tricker, B. J. K. (1969). Competition and extinction: the mice of Foula, with notes on those of Fair Isle and St. Kilda. *J. Zool., Lond.* **158**, 247–265.

Berry, R. J. and Truslove, G. M. (1968). Age and eye lens weight in the house mouse. *J. Zool. Lond.* **155**, 247–252.

Boyd, J. M. (1959). Observations on the St. Kilda field-mouse *Apodemus sylvaticus hirtensis*, Barrett-Hamilton. *Proc. zool. Soc. Lond.* **133**, 47–65.

Brown, L. N. (1965). Selection in a population of house mice containing mutant individuals. *J. Mammal.* **46**, 461–465.

Brues, A. M. (1969). Genetic load and its varieties. *Science, N.Y.* **164**, 1130–1136.

Chitty, D. (1967). The natural selection of self-regulatory behaviour in animal populations. *Proc. ecol. Soc. Aust.* **2**, 51–78.

Chitty, D. (1970). Variation and population density. *Symp. zool. Soc. Lond.* No. 26, 327–333.

Clarke, W. E. (1904). On some forms of *Mus musculus*, Linn., with description of a new sub-species from the Faeroe Islands. *Proc. R. phys. Soc. Edinb.* **15**, 160–167.

Clegg, T. M. (1963). Observations on an East Yorkshire population of the house mouse (*Mus musculus* Linn.). *Naturalist, Hull* No. 885, 39–40.

Cook, L. M. (1961). The edge effect in population genetics. *Am. Nat.* **95**, 295–307.

Corbet, G. B. (1961). Origin of the British insular races of small mammals and of the 'Lusitanian' fauna. *Nature, Lond.* **191**, 1037–1040.

Corbet, G. B. (1963). An isolated population of the bank vole (*Clethrionomys glareolus*) with aberrant dental pattern. *Proc. zool. Soc. Lond.* **140**, 316–319.

Corbet, G. B. (1964*a*). *The identification of British mammals.* London: British Museum (Natural History).

Corbet, G. B. (1964*b*). Regional variation in the bank-vole *Clethrionomys glareolus* in the British Isles. *Proc. zool. Soc. Lond.* **143**, 191–219.

Crowcroft, W. P. (1966). *Mice all over.* London: Foulis.

Crowcroft, W. P. and Jeffers, J. N. R. (1961). Variability in the behaviour of wild house mice (*Mus musculus* L.) towards live traps. *Proc. zool. Soc. Lond.* **137**, 573–582.

Crowcroft, W. P. and Rowe, F. P. (1961). The weights of wild house mice (*Mus musculus*, L.) living in confined colonies. *Proc. zool. Soc. Lond.* **136**, 177–185.

Degerbøl, M. (1942). Mammalia. *Zoology Faroes* Pt 65, 1–133.

Delany, M. J. (1964). Variation in the Long-tailed field-mouse (*Apodemus sylvaticus* (L.) in north-west Scotland. I. Comparison of individual characters. *Proc. R. Soc. Lond.* (B) **161**, 191–199.

Delany, M. J. (1970). Variation and ecology of island populations of the Long-tailed field-mouse (*Apodemus sylvaticus* (L.)). *Symp. zool. Soc. Lond.* No. 26, 283–295.

Delany, M. J. and Healy, M. J. R. (1964). Variation in the long-tailed field-mouse (*Apodemus sylvaticus* (L.)) in north-west Scotland. II. Simultaneous examination of all characters. *Proc. R. Soc.* (B) **161**, 200–207.

Deol, M. S. (1970). The determination and distribution of coat colour variation in the House mouse. *Symp. zool. Soc. Lond.* No. 26, 239–250.

Deol, M. S., Grüneberg, H., Searle, A. G. and Truslove, G. M. (1957). Genetical differentiation involving morphological characters in an inbred strain of mice. I. A British branch of the C57BL strain. *J. Morph.* **100**, 345–376.

Deol, M. S. and Truslove, G. M. (1957). Genetical studies on the skeleton of the mouse. XX. Maternal physiology and variation in the skeleton of C57BL mice. *J. Genet.* **55**, 288–312.

Dice, L. R. (1947). Effectiveness of selection by owls of deermice (*Peromyscus maniculatus*) which contrast in colour with their background. *Contr. Lab. vert. Biol. Univ. Mich.* **34**, 1–20.

Dunn, L. C. (1960). Variations in the transmission ratio of alleles through egg and sperm in *Mus musculus*. *Am. Nat.* **94**, 385–393.

Dunn, L. C., Beasley, A. B. and Tinker, H. (1960). Polymorphisms in populations of wild house mice. *J. Mammal.* **41**, 220–229.

Falconer, D. S. (1953). Selection for large and small size in mice. *J. Genet.* **51**, 470–501.

Falconer, D. S. (1960). *An introduction to quantitative genetics*. Edinburgh and London: Oliver & Boyd.

Fisher, R. A. (1936). 'The coefficient of racial likeness' and the future of craniometry. *Jl R. anthrop. Inst.* **46**, 57–63.

Ford, C. E. and Hamerton, J. L. (1970). Robertsonian variation in the Common shrew. *Symp. zool. Soc. Lond.* No. 26, 223–236.

Ford, E. B. (1964). *Ecological genetics*. London: Methuen.

Ford, H. D. and Ford, E. B. (1930). Fluctuation in numbers and its influence on variation in *Melitaea aurinia*. *Trans. ent. Soc. Lond.* **78**, 345–351.

Foster, J. B. (1964). Evolution of mammals on islands. *Nature, Lond.* **202**, 234–235.

Fullagar, P. J., Jewell, P. A., Lockley, R. M. and Rowlands, I. W. (1963). The Skomer vole (*Clethrionomys glareolus skomerensis*) and Long-tailed field mouse (*Apodemus sylvaticus*) on Skomer Island, Pembrokeshire in 1960. *Proc. zool. Soc. Lond.* **140**, 295–314.

Grant, V. M. (1964). *Architecture of the germplasm*. New York and London: Wiley.

Green, E. L. (ed.) (1966). *Biology of the laboratory mouse*. 2nd edn., New York and London: McGraw-Hill.

Gropp, A., Tettenborn, U. and Lehmann, E. v. (1969). Robertson'sche Chromosomen variation bei der Maus (*M. musculus*), der Tabakmaus (*M. poschiavinus*) und ihren Hybriden. *Cytogenetics*, **9**, 9–23.

Grüneberg, H. (1951). The genetics of a tooth defect in the mouse. *Proc. R. Soc.* (B.) **138**, 437–451.

Grüneberg, H. (1952). The genetics of the mouse. *Biblphia genet.* **15**, 1–650.

Grüneberg, H., Bains, G. S., Berry, R. J., Riles, L. E., Smith, C. A. B. and Weiss, R. A. (1966). A search for genetic effects of high natural radioactivity in South India. *Spec. Rep. Ser. med. Res. Coun.* No. 307, 1–59.

Haldane, J. B. S. (1948). The theory of a cline. *J. Genet.* **48**, 277–284.

Haldane, J. B. S. (1957). The cost of natural selection. *J. Genet.* **55**, 511–524.

Harris, H. (1966). Enzyme polymorphisms in man. *Proc. R. Soc.* (B.) **164**, 298–310.

Hellwald, H. (1931). Untersuchungen über Triebstärken bei Tieren. *Z. Psychol.* **123**, 94–141.

Hinton, M. A. C. (1914). Notes on the British forms of *Apodemus*. *Ann. Mag. nat. Hist.* (9) **14**, 117–134.

Jameson, H. L. (1898). On a probable case of protective coloration in the House Mouse (*Mus musculus*, Linn.). *J. Linn. Soc.* (*Zool.*) **26**, 465–473.

Jewell, P. A. and Fullagar, P. J. (1965). Fertility among races of the field-mouse (*Apodemus sylvaticus* (L.)) and their failure to form hybrids with the yellow-necked mouse (*A. flavicollis* (Melchior)). *Evolution, Lancaster, Pa.* **19**, 175–181.

Jones, G. (1968). *A history of the Vikings.* London: Oxford University Press.

Kettlewell, H. B. D. and Berry, R. J. (1969). Gene flow in a cline. *Heredity* **24**, 1–14.

King, H. D. (1939). Life processes in gray Norway rats during fourteen years of captivity. *Am. anat. Mem.* No. 17, 1–77.

Koller, P. C. (1941). The genetical and mechanical properties of the sex chromosomes. VII. *Apodemus sylvaticus* and *A. hebridensis. J. Genet.* **41**, 375–389.

Lack, D. (1948). The significance of litter size. *J. Anim. Ecol.* **17**, 45–50.

Lerner, I. M. (1954). *Genetic homeostasis.* Edinburgh and London: Oliver & Boyd.

Lewontin, R. C. and Hubby, J. L. (1966). A molecular approach to the study of genic heterozygosity in natural populations of *Drosophila pseudoobscura. Genetics* **54**, 595–609.

Livingstone, F. B. (1969). The founder effect and deleterious genes. *Am. J. phys. Anthrop.* **30**, 55–60.

Lloyd, H. G. (1970). Variation and adaptation in reproductive performance. *Symp. zool. Soc. Lond.* No. 26, 165–188.

Lockley, R. M. (1947). *Letters from Skokholm.* London: Dent.

Lush, I. E. (1966). *The biochemical genetics of vertebrates except man.* Amsterdam: North-Holland Publ. Co.

Lyon, M. F. and Meredith, R. (1964). The nature of the *t*-alleles in the mouse. *Heredity* **19**, 301–330.

Mather, K. (1941). Variation and selection of polygenic characters. *J. Genet.* **41**, 159–193.

Mather, K. (1953). The genetical structure of populations. *Symp. Soc. exp. Biol.* **7**, 76–95.

Mather, K. (1964). *Human diversity.* Edinburgh and London: Oliver & Boyd.

Matthews, L. H. (1952). *British mammals.* London: Collins.

Mayr, E. (1954). Change of genetic environment and evolution. In *Evolution as a process:* 157–180. Huxley, J., Hardy, A. C., Ford, E. B. (eds). London: Allen & Unwin.

Meylan, A. (1970). Chromosomal polymorphisms in mammals. *Symp. zool. Soc. Lond.* No. 26, 211–222.

Miller, G. S. (1912). *Catalogue of the mammals of western Europe.* London: British Museum (Natural History).

Muller, H. J. (1950). Our load of mutations. *Am. J. hum. Genet.* **2**, 111–176.

O'Gorman, F. A. and Von Rizzori, A. (1969). Colour variation in the House mice of the Bull Island, Dublin Bay. Paper read at Symposium on Mammalian Variation, Zoological Society of London, November, 1969.

Pantelouris, E. M. and Arnason, A. (1967). Serum proteins of *Apodemus sylvaticus* and *Mus musculus. Comp. Biochem. Physiol.* **21**, 535–539.

Parsons, P. A. (1967). *The genetic analysis of behaviour.* London: Methuen.

Petras, M. L. (1967). Studies of natural populations of *Mus.* II. Polymorphism at the T locus. *Evolution, Lancaster, Pa,* **21**, 466–478.

Petras, M. L., Reimer, J. D., Biddle, F. G., Martin, J. E. and Linton, R. D. (1969). Studies of natural populations of *Mus.* V. A survey of nine loci for polymorphisms. *Canad. J. Genet. Cytol.* **11**, 497–573.

Rasmussen, D. I. (1964). Blood group polymorphism and inbreeding in natural populations of the Deer Mouse, *Peromyscus maniculatus. Evolution, Lancaster, Pa.* **18**, 219–229.

Rensch, B. (1959). *Evolution above the species level.* New York: Columbia University Press.

Roberts, R. C. (1965). Some contributions of the laboratory mouse to animal breeding research. *Anim. Breed. Abstr.* **33**, 339–353; 515–526.

Sadleir, R. M. F. S. (1965). The relationship between agonistic behaviour and population changes in the deermouse, *Peromyscus maniculatus* (Wagner). *J. Anim. Ecol.* **34**, 331–352.

Schlager, G. and Dickie, M. M. (1966). Spontaneous mutation rates at five coat-colour loci in mice. *Science, N.Y.* **151**, 205–206.

Searle, A. G., Berry, R. J. and Beechey, C. V. (1970). Cytogenetic radiosensitivity and chiasma frequency in wild-living male mice. *Mutation Res.* **9**, 137–140.

Selander, R. K. (1970). Biochemical polymorphism in populations of the House mouse and Old-field mouse. *Symp. zool. Soc. Lond.* No. 26, 73–91.

Selander, R. K., Hunt, W. G. and Yang, S. Y. (1969). Protein polymorphisms and genetic heterozygosity in two European subspecies of the house mouse. *Evolution, Lancaster, Pa.* **23**, 379–390.

Silberberg, M. and Silberberg, R. (1941). Age changes of bones and joints in various strains of mice. *Am. J. Anat.* **68**, 69–95.

Smith, J. M. (1968). "Haldane's dilemma" and the rate of evolution. *Nature, Lond.* **219**, 1114–1116.

Smyth, M. (1966). Winter breeding in woodland mice, *Apodemus sylvaticus*, and voles, *Clethrionomys glareolus* and *Microtus agrestis*, near Oxford. *J. Anim. Ecol.* **35**, 471–485.

Southern, H. N. (ed.) (1954). *Control of rats and mice.* **3**. *House mice.* Oxford: Clarendon Press.

Southwick, C. H. (1958). Population characteristics of house mice living in English corn ricks: density relationships. *Proc. zool. Soc. Lond.* **131**, 163–175.

Spickett, S. G. and Thoday, J. M. (1966). Regular responses to selection. 3. Interaction between located polygenes. *Genet. Res.* **7**, 96–121.

Steven, D. M. (1953). Recent evolution in the genus *Clethrionomys. Symp. Soc. exp. Biol.* **7**, 310–319.

Sumner, F. B. (1930). Genetic and distributional studies of three subspecies of *Peromyscus. J. Genet.* **23**, 275–376.

Sved, J. A. (1968). Possible rates of gene substitution in evolution. *Am. Nat.* **102**, 283–294.

Thoday, J. M. (1955). Balance, heterozygosity, and developmental stability. *Cold Spring. Harb. Symp. quant. Biol.* **20**, 318–326.

Turner, D. T. L. (1965). A contribution to the ecology and taxonomy of the vole *Microtus arvalis* on the island of Westray, Orkney. *Proc. zool. Soc. Lond.* **144**, 143–150.

Van Valen, L. (1965–1966). Selection in natural populations. IV. British house mice (*Mus musculus*). *Genetica* **36**, 119–134.

West, R. G. (1968). *Pleistocene geology and biology.* London: Longmans.

Symp. zool. Soc. Lond. (1970) No. 26, 27–39.

THE NATURE AND SIGNIFICANCE OF VARIATION IN WILD POPULATIONS

KENNETH MATHER

University of Southampton, Southampton, England

SYNOPSIS

The genetical constitution and changes of a wild population are the resultant of the variation that arises in it and the forces of natural selection that impinge on it. Variation may affect the number of the chromosomes and their structure, or it may take the form of gene changes which are cytologically indetectable, revealing their occurrence only by their effects on the phenotypes of individuals carrying them. All these types of variation are known to occur in man.

Changes in chromosome number may involve whole sets of chromosomes, as in polyploidy, or individual chromosomes, as in trisomy. Under some circumstances, especially in plants, polyploids may be favoured by natural selection; but polyploidy commonly leads to a greater or lesser degree of sterility and so is disfavoured. Trisomy distorts development, often to a lethal degree, and so reduces fitness. There is evidence in man of the differential elimination of foetuses with variant numbers of chromosomes. Thus most numerical variants immediately reduce fitness and have little prospect of continuation in the population. Their presence attests only their continual recurrence—that cytological accidents continue to happen.

Certain structural variants of the chromosomes do not affect the phenotype and do not reduce fitness in a serious way. They may be retained in the population because of the control they facilitate over the recombination of the genes carried in the segments of the chromosome that they affect.

Gene changes affect every character of an organism to an extent ranging from that so drastic as to be lethal to the barely perceptible. Some appear to be unconditionally disadvantageous: like certain chromosome variants, they are found in populations only because they continually recur, and they attest that genic, like cytological, accidents continue to happen. Other genic variants form the basis of polymorphisms, the number of which known in human and other populations is growing steadily. The maintenance of these polymorphisms is discussed in relation to heterozygous advantage, the notion of genetic load, and co-operative relations between unlike phenotypes.

The most common type of variation in populations is the so-called continuous variation which is shown by virtually all characters of all organisms and the heritable part of which is mediated by polygenic systems. A great proportion of the variation stemming from such a system lies concealed in the genotype as opposed to being freely expressed in the phenotype. Natural selection makes conflicting demands for uniformity of phenotype at any given time and for change of phenotype to adjust to trends in the environment. The pool of concealed variability makes possible the reconciliation in large measure of these opposing requirements. Continuous variation allows fine adjustment to the demands of the environment and is the prime variation of adaptation. The population as we see it reflects both the immediate impact of selection and its changes and also, in the organization of the system, the impact of selection on past generations.

INTRODUCTION

Populations of living organisms as we see them in the wild result from the interplay of variation and natural selection. If we are to understand the populations that we see, and also changes that we might observe to be in progress, we must seek to understand the nature and extent of the variation that they carry and the significance of this variation in relation to natural selection.

We are most familiar with variation in our own species, *Homo sapiens*, where indeed the whole of our system of society is based on the assumption not only that no two of us are alike but that we can be readily distinguished from each other by simple ocular inspection. We know, of course, that the differences visible to the eye are but a part of the picture; that each of us has his biochemical individuality (with all the trouble that it gives to the medical profession in the treatment of disorders, the administration of anaesthetics and the transplantation of tissues and organs) and his own psychological individuality (with all the trouble it gives to educationalists in seeking to combine appropriate education for each with the provision of equal opportunity for all), as well as his own morphological individuality. We know, too, that this variation can have its consequences for Darwinian fitness—for our prospects of contributing to posterity. In the grosser cases we agree on what these consequences are, though in less extreme cases we dispute them and the effects of any alleviating or remedial measures that we deem it wise or humane to apply to variant individuals.

We are in general less familiar with variation in other species. Indeed, initially the difficulty may be that of detecting any variation at all, but the more one becomes familiar with another species the more one comes to recognize variation and individuality. The types and incidence of variation in wild populations has been most thoroughly analysed in species of *Drosophila*, particularly *D. pseudoobscura*; but we now have quite extensive information about a number of species, both plant and animal, and a rapidly growing body of knowledge about human variation. All the types of variation which we have come to know from laboratory studies are found side by side in wild populations of species that have been adequately examined, and few geneticists would doubt that similar examination would reveal them in species where they have yet to be recorded.

TYPES OF VARIATION

The organism, as we see it, is the resultant of its genotype and its environment. It may vary therefore as a result of change in the environment as well as change in the hereditary materials. Variations traceable to differences in the environment are of course in general non-heritable and as such are of less interest to us, but we should note that certain external agencies can and do stimulate heritable change. The best known of these are ionizing radiations which raise the mutation rate in a relatively unspecific way. Species living in areas where background radiation levels are unusually high, such as Kerala in Southern India might therefore be expected, *prima facie*, to show more variation. A search for such enhanced variation among rats in Kerala has, however, failed to reveal it, at any rate in significant amount (Grüneberg, 1964), though there may be other reasons for this as we shall see later.

Certain chemicals have mutagenic effects, similar to those of ionizing radiation, inducing heritable changes that are broadly, if not wholly, non-specific in their effects on the phenotype. More remarkable, however, are the directed changes which have been induced in certain plants, notably flax (Durrant, 1962). Here the imposition of a particular environmental regime, defined usually in terms of fertilizers, always produces a specific effect on the growth and size of the plant, which is transmissible from parent to offspring for a number of generations. Cases of such successful "conditioning", as Durrant calls it, are not common in plants and, so far as I am aware, there is no established case in animals despite a number of attempts that have been made to produce them.

In these various cases heritable variants arise as a result of environmental intervention. The occurrence and extent of non-heritable variation may also depend on the genotype. In an ultimate sense this must obviously be true for, without an appropriate genotype, the organism could not display the characters whose variation we observe and record (or, for that matter, even exist). In a narrower and more helpful sense, however, we might observe that differential responses to changes in the environment, as when a plant's leaves change form according to whether they are submerged or not, or a mammal changes its pelage according to its environmental background, depend on the possession of an appropriate genotype. Such variation is presumably adaptive; but the extent of non-adaptive fluctuation also depends on the genotype. Within a species, individuals with certain genotypes appear to be "pushed around" by their environments to a very much greater extent than their fellows with another genotype. In naturally cross-breeding species this property of developmental stability, or homeostasis is greater in individuals with more heterozygous loci than it is in homozygotes (Mather, 1946; Lerner, 1954). It is not, however, a property of heterozygosity *per se* (Jinks and Mather, 1955) and stability of development has been reduced by selection in *Drosophila* (Mather, 1953).

Variations in the hereditary materials themselves range from gross changes in the number of chromosomes to small changes within the chromosomes, indetectable by cytological inspection and possibly in some cases involving no more than a few nucleotides, which reveal themselves only by their effects on the phenotype. These latter are the gene mutations from whose effects genes are inferred. They may affect any character of the organism and their effects may either be so gross as to be incompatible with the full development of the individual or so small as to escape detection other than by the most refined genetical and statistical techniques.

Changes in chromosome number may involve complete sets of chromosomes, producing polyploids, or they may be replications of single chromosomes, producing trisomics. Both types of change are now known in man, but their consequences were worked out thoroughly in plants and in *Drosophila* some decades ago. Polyploidy seldom produces any striking change in development but it has a profound effect on fertility: it has been important in plant evolution because of the capacity of tetraploids to combine two parental sets of chromosomes in a fertile, true-breeding hybrid. Triploids, which are probably the commonest polyploids to arise in nature are virtually completely sterile. Their occurrence in a population therefore represents no more than the outcome of an unfortunate cytological accident and is of virtually no significance for the population's future.

The trisomic condition has less effect on fertility, but a greater effect on the somatic phenotype which is distorted by a syndrome of abnormalities characteristic of the chromosome in question. Thus trisomy for chromosome 21 in man produces mongolism. Trisomy for larger chromosomes may well be lethal in man, as is known to be the case in *Drosophila*. Thus, like triploids, trisomics are unfortunate cytological accidents with no significance for the population's future. Their disabling effects are illustrated in another way by the high incidence of their occurrence in human foetuses which have failed to reach term (Polani, 1967). Evidently because of their profound effect on development and hence fitness they are selectively eliminated during foetal life. Curiously enough triploids, which do not show an especially abnormal somatic phenotype, also appear to be selectively eliminated during foetal life (Polani, 1967).

In addition to changes of number in the chromosomes, changes of their structure also occur and certain of these, notably inversions and interchanges (or translocations as they are often loosely called) may have no effect on the phenotype, and reduce fertility to no more than an acceptable extent. They may therefore persist in the populations, and indeed play an adaptive role by reducing effective recombination and so holding together valuable constellations of linked genes. Structural change, notably by unequal interchange, may also play its part in bringing about viable changes in the number of chromosomes such as commonly distinguish one species from another.

We have seen that most changes in chromosome number are no more than the unfortunate consequences of accidents to pairing, disjunction and spindle behaviour during cell divisions. They do not persist in a population, and insofar as they are found in each generation, they bear testimony to the continuing occurrence of these cytological accidents. Genic accidents also happen and give rise to gene mutations, perhaps by failure of perfect replication during cell division. Since the gene is a highly integrated entity closely adjusted to doing its biochemical work, these mutant genes are commonly less able to carry out the normal function in development. Individuals may show a consequent change when bearing the mutant gene in the homozygous condition, or less often, when in the heterozygous condition; but whether shown in homozygous or heterozygous state the change is usually disabling to some degree. These are the familiar mutations of the laboratory. Individuals displaying their effects are seldom observed directly in wild populations, presumably because, being in some measure disabled, they seldom survive long under wild conditions. However, breeding experiments, chiefly of *Drosophila* spp., have shown that individuals from wild populations regularly carry in heterozygous condition recessive mutant genes whose effects range over the full spectrum of change seen in the laboratory mutations, including lethality. The occurrence of, for example, albinos in artificial breeding herds of such creatures as lions and kangaroos testifies to the same thing.

These mutations, producing major effects, will not be eliminated at once like certain cytological changes, if only because they are commonly recessive and hence will be sheltered from the action of natural selection behind the cloak of their normal allele in heterozygotes. Yet natural selection will always be tending to reduce the frequency of these genes in wild populations and they will maintain themselves only because of their continuing recurrence by mutation. Indeed, if such a gene reduces the fitness of an individual displaying its effects to $1-s$, where the fitness of a normal individual is 1, and μ is the rate at which the gene is arising by mutation, the frequency of individuals displaying the effects of the gene in a population at equilibrium will be proportional to μ/s. Since mutation rates are small, commonly of the order of 10^{-5} or 10^{-6}, whereas s may be as high as 1, (as with a lethal or grossly disabling gene) μ/s usually has a very low value. So the proportion of individuals displaying the effects of such a gene on a population will be small: and this we know to be the case in man. At the same time there are so many loci at which such mutations (each with its own effect: anatomical, physiological or psychological) can arise that a not inconsiderable proportion of the newly formed zygotes potentially display the effects of one or other of these genes. Thus in man it has been estimated that between 1% and 4% of live births are of individuals showing some genetically determined abnormality of form or function—and this despite such elimination as may have taken place during foetal life.

This type of variation can be of no adaptive significance. It is presumably the price that must be paid for having genes, with the risk of their going wrong, just as many of the cytological variants represent no more than the price that must be paid for having a chromosome apparatus, with its risk of going wrong. In both cases, however, the price can be kept under control since both the rate of mutation of genes and the functioning of the chromosome apparatus are under genetical control and so subject to adjustment by selection (Darlington, 1958; Rees, 1961).

POLYMORPHISM AND GENETIC LOAD

Most of the variation to which reference has so far been made is immediately debilitating and prospectively profitless. It is variational detritus, reflecting the breakdown of materials and mechanisms rather than their adjustment, and is held in the population only by its recurrence. This is not true of polymorphisms which divide populations into groups of genetically and phenotypically distinct individuals the rarest of which is too common to spring from the pressure of mutation alone.

Polymorphisms are well known in animals, especially perhaps in certain groups of insects, including the Lepidoptera where they are not uncommonly associated with mimicry. The classical polymorphism in man is that of the A-B-O series of blood groups, to which, of course, have been added an extensive series of other blood group polymorphisms: M-N-S, Rhesus, and so on. The number of other blood character polymorphisms is now being extended rapidly: indeed whenever a search is made for biochemical or immunological variation (as for example, by Harris, 1966) more cases are found, and we are left with the impression that the number of loci at which there exist two or more alleles producing a polymorphism must be large. The same is true of other species, including *Drosophila*, and examples in mammals are reviewed in this symposium by Lush (1970).

What is the significance of these polymorphisms? The classical interpretation of polymorphism is due to Fisher (1927) who pointed out that if a heterozygote has the advantage of greater fitness than both the two corresponding homozygotes, the two alleles will be held in the population by natural selection and a polymorphism will result. This is how the polymorphism for sickle-cell anaemia is maintained in certain areas: the heterozygote is free from the anaemia which characterizes one homozygote and from the susceptibility to malaria which characterizes the other (Allison, 1955). How far this interpretation can be applied to the extensive series of polymorphisms that are being discovered has, however, been a matter of dispute. The sickle-cell polymorphism is maintained by the mortality arising from the anaemia on the one hand and malaria on the other. It thus imposes a "genetic load" of unfitness on the population and if it did not impose this load it would not be maintained. Such a load (which with sickle-cell anaemia may in an extreme case amount to an average reduction in fitness of 10%–15% across the population as a whole) is perhaps acceptable for a single polymorphism or, even if less extreme, several polymorphisms. But if the loads imposed by a large number of polymorphisms were cumulative their maintenance would appear to require that the cumulative loss of fitness would be such as to deny the population the ability to maintain itself as a continuing entity: the total load would be too great.

Several counter arguments have been advanced. It has been argued that perhaps the polymorphisms in many cases do not represent a balanced situation depending on opposing selective forces as with sickle-cell anaemia, but reflect no more than the chance distribution of alleles which are selectively neutral in relation to one another. The acceptability of this argument depends on the credibility of selective neutrality. To some of us Fisher's (1930) demonstration that the effect on fitness of a gene difference must be no more than of the order of the reciprocal of the number of individuals in the population if it is to be selectively neutral, makes this interpretation of polymorphism difficult to accept, except possibly for a few special cases.

It has been pointed out by Sved, Reed and Bodmer (1967) that the loads springing from a number of different loci may not be simply cumulative and that the total load springing from a large number of loci may thus not be excessive. This is in fact likely to be the case where competition enters in as a selective force. Indeed, the argument that the total load from a number of polymorphisms would become excessive assumes implicitly that the differences in fitness of the various genotypes are unconditional. Now selection through differences in ability to compete with other individuals leads to exactly the same genetical equilibria as do unconditional selective forces (Mather, 1969). At the same time competitive selection eases in its rigour if the density of the population falls. Thus over-heavy selection resulting in a fall in population size carries its own remedy in that the fall in number leads to reduction in the loss of fitness, and the argument from load falls to the ground. In fact, when selection is competitive—and most selection must have a competitive element in it—the concept of load itself fails (Mather, 1969).

The final point to make about polymorphism is that it need not depend for its maintenance on the advantage of heterozygotes over the corresponding homozygotes. In fact, the most common dimorphism in the animal kingdom—the sexual dimorphism—manifestly cannot depend on heterozygous advantage. Polymorphism can arise wherever selection is simultaneously favouring two or more types of individual which are bound together in a symbiotic relation in respect of some function, as are males and females in reproduction, or the different form of mimic in Batesian mimicry (Mather, 1955). We do not know the extent to which the many polymorphisms that are coming to be recognized in man and other mammals are to be interpreted in this way; but it at least seems possible that some of the blood group series, including A-B-O, might represent this kind of situation where the adjusted diversity of the phenotype is itself advantageous. More evidence is clearly needed.

CONTINUOUS VARIATION

There remains one further type of variation to consider, the most common type of all, by which individuals are not classifiable into discrete, sharply distinct classes by simple genetic differences, but show every grade of character expression between wide limits, the central phenotypes being most common with the more extreme individuals becoming progressively rarer. This is not the place to give a detailed account of the genetic control of such continuous variation: many such accounts exist in the literature. It is sufficient to say that in general variation of this kind is controlled by polygenic systems whose member genes have small, similar and supplementary effects. The consequence of this is that the expression of a character depends more on the number of increasing and decreasing alleles the individual carries at the various loci involved in the system, than on which precise genes are present. Some of the properties of this type of variation must, however, be recapitulated if we are to understand the variation we see in wild populations and the reaction of these populations to selection.

Continuous variation is ubiquitous and is, indeed, the commonest type of variation to be found in living organisms. It is shown by all the characters that can be recognized in any organism, the variation sometimes being meristic rather than truly continuous, as for example where we are considering the number of vertebrae an individual possesses. Such meristic variation, though necessarily discrete at the phenotypic level, is most readily interpretable in terms of an underlying, continuously distributed, potential, and hence has been appropriately described as "quasi-continuous" (Grüneberg, 1952). Typically, continuous variation as observed in a population is partly non-heritable and partly heritable, the proportion which is heritable often being termed the "heritability" of the character in that population.

Any single character is subject to both continuous variation and discontinuous of the type seen when single genes of large effect are segregating. Thus, for example, stature in man may display the effects of genes such as that for achondroplasia, the achondroplasic dwarfs being clearly recognizable as abnormal, and at the same time show the continuous variation, as a result of which "normal" individuals differ one from another in stature and which we have known since Galton's time to be very largely heritable. Continuous variation is the variation which makes possible fine adjustment of a character by the action of selection, while yet allowing larger differences to be built up by the accumulation of an appropriate number of small ones. It is in fact the very variation to which Darwin in the main referred adaptative and evolutionary change.

One of the most important properties of the polygenic systems which mediate continuous variation is their capacity for storing, concealed within the genotype, very large amounts of the variation that they are potentially able to produce. Such concealed or latent variation is of course a property of all heterozygotes: thus, for example, Mendel's F_1 peas, though varying but little among themselves, carried concealed within their genotypes the capacity for reproducing the extreme parental differences in F_2. In a randomly breeding population up to half the variation that can be produced by a pair of alleles may be concealed in the heterozygotes, and, if there are k alleles at the locus, up to $(k-1)/k$ of the variation may be so concealed. The potentialities with polygenic variation are however even greater, since in them not only alleles but also non-allelic genes may balance one another's effects and so bring about the concealment of variation. Again with a system of k loci, a proportion up to $(k-1)/k$ of the variation may be concealed by the balancing of non-allelic genes. Thus with a polygenic system comprising 10 loci each with a pair of alleles, up to 50% of the variation may be concealed by heterozygosity and 90% of the remainder by the balancing action of non-allelic genes, i.e. up to 95% of the variation the system is capable of producing is concealed, no more than 5% of it being displayed as observable phenotypic differences between the individuals of the population. With continuous variation, therefore, the variation that is freely observable in the phenotype of a population is but the small tip of a large iceberg, much the greater part of which is hidden below the surface. We may note too that while heterozygosity can conceal variation only in diploid organisms, the balancing action of non-alleles can conceal it in haploid organisms or in chromosomes which are monosomic, such as certain sex-chromosomes in an otherwise diploid species.

The importance of concealed variation appears when we consider the action of selection. In general, unless the environment is changing—and changing moderately rapidly—we should expect the mean expression of a character in a population to be close to the optimal expression required by the environment, for otherwise the expression would shift until this state had been reached. When the mean is near the optimum, individuals with the more extreme expressions of the character will be penalized by natural selection, and the more extreme the expression, the more they will be penalized, as is well brought out by the occurrence of neonatal mortality in relation to birth-weight in man (Karn and Penrose, 1952). Selection is then stabilizing and it will tend to reduce the spread of the frequency distribution of expression round the mean. Many examples of this have been observed, among recent ones being that of sternopleural chaetae in *Drosophila melanogaster* reported by Kearsey and Barnes (1970) and that of the Dog-whelk (*Nucella lapillus*) where Berry and Crothers (1968) were able to relate the strength of the selection to the physical conditions of the environment, which in extreme cases reduced the variation by 90%.

Under such stabilizing selection, departure from the mean is a handicap and variation is thus disadvantageous. But variation is essential if the population is to adjust to the directional selection imposed by the demands of a changing environment. This is where concealed variation is significant. Being concealed, it does not affect the immediate behaviour of the population under stabilizing selection, and a high level of adjustment to the demands of the temporary environment can be achieved. At the same time, in a normally cross-breeding species, the concealed variation is gradually released by segregation and recombination over the generations, and so makes possible change to meet the demands of a changing environment. The changes that can be brought about by directional selection have been extensively investigated in the laboratory by authors too numerous to be listed here. The chief feature of these experiments from our present point of view is the amount by which the character under selection can be altered. A dozen or so generations of selection will commonly see the mean expression of the character pushed well beyond the limits of expression detectable in the original population. The range of expression initially detectable reflects only the free variation of the population, while the ultimate response to directional selection depends on the total variation, including that concealed in the genotypes of the initial population. The great changes achieved in these experiments therefore bear eloquent testimony to both the amount and the importance of concealed variation in populations.

One last point remains to be made. The proportions of the variability that are on the one hand freely expressed and on the other concealed in a population, together with the mechanisms of crossing, segregation and recombination by which concealed variability is released, are themselves also adaptive features of a species, capable of adjustment and readjustment by selection. Thus not only the compromise between response to the demands of the short-term environment and those of the long-term environment, but also the mechanism by which this compromise is reached, are the products of adaptive change. Indeed, it is only in terms of the flow of variability that we can understand the breeding system and chromosome system of a species and the ways in which these systems vary from one species to another.

CONCLUSION

In this brief survey we have seen that populations of living things carry many types of variation, reflecting a variety of kinds of change in the hereditary materials, the significance of which lies in their effects on the fitness of the individuals carrying them. Some of these changes have an immediately and more or less unconditionally adverse effect on fitness and they are found in the population only because of their constant reappearance through recurrent, even if relatively rare, breakdown of the chromosome system or mutation of the genes. In other cases, either because of the selective advantage of heterozygotes or symbiotic relation between unlike types of individual, selection preserves the variation and polymorphism results. In still further cases the variation is continuous and is mediated by polygenic systems which have a great capacity for the concealment of variation through balancing action between both alleles at the same locus, and non-allelic genes. The concealment of variation allows adjustment to meet the conflicting needs of stabilizing selection imposed by the immediate environment and directional selection imposed by longer term changes in environmental circumstances, and the genetical mechanisms by which this variation is controlled are themselves adjusted by selection to the biological needs and circumstances of the species.

The population as we see it, and in particular the variation that it carries and displays, reflects not only the types of variation that can arise but also the action of selection on both the new variants themselves and the system of variation as a whole. Variation therefore is to be understood only in terms of selection, just as the consequences of selection are to be understood only in terms of the variation on which it is acting: variation and selection cannot in fact be divorced in our consideration of the population.

One last point should be made. Apart from man, little reference has been made to mammals in this survey. The information and evidence has been drawn from many species, plants as well as animals, *Drosophila* as well as man. But this reflects only our relative lack of information about mammalian species in the wild. That they will conform to the same principles we cannot doubt. Nor can we doubt that their study will illuminate these principles by displaying their application under the special circumstances of mammalian life.

References

Allison, A. C. (1955). Aspects of polymorphism in man. *Cold Spring Harb. Symp. Quant. Biol.* **20**, 239–255.

Berry, R. J. and Crothers, J. H. (1968). Stabilizing selection in the dog-whelk (*Nucella lapillus*). *J. Zool., Lond.* **155**, 5–17.

Darlington, C. D. (1958). *Evolution of genetic systems*, 2nd edn. Edinburgh: Oliver & Boyd.

Durrant, A. (1962). The environmental induction of heritable change in *Linum*. *Heredity* **17**, 27–61.

Fisher, R. A. (1927). On some objections to mimicry theory: statistical and genetic. *Trans. R. ent. Soc. Lond.* **75**, 269–278.

Fisher, R. A. (1930). *The genetical theory of natural selection.* London: Oxford University Press.

Grüneberg, H. (1952). Genetical studies on the skeleton of the mouse. IV. Quasi-continuous variation. *J. Genet.* **51**, 95–114.

Grüneberg, H. (1964). Genetical research in an area of high natural radioactivity in South India. *Nature, Lond.* **204**, 222–224.

Harris, H. (1966). Enzyme polymorphisms in man. *Proc. R. Soc.* (B.) **164**, 298–310.

Jinks, J. L. and Mather, K. (1955). Stability of development of heterozygotes and homozygotes. *Proc. R. Soc.* (B.) **143**, 561–577.

Karn, M. W. and Penrose, L. S. (1952). Birth weight and gestation time in relation to maternal age, parity and infant survival. *Ann. Eugen.* **16**, 147–164.

Kearsey, M. J. and Barnes, B. W. (1970). Variation for metrical characters in *Drosophila* populations. II. Natural Selection. *Heredity* **25**, 11–21.

Lerner, I. M. (1954). *Genetic homeostasis*. Edinburgh: Oliver & Boyd.

Lush, I. E. (1970). The extent of biochemical variation in mammalian populations. *Symp. zool. Soc. Lond.* No. 26, 43–71.

Mather, K. (1946). The genetical requirements of bio-assay with higher organisms. *Analyst* **71**, 401–411.

Mather, K. (1953). Genetical control of stability in development. *Heredity* **7**, 297–336.

Mather, K. (1955). Polymorphism as an outcome of disruptive selection. *Evolution* **9**, 52–61.

Mather, K. (1969). Selection through competition. *Heredity* **25**, 529–540.

Polani, P. E. (1967). Occurrence and effect of human chromosome abnormalities. *Eugen. Soc. Symp.* **3**, 3–19.

Rees, H. (1961). Genotypic control of chromosome form and behaviour. *Bot. Rev.* **27**, 288–318.

Sved, J. A., Reed, T. E. and Bodmer, W. F. (1967). The number of balanced polymorphisms that can be maintained in a natural population. *Genetics* **55**, 469–481.

CRYPTOMORPHISMS

Symp. zool. Soc. Lond. (1970) No. 26, 43–71.

THE EXTENT OF BIOCHEMICAL VARIATION IN MAMMALIAN POPULATIONS

I. E. LUSH

Department of Biology, Royal Free Hospital School of Medicine, London, England

SYNOPSIS

A wide variety of techniques is used to demonstrate biochemical variation within mammalian species. These include the measurement of enzyme activity, kinetic characteristics, and resistance to heat inactivation. Immunological techniques are sometimes useful, but the most popular routine method of demonstrating variation is electrophoresis, particularly starch gel electrophoresis. Using this technique, 6 out of 18 human erythrocyte enzymes have been shown to be polymorphic. Other mammals are probably no less variable.

The role of selection in maintaining biochemical polymorphisms is uncertain. However, *prima facie* evidence of the action of selection on some polymorphisms in mammals and fish is available.

Biochemical variants can be used as tools for the analysis of the structure of natural populations; they are also sometimes of economic importance.

The electrophoretic mobility of a protein is affected by amino acid substitutions which alter the charge of the protein molecule. The presence of undetected variation due to amino acid substitutions which do not involve an alteration in charge should be borne in mind when interpreting results.

INTRODUCTION AND TECHNIQUES

I have interpreted my function in this Symposium as not to present any new and unpublished information, but rather to summarize some aspects of biochemical variation in mammals so that those who are unfamiliar with the subject will have a general picture in which they can place the contributions of later speakers who will discuss particular examples of biochemical variation in more detail. The definition of biochemical variation meets the same difficulty as the definition of molecular biology, because presumably all biological phenomena are the results of the activities of molecules, and therefore all biology is in that sense molecular biology. Similarly, all genetical variation, whether of coat colour or skeletal morphology, is presumably due to biochemical differences. Nevertheless I shall use the term biochemical variation to mean Mendelian variation in which individuals are classified principally by means of some biochemical measurement, such as the level of a hormone or metabolite or the characteristics of an enzyme or the electrophoretic mobility of a protein.

The earliest example of biochemical variation in mammals, other than man, was described by Fleischmann (1910) who found that some individual rabbits had in their blood an enzyme which could hydrolyse the drug atropine, whereas other rabbits lacked this ability. It has been shown (see Lush, 1966) that there is a simple Mendelian explanation for this variation in terms of 2 alleles at an autosomal locus. One allele, *as*, is a silent allele in that it produces no detectable enzyme. The alternative allele, *As*, determines the presence of the atropinesterase and shows a dosage effect in that the plasmas of homozygous, *As/As*, rabbits have a higher mean enzyme activity than the plasmas of heterozygotes by a factor of between 2 and 3. The variation of plasma atropinesterase in rabbits is now just one of about 25 examples which have since been found of variation of enzyme activity between individuals within mammalian species. Rabbit atropinesterase is an extreme example in that in the *as/as* homozygote the enzyme is totally absent. In other examples one allele may determine a *high* activity and the other allele a *low* activity. Usually the alleles appear to act independently so that the heterozygote has an intermediate activity. Other examples of enzyme variation may involve differences in resistance to inactivation by heat (Walsh, Ericsson and Neurath, 1966) or differences in kinetic characteristics (Huff and Chaykin, 1967). No doubt there are some enzymes which function so critically that almost any genetical alteration of this kind would be harmful or even lethal, but it is clear that there is an unpredictably large number of enzymes which can vary genetically in their activity, etc., in individuals which are apparently unharmed.

Variation in protein structure can sometimes be demonstrated by immunological techniques. The name "allotypes" is given to such variants, and although this approach can in principle be applied to any protein it has been particularly valuable in defining variants of immunoglobulins in the rabbit and the mouse. However, in the last decade starch gel electrophoresis has become by far the most popular technique for demonstrating protein variation. The success of this technique depends upon the fact that at all pH values except its isoelectric pH, each protein carries either a positive or a negative charge. This charge is the net result of the charges on all the individual amino acid residues, some of which are positive, some negative and some uncharged. Because of its net charge, a protein can be made to move through a suitable supporting medium such as a starch gel if a potential gradient is maintained along the gel. It so happens that if even one amino acid residue in a protein is replaced by another of different charge, then the effect on the net charge is usually sufficient to make a relatively large change in the rate at which the protein moves through the gel, and therefore a visible difference to the position of the protein in the gel at the end of the period of electrophoresis. In other words, even the smallest genetical change in the structure of a protein can be detected if it involves amino acid residues of different charge. The value of this technique has been greatly increased by specific staining methods which can make visible the final position of some enzymes in a starch gel so that it is no longer necessary to isolate them in a pure state before electrophoresis. Even crude aqueous extracts can be used with success, and at the present time the wider application of the technique is limited mainly by the need to find specific staining methods for more enzymes. The increasing use of starch gel electrophoresis to demonstrate biochemical variation is shown in Fig. 1 where the total accumulated number of biochemical loci known in all vertebrates except man at the end of each year is compared with the accumulated number of loci which were discovered by techniques other than starch gel electrophoresis. The difference between A and B is the number of loci identified by starch gel electrophoresis and it is easy to see how successful this has been. The corresponding histograms for mammalian loci are very similar in shape to those in Fig. 1, the total number of mammalian loci recorded at the end of 1968 being 210. I have listed the known biochemical variants in mammals in the Appendix.

POPULATION DATA AND THE OCCURRENCE OF SELECTION

With all these rapidly accumulating data at our disposal, the population biologist is entitled to ask for an estimate of what proportion of all the different proteins (including enzymes) in a given species can be expected to be genetically variable. The mammal for which the best estimate is available is man himself. Professor Harry Harris and his colleagues (Harris, 1969) have used starch gel electrophoresis to survey large populations with respect to 18 enzymes (or isoenzymes) which are present in the human erythrocyte. They found that in both Europeans and Negroes, about one third of these 18 erythrocyte enzymes are polymorphic in the sense that the rarer alleles have frequencies of not less than 0·01. They also confirmed that with most, perhaps all, enzymes it is possible to find a few very rare variants in the population, provided the survey is sufficiently large. It must be borne in mind that electrophoresis usually detects variation only when it involves a change in the net charge of a protein. Most amino acid substitutions caused by a single base change in DNA do not involve a change in charge, and therefore the population must contain many more variants (both

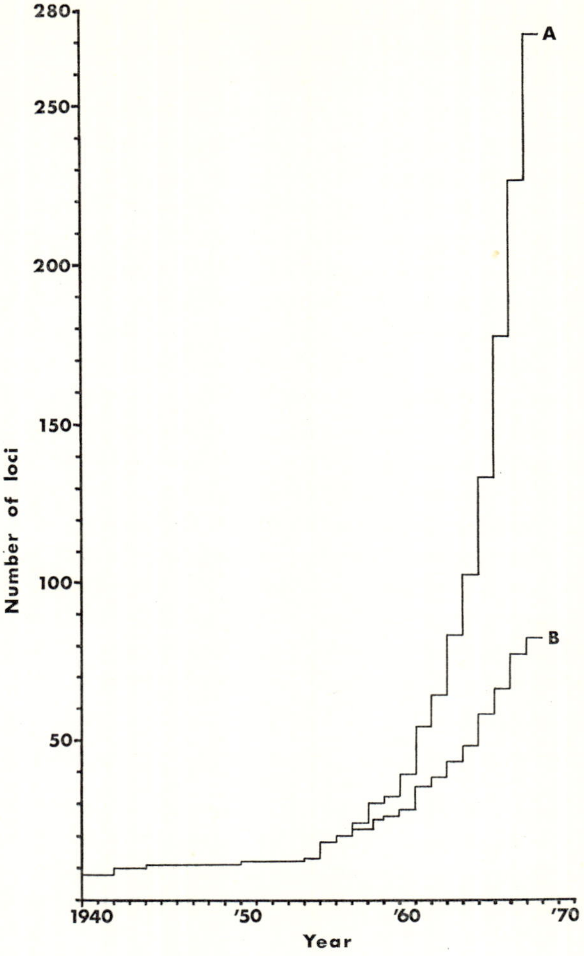

Fig. 1. Accumulated totals of biochemical loci known in vertebrates, except man, at the end of each year from 1940 to 1958. Histogram A refers to all loci. Histogram B to those identified by techniques other than starch gel electrophoresis.

common and rare) than we can at present detect with this technique. A very similar proportion of polymorphic loci has been found in wild populations of mice by Selander, Hunt and Yang (1969).

If the erythrocyte enzymes studied by Harris (1969) are typical of all mammalian enzymes, then it follows that each species probably contains several thousand polymorphisms and every individual within each species may be heterozygous at several hundred loci. At first sight it might seem that if all these polymorphisms were maintained by selection in favour of heterozygotes, then the load of selection against homozygotes would make it impossible for mammalian species, with their small family size, to survive. The fact that they do survive would therefore mean that most of the polymorphisms are selectively neutral. However when it is remembered that the unit which is actually selected by selection is the individual rather than the phenotype at each locus and that it is therefore unrealistic to consider each locus independently, it becomes possible to show that a large number of polymorphisms can in fact be maintained in a population without the necessity of an unrealistic load of selection (Milkman 1967; Sved, Reed and Bodmer, 1967; King, 1967).

An alternative point of view, which has some support from biological observation, is that each polymorphism may have a certain gene frequency at which the polymorphism is selectively neutral. Only when the gene frequency in a population strays from this value will selection come into force to restore the *status quo* (Clarke, 1962). This gene frequency-dependent mechanism is therefore also compatible with the existence of a large number of polymorphisms within a species. Even apart from frequency-dependent selection, it should be remembered that selective forces may change or disappear and that a polymorphism which was originally established by heterozygote advantage may now be selectively neutral. This is presumably particularly true of man, as a result of advances in medicine. The situation at the moment seems to be that, while the existence of widespread biochemical polymorphism in animals has been firmly established, there is still disagreement about the applicability of various theoretical explanations, in particular as to the importance of selection (O'Donald, 1969). It may be worthwhile, therefore, to consider here the kinds of data which could be taken as *prima facie* evidence of the action of selection on a particular polymorphism.

The longer a polymorphism exists the more likely it is to be maintained by selection, in other words the more likely it is to be balanced rather than transient. We cannot examine the protein polymorphisms of fossils, but if we find 2 closely-related species which have the same polymorphism, then it is possible that they have both inherited it from their common ancestral species, which would be evidence of its great age. To be certain that this has happened in a given case it is necessary to establish, firstly, that the biochemical difference which is the basis of the polymorphism is exactly the same in each species and, secondly, that the polymorphism has not arisen in each species independently. The first criterion could in principle be satisfied by detailed biochemical investigation. The best evidence against the possibility that the polymorphism had originated independently in each species would be if the biochemical basis of the polymorphism were found to be so complex and unusual that it seemed unlikely to have arisen more than once. For example the common haptoglobin polymorphism in man, which is due to a particular partial duplication of the haptoglobin α-chain, is probably the result of an unique mutational event (Dixon, 1966). However although this polymorphism is now present throughout the races of mankind (Kirk, 1968) it has not been convincingly demonstrated in any other primate and presumably originated during the early evolution of the human species. A polymorphism which is present in man and also, apparently in many other primate species is the variation in the ability to taste phenylthiourea (PTC). Unfortunately the biochemical basis of this polymorphism is unknown and so one cannot exclude the possibility of its independent origin in each species, although according to Chiarelli (1963) it is present in an astonishingly large number of different primates. Polymorphism for some of the allotypic variants of the human immunoglobulin Gm polymorphism is also found in the chimpanzee and orangutan (see Appendix) but the available evidence suggests that the biochemical difference between the variants may be small, and therefore could be of independent origin in the different species (Wang and Fudenberg, 1969). Similarly the vasopressin variants in different species of Suiformes (see Appendix) differ by only a single amino acid replacement (Table I) and could also be of independent origin. It is of interest that the same vasopressin polymorphism may occur in the mouse, *Mus musculus* (Stewart, 1968).

Selection is presumed to be at work in those polymorphisms which show differences in gene frequency related to differences in habitat. For example, domestic sheep are polymorphic with respect to both haemoglobin and also erythrocyte potassium concentration (see Appendix). There are 2 alleles at the haemoglobin locus, Hb^{A} and Hb^{B}, and 2 alleles at the locus which controls erythrocyte potassium concentration, K^{H} and K^{L}. Evans, Harris and Warren (1958) have shown that in Britain the Hb^{A} and K^{H} alleles are regularly found at higher frequencies in hill breeds of sheep which inhabit exposed and mountainous terrain than in lowland breeds which live under more sheltered conditions (Fig. 2). The physiological characteristics of the different phenotypes which may have led to this distribution are not known, but with regard to haemoglobin it may be significant that when sheep which possess the Hb^{A} allele become anaemic, they are able to synthesize an additional form of haemoglobin, HbC, which is never found in sheep homozygous for Hb^{B}. The kind of correlation between gene frequency and habitat which may soon become commonplace in

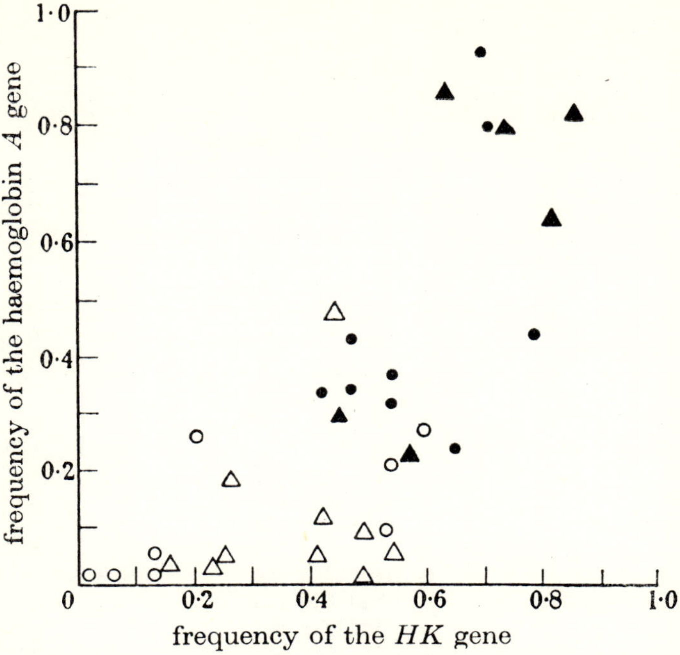

FIG. 2. Correlation between frequency of $K^H(HK)$ allele and frequency of haemoglobin *A* allele in 33 different British breeds of sheep.
○, Lowland Longwool breeds; △, Lowland Shortwool breeds; ●, Hill breeds; ▲, Blackface Hill breeds (from Evans, Harris and Warren, 1958).

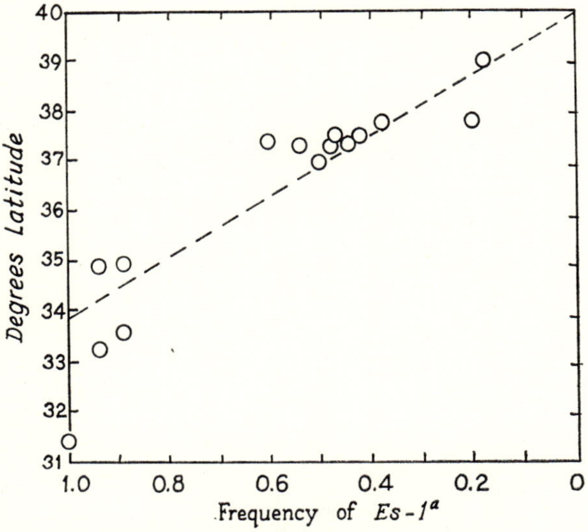

FIG. 3. Latitudinal distribution of the frequency of the esterase allele $Es\text{-}1^a$ in 15 populations of *Catostomus* (*P.*) *clarki*, in the tributaries of the lower Colorado River basin, U.S.A. (redrawn from Koehn and Rasmussen, 1967).

TABLE I

Differences in primary structure between variants of twelve polymorphic proteins in mammals

Within each polypeptide the amino acids listed in each column probably occupy the same position in the chain. Where the position is numbered it is counted from the N-terminal end.

Genetic variants	Amino acid replacements	References
Cattle		
Haemoglobin β-chain		128
A	Gly Lys Lys	
B	Ser His Asn	
position in chain:	15 18 119	
Pancreatic carboxypeptidase A		107
Val-type	Ile Ala Val	
Leu-type	Val Glu Leu	
position in chain:	179 228 305	
β-lactoglobulin		22, 89
A	Asp Val Gln Glu	
B	Gly Ala Gln Glu	
C	Gly Ala His Glu	
D	Gly Ala Gln Gln	
α_{S1}-casein		81, 146
A*	Glu Ser Lys	
B	Glu Ser Asn	
C	Gly Ser Asn	
D	Glu Pro Asn	
β-casein		81
A	Glu Gln	
B	Glu Arg	
C	Lys Gln	
κ-casein		81
A	Asp Thr	
B	Ala Ile	
Sheep		
Haemoglobin α-chain		68
A	Gly	
B	Asp	
position in chain:	15	
Haemoglobin β-chain		10, 16, 32, 155
A	Ser Ala Val Glu Ser Glx Arg	
B	Asx Pro Met Lys Asn Asp Lys	
position in chain:	49 57 74 75 119 128 143	

Genetic variants	Amino acid replacements	References
Goat		
Haemoglobin α-chain		67
	Asp	
	Tyr	
position in chain:	75	
Rabbit		
Haemoglobin α-chain		69, 133
Val-type	Val Phe Thr	
Leu-type	Leu Leu Ser	
position in chain:	29 48 49	
Mouse		
Haemoglobin α-chain		60
C57BL	Gly Val Asn	
NB	Val Ile Ser	
BALB/c	Gly Val Ser	
	Gly Val Thr	
C3HB	Gly Val Ser	
	Gly Val Asn	
C3H	Val Ile Ser	
	Gly Val Asn	
position in chain:	25 62 68	
Orangutan		24
Haemoglobin α-chain		
A	Asp	
B	Gly	
Warthog, and other Suiformes (see Appendix)		38, 39
Vasopressin		
Lys-type	Lys	
Arg-type	Arg	
position in chain:	8	

*Variant A of α_{s1}-casein differs from the other 3 variants in that it contains 1 less arginine, alanine and valine, 2 less phenylalanines and 3 less leucines.

mammalian ecology is well shown in the data of Koehn and Rasmussen (1967) on the freshwater fish *Catostomus clarki*. Populations of this species in the tributaries of the lower Colorado River basin, U.S.A., were sampled over a wide area, and it was found that the allele frequencies of one particular serum esterase polymorphism showed a steady change (i.e. a cline) correlated with the latitude of the population sampled (Fig. 3). Many of these populations are genetically separate, and it seems clear that there is a selective influence which varies with latitude and which determines the optimum gene frequency in each locality. Koehn (1969) has shown that the selective influence is probably environmental temperature. He found that the *Es*-1*b* variant, which is common in the northern part of the area, is much more active at low temperatures than the *Es*-1*a* variant, which is common in the southern part of the species' habitat.

Another way to look for evidence of the action of selection on a biochemical polymorphism is to see if the allele frequencies alter in a systematic way as the population ages. The only mammalian example known to me is an unconfirmed report by Osterhoff (1964) that in South African dairy cattle the transferrin allele Tf^{A} becomes more frequent in older animals at the expense of the alternative allele, Tf^{B}. Fujino and Kang (1968), working on a fish, the Hawaiian skipjack tuna (*Katsuwonus pelamis*), found that the transferrin is polymorphic with two common alleles, $T_{SJ}{}^{2}$ and $T_{SJ}{}^{3}$ which determine the homozygous phenotypes "2" and "3" and the heterozygous phenotype "2–3". Fujino and Kang (1968) found that, in the general population, heterozygotes occurred more frequently than would be expected on the basis of the Hardy-Weinberg distribution. This effect was greatest in young fish and gradually decreased in older age groups (Fig. 4). It seems that there must be selection in favour of heterozygotes at some very early stage in the life of the tuna and that this is counteracted by selection against heterozygotes in the adult years. Possibly the initial selection occurs at the stage of fertilization, but whatever the mechanisms may be the total result is no change in the allele frequencies.

The action of selection can be seen if the allele frequencies of a polymorphism fluctuate in a regular way with the seasons of the year. Berry (1970) found this in his work on the mice of Skokholm. The density of individuals in a population may also be an important selective influence in some mammals. For example, Semeonoff and Robertson (1967) found a serum esterase polymorphism in populations of the field vole (*Microtus agrestis*) which inhabit the Carron Valley in Stirlingshire, Scotland. One allele, which is dominant, determines the presence of the esterase and the other allele is silent and produces no esterase. The evidence seems to show that during a normal winter there is a decrease in the proportion of esterase-negative voles in each population. However, this species is peculiarly prone to an occasional build up of numbers followed by a catastrophic fall, or crash. When this happened in the winter of 1963/64 the proportion of esterase-negative individuals increased sharply, indicating that under these conditions this phenotype may be at an advantage. It may be that seasonal and density-

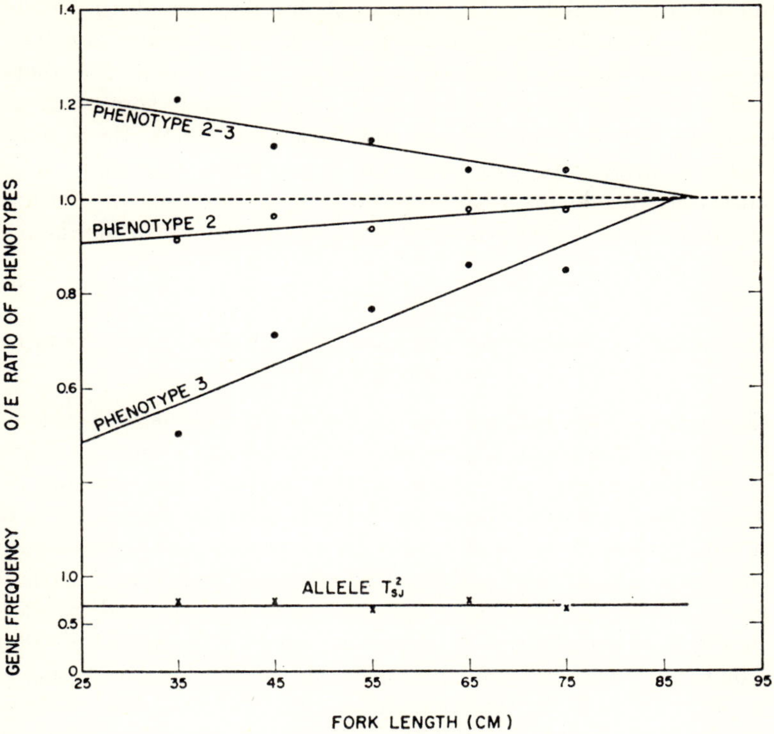

FIG. 4. Relationship between transferrin phenotype and fork length in Hawaiian skipjack tuna. Fork length (from snout to tail fork) is a measure of age. The ratios of the observed numbers of each phenotype to the numbers expected on the basis of the Hardy-Weinberg distribution (O/E ratios) can be seen to approach unity in the older fish (from Fujino and Kang, 1968).

dependent selection are 2 opposing factors which act together to keep this polymorphism in existence.

THE USE OF VARIANTS AS BIOLOGICAL TOOLS

Apart from their intrinsic interest, one can regard biochemical polymorphisms as convenient tools for the analysis of the breeding structure of mammalian populations. The question whether 2 neighbouring populations do in fact interbreed can be answered in the negative if their allele frequencies for several polymorphisms are very different. Some work along these lines is being done with macaque monkeys in south-east Asia (Goodman, 1968) and chimpanzees in Africa (Goodman and Tashian, 1969). Naevdal (1965) has compared the frequency of transferrin variants in 3 populations of the harp seal (*Pagophilus groenlandicus*) although with inconclusive results. The process of the drifting apart of populations in terms of the allele frequencies at their polymorphic loci is stimulating some interesting theoretical work (Cavalli-Sforza and Edwards, 1967). Economic considerations enter into the work on biochemical variation in cattle because polymorphisms with suitable gene frequencies can be used to check on the parentage of calves when the identity of the sire is in some doubt.

DIFFERENCES BETWEEN PROTEIN VARIANTS AT THE LEVEL OF PRIMARY STRUCTURE

One aspect of the subject which remains to be discussed is the extent of the primary structural differences between protein variants. Thirteen different polymorphic proteins in mammals have been studied at this level and the results are shown in Table I. Some of the variants differ from each other by a single amino acid replacement. For example, bovine β-lactoglobulins C and D can be derived from the B variant by single replacements, and α_{SI}-caseins C and D can be derived from α_{SI}-casein B in the same way. However, some of the variants differ by 2, 3, or even more replacements, as in the cattle k-caseins and carboxypeptidases and the sheep haemoglobin β-chains. The differences between rabbit immunoglobulins of different allotypes also involve several replacements, but the published analyses are not yet complete (Wilkinson, 1969). α_{SI}-casein A is a rare variant which is unusual in that it seems to be derived from α_{SI}-casein B by one replacement and the total loss of a block of eight other amino acids. A rare bovine β-lactoglobulin variant named β-lactoglobulin Droughtmaster has been found in the Droughtmaster breed of cattle by Bell, McKenzie and Murphy (1966). This variant has an amino acid composition identical to that of β-lactoglobulin A, but also contains 9·8% covalently linked carbohydrate (sialic acid, N-acetyl hexosamine and hexoses). The mouse haemoglobin α-chain locus is duplicated in some strains and genetical variation occurs at both α-chain loci. The goat haemoglobin α-chain locus is also duplicated, but in this species variation has so far been found at only one of the two loci.

I think it should be realized that the differences between protein variants are likely to be less straightforward than they seem to be in Table I. Let us consider the hypothetical polypeptide shown under stage I in Fig. 5 and restrict our attention to 3 positions in this polypeptide at which the amino acids a, p, and x are originally present

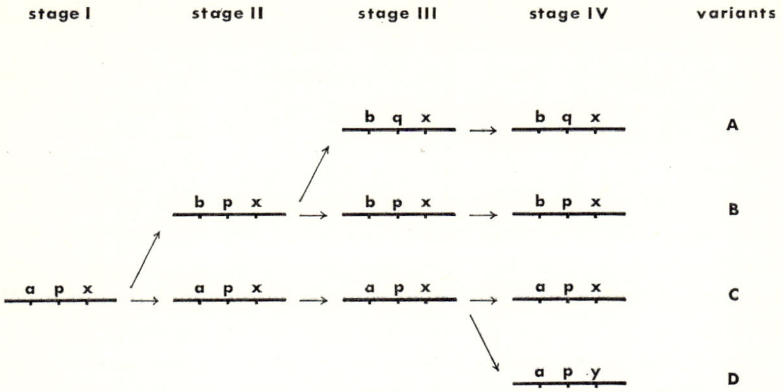

FIG. 5. Stages in the divergence of genetic variants of a polypeptide as a result of amino acid replacements at 3 different positions.

throughout a population of animals. Let us suppose that a point mutation gives rise to a variant polypeptide, in which *a* is replaced by *b* (stage II) thus giving rise to a polymorphism in the population in which the variants differ by a single amino acid replacement. Several examples of this are shown in Table I. Now if another replacement (*p* to *q*) arises at a different position in one of the variants a more complex polymorphism could be established (stage III) consisting of 3 variants, A, B, and C, of the kind seen in the bovine β-caseins. If at this stage variant B were lost from the population, either by selection or chance or by a combination of both, the polymorphism would consist of 2 polypeptides which differed by 2 amino acid replacements, as in the bovine κ-caseins. However if variant B remained in existence and a third replacement (*x* to *y*) occurred in one of the polypeptides, a polymorphism such as that shown at stage IV could arise. At this stage the loss of variants B and C would leave 2 variants, A and D, which differ by 3 replacements, as do the bovine carboxypeptidases. Continuation of this process through more stages would give rise to 2 variants which differed by more replacements, as is found in the sheep haemoglobin β-chain polymorphism.

It may seem that hypothetically losing unwanted variants in this rather carefree way is not likely to reflect what actually happens in real populations. Nevertheless the loss of variants by chance is certainly made more likely by the artificial population structure which is a feature of farm animals such as cattle and sheep. For example the use of only a small number of bulls as sires for the artificial insemination of very large numbers of cows must lead to large fluctuations in the frequencies of polymorphic variants in cattle, and this could bring about the temporary loss of a variant from some herds. Similarly the genetical bottleneck represented by the small foundation population of a new breed might by chance lack some variants of a particular polymorphic protein and consequently those variants would be permanently absent from that breed.

The reason why Table I may present a somewhat simplified version of the extent of polymorphism in those proteins is that electrophoretic mobility was in fact the only criterion used to identify the majority of the genetic variants listed there. Electrophoresis does not usually detect variants which differ by amino acid replacements which do not involve a difference of charge. To illustrate this point let us suppose that in Fig. 5 the replacement of *a* by *b* is the only one which involves a change of charge. At stage III the variants A and B would then be indistinguishable from each other and there would appear to be only 2 variants. However the chemical analysis of the A + B "variant" would give variable results for the amino acid at the position occupied by *p* or *q*, and the results would depend on the proportion of A and B in the sample analysed. To use a sample from one animal would be no way out of the difficulty because it might be an *AB* heterozygote. Even if, by some happy chance, the animal were, say, an *A* homozygote the next sample might come from a *B* homozygote. The same difficulties, only worse, would attend the amino acid analysis of a stage IV polymorphism in which electrophoresis would still only demonstrate the 2 "variants" A + B and C + D. Still more confusion might be introduced by intra-allelic crossing-over which could theoretically produce the other 2 possible variants not shown in Fig. 5, i.e. *b p y* and *a q x*.

With all the above considerations in mind, it is not difficult to imagine a situation in which 2 variants which are defined solely by the criterion of electrophoresis differ in a consistent and repeatable way with regard to amino acid replacements at some positions, but in an inconsistent and ambiguous way at other positions. This may appear to be conjuring up difficulties where none in fact has been found, but these considerations should be borne in mind when independent analyses of the same electrophoretic variants are found to disagree (Wilson *et al.*, 1966; Boyer *et al.*, 1967).

The same kind of difficulties that apply to the use of electrophoresis as the sole technique for defining certain protein variants also apply to any other partial criterion used by itself, for example allotypy. If the allotypic classification of, for example, the immunoglobulin heavy chains of the rabbit depends on amino acid replacements at certain antigenically important positions of the molecule, then replacements at other positions will be undetected and may cause confusion when complete amino acid analyses of the allotypes are attempted (Wilkinson, 1969). The only solution to the problem of the *definition* of polypeptide variants in animals which cannot be highly inbred is to define each variant initially as one of the variants which are produced by a particular heterozygous individual. For example, in a stage III polymorphism in Fig. 5 in which the variants are initially typed only by electrophoresis, one would take as a reference point a particular animal which is demonstrably a heterozygote. The C variant would be *defined* as the polypeptide produced by the *C* allele present in that particular animal or in its heterozygous descendants. The A+B "variant" would be similarly defined until such time as the p/q variation was discovered, then *A* and *B* alleles would be transmitted to an *AC* and *BC* heterozygote respectively and the A and B variants *defined* as the variants present in these individuals or in their *AC* or *BC* descendants. By keeping each reference allele and its descendants always in the heterozygous state, different laboratories can be certain of dealing with alleles which are identical by descent, although the remote possibility of a very rare change through intra-allelic crossing-over or mutation should not be overlooked. These genetic precautions may be inconvenient and expensive but in many cases they may prove to be unavoidable.

References

1. Adams, H. R., Wrightstone, R. N., Miller, A. and Huisman, T. H. J. (1969). Quantitation of haemoglobin α-chains in adult and fetal goats; Gene duplication and the production of polypeptide chains. *Arch. Biochem. Biophys.* **132**, 223–236.
2. Albers, J. J. and Dray, S. (1969). Identification and genetic control of 2 new low-density lipoprotein allotypes: phenotypes at the *Lpq* locus. *J. Immunol.* **103**, 155–162.
3. Alepa, F. P. (1968). Antigenic factors characteristic of human immunoglobulin G detected in the sera of non-human primates. *Primates in Medicine*, **1**, 1–9.
4. Ashton, G. C. and Carr, W. R. (1965). Serum transferrins in some African antelopes. *Rhod. Zamb. Malaw. J. Agr. Res.* **3**, 109–111.
5. Augustinsson, K-B. and Henricson, B. (1966). A genetically controlled esterase in rat plasma. *Biochim. biophys. Acta* **124**, 323–331.
6. Barnicot, N. A., Jolly, C. J., Huehns, E. R. and Dance, N. (1965). Red cell and serum protein variants in baboons. In *The baboon in medical research.* 1–15. H. Vagtborg (ed.). Univ. Tenes. Press.
7. Barnicot, N. A., Huehns, E. R. and Jolly, C. J. (1966). Biochemical studies on haemoglobin variants of the irus macaque. *Proc. R. Soc.* (B), **165**, 224–244.
8. Barnicot, N. A. and Jolly, C. J. (1966). Haemoglobin polymorphism in the Orangutan, an animal with four major haemoglobins. *Nature, Lond.*, **210**, 640–642.

9. Beale, D. (1966). Differences in amino acid sequence between sheep haemoglobins A and B. *Biochim. biophys. Acta* **127**, 239–241.
10. Beale, D. (1967). A partial amino acid sequence for sheep haemoglobin A. *Biochem. J.* **103**, 129–140.
11. Bell, K. and McKenzie, H. A. (1967). The whey proteins of ovine milk: β-lactoglobulins A and B. *Biochim. biophys. Acta* **147**, 123–134.
12. Bell, K., McKenzie, H. A. and Murphy, W. H. (1966). Isolation and properties of β-lactoglobulin Droughtmaster. *Aust. J. Sci.* **29**, 87.
13. Ben-Efraim, S. and Liacopoulos, P. (1969). The competitive effect of DNP-poly-L lysine in responder and non-responder guinea pigs. *Immunology*, **16**, 573–580.
14. Berry, R. J. (1970). Covert and overt variation, as exemplified by British mouse populations. *Symp. zool. Soc. Lond.* No. 26, 3–26.
15. Biddle, F. G. and Petras, M. L. (1967). The inheritance of a non-haemoglobin erythrocyte protein in *Mus musculus*. *Genetics, Austin* **57**, 943–949.
16. Boyer, S. H., Hathaway, P., Pascasio, F., Bordley, J., Orton, C. and Naughton, M. A. (1967). Differences in the amino acid sequences of tryptic peptides from 3 sheep haemoglobin β-chains. *J. biol. Chem.* **242**, 2211-2232.
17. Braend, M. (1966). Serum transferrins of dogs. *Proc. Xth Conf. Europ. Soc. Blood Group Research*, 319–322.
18. Braend, M. (1968). Genetic variation of horse haemoglobin. *Hereditas* **58**, 385–392.
19. Brdicka, R. (1966). Genetics of the rat haemoglobin. *Proc. Xth Conf. Europ. Soc. Blood Group Research*, 407–411.
20. Brew, K., Vanaman, T. C. and Hill, R. L. (1968). The role of α-lactalbumin and the A protein in lactose synthetase: a unique mechanism for the control of a biological reaction. *Proc. natn. Acad. Soc. U.S.A.* **59**, 491–497.
21. Brewer, G. J., Eaton, J. W., Knutsen, C. S. and Beck, C. C. (1967). A starch gel electrophoretic method for the study of diaphorase isoenzymes and preliminary results with sheep and human erythrocytes. *Biochem. biophys. Res. Commun.* **29**, 198–204.
22. Brignon, G., Ribadeau-Dumas, B., Garnier, J., Pantaloni, D., Guinand, S. and Jay, J. (1969). Chemical and physico-chemical characterization of genetic variant D of bovine β-lactoglobulin. *Arch. Biochem. Biophys.* **129**, 720–727.
23. Buettner-Janusch, J. (1963). Hemoglobins and transferrins in baboons. *Folia primatol.* **1**, 73–87.
24. Buettner-Janusch, J., Buettner-Janusch, V. and Mason, G. A. (1969). Amino acid compositions and amino-terminal end groups of α and β chains from polymorphic haemoglobins of *Pongo pygmaeus*. *Arch. Biochem. biophys.* **133**, 164–170.
25. Cattanach, B. M. and Perez, J. N. (1969). A genetically determined variant of the A-subunit of lactic dehydrogenase in the deer mouse. *Biochem. Genet.* **3**, 499–506.
26. Cavalli-Sforza, L. L. and Edwards, A. W. F. (1967). Phylogenetic analysis: models and estimation procedures. *Evolution, Lancaster, Pa.* **21**, 550–570.
27. Chiao, J. W. and Dray, S. (1969). Identification and genetic control of rabbit haptoglobin allotypes. *Biochem. Genet.* **3**, 1–13.
28. Chiarelli, B. (1963). Sensitivity to PTC (phenylthiocarbamide) in primates. *Folia primatol.* **1**, 88–94.

29. Clarke, B. (1962). Balanced polymorphism and the diversity of sympatric species. *Syst. Ass. Publs* No. 4, 47–70.
30. Conway, T. P., Dray, S. and Lichter, E. A. (1969). Identification and genetic control of 3 rabbit γA immunoglobin allotypes. *J. Immunol.* **102**, 544–554.
31. Crawford, M. H. (1966). Haemoglobin polymorphism in *Macaca nemestrina*. *Science, N.Y.*, **154**, 398–399.
32. Dayhoff, M. O. and Eck, R. V. (1968). *Atlas of protein sequence and structure.* U.S.A. Nat. Biomed. Res. Fdn.
33. DeLorenzo, R. I. and Ruddle, F. H. (1969). Genetic control of two electrophoretic variants of glucosephosphate isomerase in the mouse, (*Mus musculus*). *Biochem. Genet.* **3**, 151–204.
34. Dixon, G. H. (1966). Mechanisms of protein evolution. In *Essays Biochem.* **2**, 147–204.
35. Ellory, J. C. and Tucker, E. M. (1969). Stimulation of the potassium transport system in low potassium type sheep red cells by a specific antigen antibody reaction. *Nature, Lond.* **222**, 477–478.
36. Evans, J. V., Harris, H. and Warren, F. L. (1958). The distribution of hemoglobin and blood potassium types in British breeds of sheep. *Proc. R. Soc.* (B.), **148**, 249–262.
37. Feinstein, R. N., Howard, J. B., Braun, J. T. and Seaholm, J. E. (1966). Acatalasemic and hypocatalasemic mouse mutants. *Genetics, Austin* **53**, 923–933.
38. Ferguson, D. R. (1969). The genetic distribution of vasopressin in the Peccary (*Tayassu angulatus*) and Warthog (*Phacochoerus aethiopicus*). *Gen. Comp. Endocrin.* **12**, 609–613.
39. Ferguson, D. R. and Heller, A. (1965). Distribution of neurohypophysial hormones in mammals. *J. Physiol.* **180**, 846–863.
40. Finlayson, J. S., Mushinsky, J. F., Hudson, D. M. and Potter, M. (1968). Components of the major urinary protein complex of inbred mice: separation and peptide mapping. *Biochem. Genet.* **2**, 127–140.
41. Fleischmann, P. (1910). Atropinengiftung durch Blut. *Arch. exp. Path. Pharmak.* **62**, 518–526.
42. Foreman, C. W. (1966). Inheritance of multiple haemoglobins in *Peromyscus*. *Genetics, Austin* **54**, 1007–1012.
43. Fujino, K. and Kang, T. (1968). Transferrin groups of tunas. *Genetics, Austin*, **59**, 79–91.
44. Gahne, B. (1966). Studies on the inheritance of electrophoretic forms of transferrins, albumins, prealbumins and plasma esterases of horses. *Genetics, Austin* **53**, 681–684.
45. Gahne, B. (1967). Alkaline phosphatase isoenzymes in serum, seminal plasma and tissues of cattle. *Hereditas* **57**, 100–114.
46. Ganschow, R. and Paigen, K. (1967). Separate genes determining the structure and intracellular location of hepatic glucuronidase. *Proc. natn. Acad. Sci. U.S.A.* **58**, 938–945.
47. Ganschow, R. E. and Schimke, R. T. (1969). Independent genetic control of the catalytic activity and the rate of degradation of catalase in mice. *J. biol. Chem.* **244**, 4649–4658.
48. Garrick, M. D. and Charlton, J. P. (1969). Inheritance of structural alleles for goat haemoglobins: site duplication and limited structural divergence for the alpha-chain locus. *Biochem. Genet.* **3**, 393–402.

49. Gasparski, J. and Stevens, R. W. C. (1968). Bovine serum amylase isozymes in several breeds of domestic cattle. *Can. J. Genet. Cytol.* **10**, 148–151.
50. Gilman, J. G. and Smithies, O. (1968). Fetal hemoglobin variants in mice. *Science N.Y.*, **160**, 885–886.
51. Glasnak, V. (1968). Inter- and intra-specific differences in milk proteins of cattle and swine. *Comp. Biochem. Physiol.* **25**, 355–357.
52. Gluecksohn-Waelsh, S., Greengard, P., Quinn, G. P. and Teicher, L. S. (1967). Genetic variations of an oxidase in mammals. *J. biol. Chem.* **242**, 1271–1273.
53. Goodman, M. (1968). Evolution of the catarrhine primates at the macromolecular level. *Primates in Med.* **1**, 10–26.
54. Goodman, M. and Poulik, E. (1961). Serum transferrins in the genus *Macaca:* species distribution of 19 phenotypes. *Nature, Lond.* **191**, 1407–1408.
55. Goodman, M. and Tashian, R. E. (1969). A geographic variation in the serum transferrin and red cell phosphoglucomutase polymorphism of chimpanzees. *Human Biol.* **41**, 237–249.
56. Groves, M. L. and Kiddy, C. A. (1968). Polymorphism of γ-casein in cow's milk. *Arch. Biochem. Biophys.* **126**, 188–193.
57. Grunder, A. A. (1966). Inheritance of a heme-binding protein in rabbits. *Genetics, Austin* **54**, 1085–1093.
58. Harris, H. (1969). Enzyme and protein polymorphism in human populations. *Br. Med. Bull.* **25**, 5–13.
59. Heston, W. E., Hoffman, H. A. and Rechcigl, M. (1965). Genetic analysis of liver catalase activity in 2 substrains of C57BL mice. *Genet. Res.* **6**, 387–397.
60. Hilse, K. and Popp, R. A. (1968). Gene duplication as the basis for amino acid ambiguity in the alpha-chain polypeptides of mouse hemoglobins. *Proc. natn. Acad. Sci. U.S.A.* **61**, 930–936.
61. Hof, J-Opt (1969). Isoenzymes and population genetics of sorbit dehydrogenase, (E.C.1.1.1.14) in swine (*Sus scrofa*). *Humangenetik* **7**, 258–259.
62. Hoffman, H. A. and Gotleib, A. J. (1968). Hemoglobins of chimpanzees and gibbons. *Primates in Med.* **1**, 27–34.
63. Hooton, B. J. and Watts, D. C. (1966). Adenosine 5′-triphosphate-creatine phosphotransferase from dystrophic mouse skeletal muscle. A genetic lesion associated with the catalytic thiol group. *Biochem. J.* **100**, 637–646.
64. Hope, R. M. and Godfrey, G. K. (1968). Transferrin polymorphism in the Australian marsupial mouse *Sminthopsis crassicaudata* (Gould). *Aust. J. biol. Sci.* **21**, 587–591.
65. Huff, S. D. and Chaykin, S. (1967). Genetic and androgenic control of N-methylnicotinamide oxidase activity in mice. *J. biol. Chem.* **242**, 1265–1270.
66. Huisman, T. H. J., Dasher, G. A., Moretz, W. H., Dozy, A. M. and Wilson, J. B. (1968). Studies of hemoglobin types in Barbary sheep (*Ammotragus lervia*). *Biochem. J.* **107**, 745–751.
67. Huisman, T. H. J., Brandt, G. and Wilson, J. B. (1968). The structure of goat hemoglobins II. Structural studies of the α-chains of the hemoglobins A and B. *J. biol. Chem.* **243**, 3675–3686.
68. Huisman, T. H. J., Dozy, A. M., Wilson, J. B., Efremov, G. D. and Vasko, B. (1968). Sheep hemoglobin D, an α-chain variant with one apparent amino acid substitution (α15 Gly$\rightarrow$Asp). *Biochim. biophys. Acta* **160**, 467–469.

69. Hunter, T. and Monro, A. (1969). Allelic variants in the amino acid sequence of the α-chain of rabbit haemoglobin. *Nature, Lond.* **223**, 1270–1272.
70. Kelus, A. S. (1969). Allotypic marker $GP\gamma_2-1$ of guinea pig immunoglobulins. *Nature, Lond.* **223**, 398–399.
71. King, J. L. (1967). Continuously distributed factors affecting fitness. *Genetics, Austin* **55**, 483–492.
72. King, J. W. B. (1966). The caseins of sheep's milk. *Proc. Xth Conf. Europ. Soc. Blood Group Research.* 427–431.
73. King, J. W. B. (1969). The distribution of sheep β-lactoglobulins. *Anim. Prod.* **11**, 53–58.
74. Kirk, P. L. (1968). *The haptoglobin groups in man.* Basel: Karger.
75. Kirsch, J. A. W. and Poole, W. E. (1967). Serological evidence for speciation in the grey kangaroo, *Macropus giganteus*, Shaw 1790 (Marsupialia: Macropodidae). *Nature, Lond.* **215**, 1097–1098.
76. Kitchen, H., Putnam, F. W. and W. J. Taylor (1967). Haemoglobin polymorphism in white-tailed deer: subunit basis. *Blood* **29**, 867–877.
77. Knight, K. L. and Dray, S. (1968). Identification and genetic control of two rabbit α_2-macroglobulin allotypes. *Biochemistry* **7**, 1165–1171.
78. Knight, K. L., Valenzuela, L. and Dray, S. (1969). Allotypes and isoenzymes of rabbit alpha-1 aryl esterase. *Fedn. Proc. Fdn Am. Socs. exp. Biol.* **28(2)**, 435.
79. Koehn, R. K. (1969). Esterase heterogeneity dynamics of a polymorphism. *Science, N.Y.* **163**, 943–944.
80. Koehn, R. K. and Rasmussen, D. I. (1967). Polymorphic and monomorphic serum esterase heterogeneity in catostomid fish populations. *Biochem. Genet.* **1**, 131–144.
81. Koning, P. J. de (1967). *Studies on rennin and the genetic variants of casein.* Wageningen: H. Veenman & Zonen, N.V.
82. Kristjansson, F. K. (1966). Fractionation of serum albumin and genetic control of two albumin fractions in pigs. *Genetics, Austin* **4**, 675–679.
83. Lai, L. Y. C. (1967). Polymorphism in red cell acid phosphatase of *Macaca iris*. *Acta genet. Statist. med.* **17**, 104–111.
84. Lange, V. and Schmitt, J. (1963). Das Serumeiweissbild der Primaten unter besonderer Berucksichtigung der Haptoglobine und Transferine. *Folia primatol.* **1**, 208–250.
85. Laris, P. C. (1967). Variation in fructose transport among erythrocytes. *J. cell. Physiol.* **70**, 1–6.
86. Lewis, W. H. P. and Truslove, G. M. (1969). Electrophoretic heterogeneity of mouse erythrocyte peptidases. *Biochem. Genet.* **3**, 493–498.
87. Lieberman, G., Edelstein, G., Dannis, I. M., Datta, S. P. and Battisto, J. R. (1964). Studies on the guinea pig serum factor that detects naturally occurring delayed iso-hypersensitivity. *Ann. N.Y. Acad. Sci.* **121**, 490–493.
88. Loghem, E. van, Shuster, J. and Fudenberg, H. H. (1968). Gm factors in non-human primates. *Vox Sang.* **14**, 81–94.
89. Lush, I. E. (1966). *The biochemical genetics of vertebrates except man.* Amsterdam: North-Holland Publ. Co.
90. Lyon, J. B., Porter, J. and Robertson, M. (1967). Phosphorylase b kinase inheritance in mice. *Science, N.Y.* **155**, 1550–1551.
91. Manwell, C. and Kerst, K. V. (1 966). Possibilities of biochemical taxonomy of bats using hemoglobin, la ctate dehyrogenase, esterases and other proteins. *Comp. Biochem. Physi ol.* **17**, 741–754.

92. Maughan, E. and Williams, J. R. B. (1967). Haemoglobin types in Deer. *Nature, Lond.* **215**, 404–405.

93. Maurer, F. W. (1967). Heritability of the plasma transferrin protein in three species of *Microtus*. *Nature, Lond.* **215**, 95–96.

94. Milkman, R. D. (1967). Heterosis as a major cause of heterozygosity in nature. *Genetics, Austin* **55**, 493–495.

95. Minna, J. D., Iverson, G. M. and Herzenberg, L. A. (1967). Identification of a gene locus for γG_1 immunoglobulin H chains and its linkage to the H chain chromosome region in the mouse. *Proc. natn. Acad. Sci. U.S.A.* **58**, 188–194.

96. Moriwaki, K., Tsuchiya, K. and Yosida, T. H. (1969). Genetic polymorphism in the serum transferrin of *Rattus rattus*. *Genetics, Austin* **63**, 193–199.

97. Nadler, C. F. (1968). The serum proteins and transferrins of the ground squirrel subgenus *Spermophilus*. *Comp. Biochem. Physiol.* **27**, 487–503.

98. Naevdal, G. (1965). Protein polymorphism used for identification of harp seal populations. *Arbok Univ. Bergen* (Mat-Nat.) **1965**(a): 1–20.

99. Neice, R. L. and Kracht, D. W. (1967). Genetics of transferrin in burros (*Equus asinus*). *Genetics, Austin* **57**, 837–841.

100. Nute, P. E. and Buettner-Janusch, J. (1969). Genetics of polymorphic transferrins in the genus *Lemur*. *Folia primatol.* **10**, 181–194.

101. Nute, P. E., Buettner-Janusch, V. and Buettner-Janusch, J. (1969). Genetics and biochemical studies of transferrins and haemoglobins of *Galago*. *Folia primatol.* **10**, 276–287.

102. O'Donald, P. (1969). "Haldane's Dilemma" and the rate of natural selection. *Nature, Lond.* **221**, 815–816.

103. Osterhoff, D. R. (1964). Recent research on biochemical polymorphism in livestock. *Jl S. Afr. vet. med. Ass.* **35**, 363–380.

104. Osterhoff, D. R. (1966). Haemoglobin, transferrin and albumin types in Equidae (Horses, mules, donkeys and zebras). *Proc. Xth Conf. Europ. Soc. Blood Group Research*, 345–351.

105. Parr, C. W. (1966). Erythrocyte phosphogluconate dehydrogenase polymorphism. *Nature, Lond.* **210**, 487–489.

106. Pelzer, C. F. (1965). Genetic control of erythrocyte esterase forms in *Mus musculus*. *Genetics, Austin* **52**, 819–828.

107. Petra, P. H., Bradshaw, R. A. Walsh, K. A., and Neurath, H. (1969). Identification of the amino acid replacements characterizing the allotypic forms of bovine carboxypeptidase. *Biochemistry*, **8**, 2762–2768.

108. Petras, M. L. and Biddle, F. G. (1968). Serum esterases in the house mouse, *Mus musculus*. *Can. J. Genet. Cytol.* **9**, 704–710.

109. Popp, R. A. (1966). Inheritance of an erythrocyte and kidney esterase in the mouse. *J. Hered.* **57**, 197–206.

110. Rae, R., Hill, G. N., Pain, R. W. and Mulhearn, C. J. (1968). Congenital goitre in the Merino sheep due to an inherited defect in the biosynthesis of thyroid hormone. *Res. vet. Sci.* **9**, 209–223.

111. Randerson, S. (1965). Erythrocyte esterase forms controlled by multiple alleles in the Deer Mouse. *Genetics, Austin* **52**, 999–1005.

112. Rapacz, J., Korda, N. and Stone, W. H. (1968). Serum antigens of cattle, I. Immunogenetics of a macroglobulin allotype. *Genetics, Austin* **58**, 387–398.

113. Rasmusen, B. A. (1965). Isoantigens of gamma globulin in pigs. *Science, N.Y.* **148**, 1742–1743.

114. Rasmussen, D. I. and Koehn, R. K. (1966). Serum transferrin polymorphism in the deer mouse. *Genetics, Austin* **54**, 1353–1357.
115. Rendel, J., Aaland, O., Freedland, R. A. and Møller, F. (1964). The relationship between alkaline phosphatase polymorphism and blood group O in sheep. *Genetics, Austin* **50**, 973–986.
116. Rother, K. (1967). Serumkomplement als möglicher Resistenzfaktor Opsonisierung und Bakterizidie. *Int. Congr. Infect. Dis.* **4**, 329–349.
117. Russell, M. A. and Semeonoff, R. (1967). A serum esterase in *Microtus agrestis*. *Genet. Res.* **10**, 135–142.
118. Ruddle, F. H., Shows, T. B. and Roderick, T. H. (1968). Autosomal control of an electrophoretic variant of glucose-6-phosphate dehydrogenase in the mouse (*Mus musculus*). *Genetics, Austin* **58**, 599–606.
119. Saison, R. (1968). A serum post-albumin system in mink. *Can. J. Genet. Cytol.* **10**, 196–197.
120. Saison, R. and E. R. Giblett (1969). 6-phosphogluconate dehydrogenase polymorphism in the pig. *Vox Sang.* **16**, 514–516.
121. Sandberg, K. (1968). Genetic polymorphism in carbonic anhydrase from horse erythrocytes. *Hereditas* **60**, 411–412.
122. Sartore, G. (1966). Richerche su un nuovo polimorfismo genetico riguardante una esteresi degli eritrociti bovini. *Atti Assoc. Genet. Ital.* **11**, 217–222.
123. Sartore, G., Bernoco, D. and Stormont, C. (1967). A new polymorphic system of cattle β-globulins. *Atti Assoc. Genet. Ital.* **12**, 366–371.
124. Sartore, G., Stormont, C., Morris, B. G. and Grunder, A. A. (1969). Multiple electrophoretic forms of carbonic anhydrase in red cells of domestic cattle (*Bos taurus*) and American Buffalo (*Bison bison*). *Genetics, Austin* **61**, 823–831.
125. Saul, G. B., Garrity, E. B., Benirschke, K. and Valtin, H. (1968). Inherited hypothalamic diabetes insipidus in the Brattleboro strain of rats. *J. Hered.* **59**, 113–117.
126. Schiff, R. and Stormont, C. (1968). Close linkage between the genes controlling two systems of rabbit esterases. *Genetics, Austin* **60**, 222.
127. Schnatz, J. D. and Cortner, J. A. (1968). Genetic variation of alkaline lipolytic activity in rabbits. *Biochem. biophys. Acta* **167**, 367–372.
128. Schroeder, W. A., Shelton, J. R., Shelton, J. B., Robberson, B. and Rabin, D. R. (1967). A comparison of amino acid sequences in the β-chains of adult bovine hemoglobin A and B. *Arch. Biochem. biophys.* **120**, 124–135.
129. Schröffel, J. (1965). Genetic determination of the serum "thread proteins" and the slow α-globulin polymorphism in pigs. *Proc. IXth Conf. Europ. Soc. Blood Group Research*, 312–329.
130. Selander, R. K., Hunt, W. G. and Yang, S. Y. (1969). Protein polymorphism and genic heterozygosity in two European subspecies of the house mouse. *Evolution, Lancaster, Pa.* **23**, 379–390.
131. Sell, S. (1966). Immunoglobulin M allotypes of the rabbit: identification of a second specificity. *Science, N.Y.* **153**, 641–643.
132. Semeonoff, R. and Robertson, F. W. (1967). A biochemical and ecological study of plasma esterase polymorphism in natural populations of the field vole, *Microtus agrestis*. *Biochem. Genet.* **1**, 205–227.
133. Shapira, G., Benrubi, M., Maleknia, N. and Reibel, L. (1969). Hemoglobine de lapin: une variante resultant d'un allelomorphisme et non d'une ambiguité. $Hb^{29\,val \rightarrow leu}$. *Biochem. biophys. Acta* **188**, 216–221.

134. Shaw, C. R. and Barto, E. (1965). Autosomally determined polymorphism of glucose-6-phosphate dehydrogenase in *Peromyscus*. *Science, N.Y.* **148**, 1099–1100.
135. Shows, T. B. and Ruddle, F. H. (1968*a*). Malate dehydrogenase: evidence for tetrameric structure in *Mus musculus*, *Science, N.Y.* **160**, 1356–1357.
136. Shows, T. B. and Ruddle, F. H. (1968*b*). Function of the lactate dehydrogenase B gene in mouse erythrocytes: evidence for control by a regulatory gene. *Proc. natn. Acad. Sci. U.S.A.* **61**, 574–581.
137. Shows, T. B., Ruddle, F. H. and Roderick, T. H. (1969). Phosphoglucomutase electrophoretic variants in the mouse. *Biochem. Genet.* **3**, 25–35.
138. Shreffler, D. C. (1967). Genetic control of cellular antigen. *Int. Congr. Hum. Genet.* **3**, 217–231.
139. Smith, J. E. and Osburn, B. I. (1967). Glutathione deficiency in sheep erythrocytes. *Science, N.Y.* **158**, 374–375.
140. Spickett, S. G., Shire, J. G. M. and Stewart, J. (1967). Genetic variation in adrenal and renal structure and function. *Mem. Soc. Endocr.* No. 15, 271–288.
141. Stewart, A. D. (1968). Genetic variation in the neuro-hypophyseal hormones of the mouse (*Mus musculus*). *J. Endocr.* **41**, XIX.
142. Sullivan, B. and Nute, P. E. (1968). Structural and functional properties of polymeric hemoglobins from Orangutans. *Genetics, Austin* **57**, 113–124.
143. Sved, J. A., Reed, T. E. and Bodmer, W. F. (1967). The number of balanced polymorphisms that can be maintained in a natural population. *Genetics, Austin* **55**, 469–481.
144. Syner, F. N. and Goodman, M. (1966). Polymorphism of lactate dehydrogenase in gelada baboons. *Science, N.Y.* **151**, 206–208.
145. Tashian, R. E., Shreffler, D. C. and Shows, T. B. (1968). Genetic and phylogenetic variation in the different molecular forms of mammalian erythrocyte carbonic anhydrases. *Ann. N.Y. Acad. Sci.* **151**, 64–77.
146. Thompson, M. P., Farrell, H. M. and Greenberg, R. (1969). α_{SI}-casein A (*Bos taurus*): A probable sequential deletion of eight amino acid residues and its effect on physical properties. *Comp. Biochem. Physiol.* **28**, 471–475.
147. Thuline, H. C., Morrow, A. C., Norby, D. E. and Motulsky, A. G. (1967). Autosomal phosphogluconic dehydrogenase polymorphism in the cat (*Felis catus*. L.). *Science, N.Y.* **157**, 431–432.
148. Tucker, E. M. (1968). Serum albumin polymorphism in sheep. *Vox Sang.* **15**, 306–308.
149. Tucker, E. M., Suzuki, Y. and Stormont, C. (1967). Three new phenotypic systems in the blood of sheep. *Vox Sang.* **13**, 246–262.
150. Valenta, M., Hyldgaard-Jensen, J. and Moustgaard, J. (1967). Three lactic dehydrogenase isoenzyme systems in pig spermatozoa and the polymorphism of subunits controlled by a third locus *C*. *Nature, Lond.* **216**, 506–507.
151. Walsh, K. A., Ericsson, L. E. and Neurath, H. (1966). Bovine carboxypeptidase A variants resulting from allelomorphism. *Proc. natn. Acad. Sci. U.S.A.* **56**, 1339–1344.
152. Wang, A-C. and Fudenberg, H. H. (1969). Genetic control of gamma chain synthesis: a chemical and evolutionary study of the Gm(a) factor of immunoglobulins. *J. molec. Biol.* **44**, 493–500.
153. Welser, C. F., Winkelmann, H. J., Cutler, E. B. and Barto, E. (1965). Albumin variations in the White-footed mouse *Peromyscus*. *Genetics, Austin* **52**, 483.

154. Wilkinson, J. M. (1969). Variation in the N-terminal sequence of heavy chains of immunoglobulin G from rabbits of different allotype. *Biochem. J.* **112**, 173–185.
155. Wilson, J. B., Edwards, W. C., McDaniel, M., Dobbs, M. M. and Huisman, T. H. J. (1966). The structure of sheep haemoglobins II. The amino acid composition of the tryptic peptides of the non-α chains of hemoglobins A, B, C and F. *Arch. Biochem. biophys.* **115**, 385–400.
156. Wistar, R. (1969). Immunoglobulin allotype in the rat: localization of the specificity to the light chain. *Immunology* **17**, 23–32.
157. Wortis, H. H. (1965). A gene locus concerned with an antigenic serum substance in *Mus musculus*. *Genetics, Austin* **52**, 267–273.

Appendix

Biochemical variants in mammals

This classification is not intended to be foolproof, but merely as an aid to discovering quickly if a particular enzyme, etc. is genetically variable in any mammal. The reference attached to each variant is sometimes to a recent article or review rather than to the original description.

Erythrocyte characters

			References
Haemoglobin			
Chimpanzee	(*Pan troglodytes*)	2 loci	62
Orangutan	(*Pongo pygmaeus*)	2 loci	8, 24, 142
Crab-eating macaque	(*Macaca irus*)		7
Pig-tailed macaque	(*Macaca nemestrina*)		31
Rabbit	(*Oryctolagus cuniculus*)		69, 133
Mouse	(*Mus musculus*)	3 loci	50, 89
Norway rat	(*Rattus norvegicus*)		19
Cotton mouse	(*Peromyscus gossypinus*)		42
Sheep	(*Ovis aries*)		89
Barbary sheep	(*Ammotragus lervia*)		66
Goat	(*Capra hircus*)	2 loci	1, 48
Cattle	(*Bos taurus* and *B. indicus*)		89
Chinese muntjac deer	(*Muntiacus reevesi*)		92
White-tailed deer	(*Odocoileus virginianus*)	2 loci	76
Zebra	(*Equus burchelli*)		104
Horse	(*Equus caballus*)		18
Non-haemoglobin protein			
Rhesus macaque	(*Macaca mulatta*)		89
Crab-eating macaque	(*Macaca irus*)		89
Mouse	(*Mus musculus*)		15
Sheep	(*Ovis aries*)		149
Potassium and sodium concentration			
Brush-tail possum	(*Trichosurus vulpecula*)		89
Sheep	(*Ovis aries*)		35, 89
Goat	(*Capra hircus*)		89

Glutathione concentration		
Sheep	(*Ovis aries*)	139
Fructose transport		
Cattle	(*Bos taurus*)	85

Plasma proteins

Plasma albumin		
Deer mouse	(*Peromyscus maniculatus bairdi*)	153
Sheep	(*Ovis aries*)	148
Pig	(*Sus scrofa*)	82
Cattle	(*Bos taurus*)	89
Zebra	(*Equus burchelli*)	104
Horse	(*Equus caballus*)	89
Pre-albumin		
Mouse	(*Mus musculus*)	89
Pig	(*Sus scrofa*)	89
Horse	(*Equus caballus*)	44
Post-albumin		
Mink	(*Lutreola lutreola* or *Mustela lutreola*)	119
Cattle	(*Bos taurus*)	89
Transferrin		
Marsupial mouse	(*Sminthopsis crassicaudata*)	64
Red kangaroo	(*Macropus rufus*)	89
Grey kangaroo	(*Macropus major*)	75
Orangutan	(*Pongo pygmaeus*)	84
Chimpanzee	(*Pan troglodytes*)	89
Gorilla	(*Pan gorilla*)	84
Rhesus macaque	(*Macaca mulatta*)	89
Pig-tailed macaque	(*Macaca nemestrina*)	54
Moor macaque	(*Macaca maura*)	53
Crab-eating macaque	(*Macaca irus*)	89
Bonnet monkey	(*Macaca radiata*)	53
Formosan macaque	(*Macaca cyclopis*)	53
Yellow baboon	(*Papio cynocephalus*)	23
Hamadryas baboon	(*Papio hamadryas*)	6
Dog-faced baboon	(*Papio doguera*)	23
Lemur	(*Lemur fulvus*)	100
Ring-tailed lemur	(*Lemur catta*)	100
Black lemur	(*Lemur macaco*)	100
Mongoose lemur	(*Lemur mongoz*)	100
Ruffed lemur	(*Lemur variegatus*)	100
Large grey bushbaby	(*Galago crassicaudatus crassicaudatus*)	101
Dwarf galago	(*Galago demidovii*)	101

Mouse	(*Mus musculus*)	89
Ground squirrel	(*Spermophilus undulatus*)	97
Ground squirrel	(*Spermophilus townsendi*)	97
Ground squirrel	(*Spermophilus richardsoni*)	97
Meadow vole	(*Microtus pennsylvanicus*)	93
Prairie vole	(*Microtus ochrogaster*)	93
Beach vole	(*Microtus breweri*)	93
Deer mouse	(*Peromyscus maniculatus*)	114
Black rat	(*Rattus rattus*)	96
Dog	(*Canis familiaris*)	17
Harp seal	(*Pagophilus groenlandicus*)	98
Sheep	(*Ovis aries*)	89
Goat	(*Capra hircus*)	89
Pig	(*Sus scrofa*)	89
Zebra	(*Equus burchelli*)	104
Grant's gazelle	(*Gazella granti*)	4
Thomson's gazelle	(*Gazella thomsoni*)	4
Eland	(*Taurotragus oryx*)	4
Reindeer	(*Rangifer tarandus*)	89
Red deer	(*Cervus elaphus*)	89
Cattle	(*Bos taurus*)	89
Horse	(*Equus caballus*)	89
Donkey	(*Equus asinus*)	99
Ceruloplasmin		
Pig	(*Sus scrofa*)	89
Haemopexin		
Rabbit	(*Oryctolagus cuniculus*)	57
Pig	(*Sus scrofa*)	89
Haptoglobin		
Rabbit	(*Oryctolagus cuniculus*)	27
Low-density lipoprotein		
Rabbit	(*Oryctolagus cuniculus*)	2
Macroglobulin		
Cattle	(*Bos taurus*)	112
α_2-macroglobulin		
Rabbit	(*Oryctolagus cuniculus*)	77
Slow α_2-protein		
Cattle	(*Bos taurus*)	89
Pig	(*Sus scrofa*)	129
β-globulin		
Guinea pig	(*Cavia porcellus*)	87

Slow β-protein			
Cattle	(*Bos taurus*)		123
Serum antigenetic substance			
Mouse	(*Mus musculus*)		157
H-2 associated protein			
Mouse	(*Mus musculus*)		138
Complement			
Rabbit	(*Oryctolagus cuniculus*)		116
Mouse	(*Mus musculus*)		89
Guinea pig	(*Cavia porcellus*)		89
Immunoglobulins			
Chimpanzee	(*Pan troglodytes*)		3, 88
Orangutan	(*Pongo pygmaeus*)		3, 88
Rabbit	(*Oryctolagus cuniculus*)	4 loci	30, 89, 131
Mouse	(*Mus musculus*)	4 loci	89, 95
Norway rat	(*Rattus norvegicus*)		156
Guinea pig	(*Cavia porcellus*)		70
Pig	(*Sus scrofa*)		113

Enzymic and metabolic variants

Acid phosphatase			
Crab-eating macaque	(*Macaca irus*)		83
Alkaline phosphatase			
Cattle	(*Bos taurus*)		46
Sheep	(*Ovis aries*)		115
Alcohol dehydrogenase			
Mouse	(*Mus musculus*)		130
Amylase			
Mouse	(*Mus musculus*)	2 loci	89
Cattle	(*Bos taurus*)		49, 89
Pig	(*Sus scrofa*)		89
β-glucuronidase			
Mouse	(*Mus musculus*)	2 loci	45, 89
Bilirubin metabolism			
Norway rat	(*Rattus norvegicus*)		89
Carbonic anhydrase			
Slow loris	(*Nycticebus coucang*)	2 loci	145
Spider monkey	(*Ateles belzebuth*)	2 loci	145
Titi monkey	(*Callicebus cupreus*)		145

			References
Capuchin monkey	(*Cebus capuchinus*)		148
Yellow baboon	(*Papio cynocephalus*)		89
Rhesus macaque	(*Macaca mulatta*)	2 loci	145
Crab-eating macaque	(*Macaca irus*)	2 loci	145
Stump-tailed macaque	(*Macaca speciosa*)		145
Orangutan	(*Pongo pygmaeus*)		145
Sheep	(*Ovis aries*)		149
Cattle	(*Bos taurus*)		124
Horse	(*Equus caballus*)		121
Carboxypeptidase			
Cattle	(*Bos taurus*)		107
Catalase			
Mouse	(*Mus musculus*)	3 loci	34, 47, 59
Dog	(*Canis familiaris*)		89
Guinea pig	(*Cavia porcellus*)		89
Creatine kinase			
Mouse	(*Mus musculus*)		63
Cystinuria			
Dog	(*Canis familiaris*)		89
δ-aminolaevulinate dehydratase			
Mouse	(*Mus musculus*)		89
Diaphorase			
Sheep	(*Ovis aries*)		21
Esterases			
Rabbit	(*Oryctolagus cuniculus*)	5 loci	78, 89, 126
Mouse	(*Mus musculus*)	5 loci	106, 108, 109
Short-tailed field vole	(*Microtus agrestis*)	3 loci	117, 132
Norway rat	(*Rattus norvegicus*)		125
Deer mouse	(*Peromyscus maniculatus*)		111
Sheep	(*Ovis aries*)		89, 149
Cattle	(*Bos taurus*)		122
Pig	(*Sus scrofa*)		89
Horse	(*Equus caballus*)		44
Glucose 6-phosphate dehydrogenase			
Mouse	(*Mus musculus*)		118
Deer mouse	(*Peromyscus m. bairdi*)		134
Haemophilias			
Dog	(*Canis familiaris*)		89
Pig	(*Sus scrofa*)		89

Immunological reactivity			
Guinea pig	(*Cavia porcellus*)		13
Indophenol oxidase			
Mouse	(*Mus musculus*)		130
Iso-citrate dehydrogenase			
Mouse	(*Mus musculus*)		89
Lactate dehydrogenase			
Little brown bat	(*Myotis lucifugus*)		91
Gelada baboon	(*Theropithecus gelada*)		144
Mouse	(*Mus musculus*)		136
Deer mouse	(*Peromyscus maniculatus*)	2 loci	25, 89
Pig	(*Sus scrofa*)		150
Lipase			
Rabbit	(*Oryctolagus cuniculus*)		127
Malate dehydrogenase			
Mouse	(*Mus musculus*)	2 loci	130, 135
N-methylnicotinamide oxidase			
Mouse	(*Mus musculus*)		65
Norway rat	(*Rattus norvegicus*)		52
Peptidase			
Mouse	(*Mus musculus*)		86
Phosphoglucomutase			
Chimpanzee	(*Pan troglodytes*)		55
Mouse	(*Mus musculus*)	2 loci	137
6-phosphogluconate dehydrogenase			
Cat	(*Felis catus*)		147
Mouse	(*Mus musculus*)		130
Norway rat	(*Rattus norvegicus*)		105
Pig	(*Sus scrofa*)		120
Phosphoglucose isomerase			
Mouse	(*Mus musculus*)		33
Phosphorylase b kinase			
Mouse	(*Mus musculus*)		90
Porphyria			
Cattle	(*Bos taurus*)		89
Pig	(*Sus scrofa*)		89
Pyrimidine catabolism			
Mouse	(*Mus musculus*)		89

Sorbitol dehydrogenase			
Pig	(*Sus scrofa*)		61
Steroid metabolism			
Mouse	(*Mus musculus*)		140
Sulfadiazine metabolism			
Rabbit	(*Oryctolagus cuniculus*)		89
Thyroid function deficiency			
Sheep	(*Ovis aries*)		110
Uric acid metabolism			
Dog	(*Canis familiaris*)		89
Vasopressin			
Norway rat	(*Rattus norvegicus*)		125
European wild boar	(*Sus scrofa*)		39
Warthog	(*Phacochoerus aethiopicus*)		38
Giant forest boar	(*Hylochoerus meinertzhageni*)		39
Collared peccary	(*Tayassu angulatus*)		38
White-lipped peccary	(*Tayassu pecari*)		39

Milk proteins

α-lactalbumin			
Cattle	(*Bos taurus*)		20, 89
β-lactoglobulin			
Cattle	(*Bos taurus*)		22, 89
Sheep	(*Ovis aries*)		12, 73
Caseins			
Cattle	(*Bos taurus*)	4 loci	56, 81, 89
Pig	(*Sus scrofa*)	2 loci	51
Sheep	(*Ovis aries*)	2 loci	72
Other milk proteins			
Pig	(*Sus scrofa*)	2 loci	51

Miscellaneous

Urinary proteins			
Mouse	(*Mus musculus*)		40

Symp. zool. Soc. Lond. (1970) No. 26, 73–91.

BIOCHEMICAL POLYMORPHISM IN POPULATIONS OF THE HOUSE MOUSE AND OLD-FIELD MOUSE

ROBERT K. SELANDER

Department of Zoology, University of Texas, Austin, Texas, U.S.A.

SYNOPSIS

Studies of electrophoretically demonstrable polymorphism in enzymes and non-enzymatic proteins in the House mouse (*Mus musculus*) and the Old-field mouse (*Peromyscus polionotus*) indicate that individuals are heterozygous at 8% of their structural gene loci. At most polymorphic loci, one or two common alleles occur throughout the range of a species, but rare alleles with very localized distributions have been detected at many loci. Patterns of geographical variation in allele frequencies suggest that much of the observed variation is adaptive, the polymorphisms being maintained by forms of balancing selection. Coadaptation of genomes in allopatrically hybridizing subspecies of House mice in Denmark is suggested by several lines of evidence, including differential introgression of alleles. For various reasons, the existence of fine-grain genetic subdivision in populations of the House mouse, as demonstrated by the analysis of protein variation, cannot be accepted as evidence that gene flow plays only a minor role in the maintenance of the genetic integrity of species. The evolutionary significance of genetic subdivision and drift in rodent populations remains to be determined.

INTRODUCTION

For several years, my associates and I have been studying the genetics of natural populations of animals, employing electrophoretic and histochemical techniques to demonstrate allelic variation at structural loci encoding polypeptides of enzymes and other proteins. Although we have worked with a variety of organisms ranging from horseshoe crabs (*Limulus*) to frogs (*Hyla*), much of our effort has been directed toward gaining an understanding of the nature and extent of genic variation in two rodents, the House mouse (*Mus musculus*), a murid of world-wide distribution, and the Old-field mouse (*Peromyscus polionotus*), a cricetid native to the south-eastern United States. Our investigations of these species, which differ in pattern of social organization, niche, extent of geographical distribution, and in numerous other aspects of their biology, have helped to provide answers to several significant questions concerning the genetic structure of wild populations. Specifically, we have been able (1) to estimate average degrees of genic heterozygosity in natural populations and to determine average numbers of alleles segregating at polymorphic loci; (2) to compare local populations, subspecies, semispecies, and species with respect to total genetic character; (3) to determine the effects of social system and other aspects of population structure on patterns of genetic variation on a microgeographical scale; and (4) to measure interpopulation variation in allele frequencies at some 30 to 40 gene loci, and to provide several lines of circumstantial evidence with bearing on the important problem of the selective neutrality or non-neutrality of allozymic variation.

In this paper I shall review some of the major findings from my laboratory relative to these four aspects of genetic variation in *Mus musculus* and *Peromyscus polionotus*. (In the following text, "*Mus*" refers to *M. musculus* and "*Peromyscus*" to *P. polionotus*, unless otherwise indicated.)

THE EXTENT OF POLYMORPHISM

Surveys of electrophoretic variation in 36 proteins in *Mus musculus* from 6 regions of the Jutland Peninsula, Denmark (Selander, Hunt and Yang, 1969*a*), and from Hallowell Farm in southern California (Selander and Yang, 1969) revealed polymorphism in 16 (44%) of the proteins. Of the 41 genetic loci controlling variation in the 36 proteins, 17 (41%) were segregating for 2 or more alleles in at least one of the 7 populations sampled. From these data, we have estimated that the average population of *Mus* is polymorphic at 26% of its loci and the average individual is heterozygous at 8·5% of its loci (Table I). Since perhaps only one-third of amino acid substitutions result in a change in the isoelectric point of polypeptides, estimates of genic heterozygosity derived from electrophoretic studies are minimal, the method detecting only part of the actual variation in protein structure. But, in any event, if the loci included in our survey are at all representative of structural genes in general, it is apparent that there must be thousands of polymorphisms in any given population, and that all individuals are heterozygous at a major proportion of their genetic loci. Moreover, we may confidently conclude that each individual has a unique protein constitution.

The House mouse is in no way unusual in regard to degree of genic variability. From an analysis of variation in 800 individuals of *Peromyscus polionotus* from 30 localities, we estimate that 53% of structural loci are polymorphic in the species as a whole, 23% of loci are polymorphic per population, and 6% of loci are heterozygous per individual. Estimates of similar magnitude have been obtained in my laboratory from studies of other rodents, including Cotton rats (*Sigmodon*) (R. K. Selander, S. Y. Yang, W. E. Johnson, Y. J. Kim and

TABLE I

Estimates of mean proportion of loci polymorphic per population and mean proportion of loci heterozygous per individual in several animal species

Species	Number of populations sampled	Number of proteins	Number of controlling loci	Proportion of loci polymorphic per population	Proportion of loci heterozygous per individual	Reference
Limulus polyphemus	4	24	25	0·250	0·057	Selander, Yang, Lewontin and Johnson (1970)
Acris crepitans	3(?)	16	20	0·14–0·23	—	Dessauer and Nevo (1969)
Mus musculus	7	36	41	0·261	0·085	Selander, Hunt and Yang (1969); Selander and Yang (1969)
Drosophila persimilis	1	24	24	0·250	0·105	Prakash (1969)
D. pseudoobscura	3	24	24	0·420	0·123	Prakash, Lewontin and Hubby (1969)
*Homo sapiens** (British population)	1	33	33	0·26	0·16	Lewontin (1967*b*)
H. sapiens (British population)	1	20	20	0·30	0·074	Harris (1969)

*Estimates based on an analysis of the time sequence of discovery of erythrocytic antigens.

M. H. Smith, unpublished), Kangaroo rats (*Dipodomys*) (W. E. Johnson, unpublished), and several Asiatic species of the genus *Mus*. It is noteworthy that our estimates for rodents are similar to those available for various species of *Drosophila* (Prakash *et al.*, 1969; Prakash, 1969; O'Brien and MacIntyre, 1969), the Horseshoe crab (Selander *et al.*, 1970), and the British human population (Lewontin, 1967*b*; Harris, 1969). Extensive protein polymorphism is a general characteristic of animals. Whether or not there is any major variation in degree of heterozygosity among phylogenetic groups will be determined only when additional species have been studied. However, we have evidence from current work suggesting that, among vertebrates, passerine birds are relatively monomorphic.

In *Mus* and *Peromyscus*, as well as in other rodents, polymorphisms are generally widespread geographically, and, at most loci, there is a strong tendency for one or two alleles to occur throughout the species range. For example, the alleles Hbb^d and Hbb^s at the β-chain hemoglobin locus are ubiquitous in populations of *Mus* in North America, South America, Hawaii, and Denmark. Similarly, in *Peromyscus*, the allele $Pgi\text{-}1^a$ at the phosphoglucose isomerase locus predominates in all mainland populations and is the allele fixed in the monomorphic populations on islands off the Florida coast (Selander, Smith, Yang and Johnson, 1970). At the transferrin locus in *Peromyscus*, the allele $Trf\text{-}1^b$ predominates in all populations.

Rodent populations generally do not have large numbers of alleles segregating at polymorphic loci. For example, in our analysis of protein variation in *Mus* from Hallowell Farm, in which we examined from 56 to 227 individuals, depending upon the locus, at no locus were more than three alleles recorded, and the average was only 2·25. And for 16 of the 17 loci that were polymorphic in 6 Danish populations of this species, only two (*Hpd*-1 and *Mdh*-2) were segregating for more than two alleles.

Among the different functional types of proteins studied, esterases are, as a group, unusually variable, both within and between species. For esterase-2 (*Es*-2) in *Mus*, we have identified 7 alleles in wild populations, and 5 alleles are recorded at the esterase-1 locus in *Peromyscus*. But in both species the number of alleles at these loci in a given population never exceeds 4 and is usually 2 or 3. We have yet to find in any rodent a situation comparable to that in *Drosophila pseudoobscura*, in which 8·3 alleles at the esterase-5 locus are found in the average population (Prakash *et al.*, 1969).

In addition to the relatively common, widely distributed alleles, a number of rare alleles which presumably arose as mutants and reached appreciable frequencies through genetic drift has been detected in single local populations. For example, 2 individuals of *Mus* heterozygous for a variant allele (*Alb*-1^b) at the albumin locus were collected in a barn on the northern Jutland Peninsula. But no additional examples of this variant or other variants at the albumin locus were found in an examination of 12 000 House mice from North America, Denmark, and Hawaii. The esterase alleles *Es*-1^d, *Es*-2^e, and *Es*-3^e in *Mus* also fall within this category, as does the phosphoglucomutase allele *Pgm*-1^c in *Peromyscus*. Rare variant alleles are likely to be detected at any given locus if large numbers of individuals are examined. But, since mutation pressure alone may be sufficient to account for their occurrence, these rare variants are of relatively minor interest compared with the common alleles accounting for polymorphism in populations.

Among rodents, particular proteins may be monomorphic in one species but highly polymorphic in another. Albumin is monomorphic in *Mus*, but most populations of *P. polionotus* are polymorphic for albumin, and comparable polymorphisms occur in *P. maniculatus* and several other species of the genus (Brown and Welser, 1968). Similarly, transferrin and lactate dehydrogenase are highly polymorphic in *Peromyscus* but monomorphic in all wild *Mus* populations we have examined. The supernatant form of NADP-dependent isocitrate dehydrogenase is monomorphic in *Peromyscus* but polymorphic in *Mus*. Certain proteins such as haemoglobin, the phosphoglucomutases, and 6-phosphogluconate dehydrogenase are polymorphic in both genera.

COMPARISON OF POPULATIONS

Similarities or differences in overall genetic character among populations may be expressed quantitatively by one of several available coefficients (Cavalli-Sforza and Edwards, 1967). Coefficients of similarity based on 17 polymorphic loci in four populations of *M.m. musculus* from the northern Jutland Peninsula and the islands east of Jutland and two populations of *M.m. domesticus* from the southern Jutland Peninsula are presented in Table II. These reflect a close genetic similarity among samples of each subspecies, and a striking difference between those of the two subspecies.

Because House mice in the southern United States have been assigned taxonomically to *M.m. brevirostris*, a subspecies native to Europe and a member of the same subspecies group (*spicilegus*) to which *M.m. domesticus* is relegated (Schwarz and Schwarz, 1943), it is of interest to compare the Hallowell Farm population in California (representing *M.m. brevirostris*) with *M.m. domesticus* from Denmark. The coefficient

TABLE II

Coefficients of genetic similarity based on 17 polymorphic loci (upper matrix) and 41 monomorphic and polymorphic loci (lower matrix)*

	Population					
	M.m. musculus				*M.m. domesticus*	
	1	2	3	4	5	6
1	X	0·89	0·85	0·86	0·31	0·31
2	0·95	X	0·92	0·91	0·35	0·35
3	0·94	0·97	X	0·94	0·32	0·33
4	0·94	0·96	0·97	X	0·36	0·37
5	0·66	0·69	0·67	0·69	X	0·99
6	0·67	0·69	0·68	0·70	0·99	X

$$*S^2 = \left(\frac{1}{L}\sum_{i=1}^{L}\left(\frac{\sum_{j=1}^{A_i} p_{ij}q_{ij}}{\sqrt{\sum_{j=1}^{A_i} p_{ij}^2 \sum_{j=1}^{A_i} q_{ij}^2}}\right)\right)^2,$$ where L = number of loci; A_i = number of alleles at locus i; p_{ij} = frequency of jth allele at the ith locus in one sample; q_{ij} = frequency of jth allele at the ith locus in another sample.

of genetic similarity between Hallowell Farm and south-eastern Jutland is 0·71. Thus, Danish populations of *M.m. domesticus* are genetically more similar to the Hallowell Farm population than to any of the neighbouring Danish populations of *M.m. musculus*. This is impressive evidence of a tight cohesion of coadapted gene pools, for, certainly, it is difficult to support the alternative proposition that there is greater similarity in external environmental selective pressures between the southern Jutland Peninsula and southern California than between the different regions of western Denmark occupied by *M.m. domesticus* and *M.m. musculus*.

At 13 of the 17 protein loci that are polymorphic in Danish mice, there are marked subspecific differences in allele frequencies, and at 6 of these loci alternate alleles are fixed or nearly so in the 2 subspecies. If we consider that the two subspecies are identical at the 4 polymorphic loci at which allele frequencies are similar and at the 24 monomorphic loci, we can say that *M.m. musculus* and *M.m. domesticus* share identical alleles at 68% of their loci. This value may be compared with that of 50% derived from a study of variation among sibling species of *Drosophila* (Hubby and Throckmorton, 1968). From our preliminary comparative studies of *Mus musculus* and the Asiatic species *M. caroli*, *M. castaneus*, and *M. cervicolor*, we are predicting that, as in *Drosophila*, species of *Mus* will exhibit allelic differences, on the average, at 50% of their loci. Elsewhere (Selander, *et al.*, 1969*a*), we have argued from the data on *Mus* and *Drosophila* that major reorganizations of gene pools normally accompany the speciation process, as long advocated by E. Mayr and T. Dobzhansky.

To account for the genetic difference between *M.m. musculus* and *M.m. domesticus* on the Jutland Peninsula, where populations of the two subspecies occupy regions differing little if at all in environmental features, we have proposed that a major part of the difference is attributable to subspecific variation in the *genetic*, rather than the environmental, background. Thus we have suggested that, for most of the polymorphic loci, a given allele has different fitness values and, hence, different equilibrium frequencies in populations of *M.m. musculus* and *M.m. domesticus* as a consequence of differential coadaptation of their genomes.

We have recently obtained supporting evidence of differential coadaptation of genomes from an analysis of patterns of introgression through a narrow band of allopatric hybridization occurring midway on the Jutland Peninsula (Ursin, 1952). If the gene pools of parental populations of *M.m. musculus* and *M.m. domesticus* are, in fact, differentially coadapted, we should expect to find varying patterns of introgression among separate loci. That is, alleles which are relatively compatible with both gene pools should show greater degrees of introgression than those having a relatively large disruptive effect on the coadapted complexes. In fact, the patterns of introgression are very different for various polymorphic loci, as illustrated for two esterase loci in Fig. 1. Hybridization obviously has not resulted in a simple "diffusion" of alleles from one parental form to the other. Despite free interbreeding of the parental types, with the production of fertile progeny, gene exchange is not extensive (W. G. Hunt and R. K. Selander, unpublished).

POPULATION STRUCTURE AND SUBDIVISION

Ecologists and population geneticists have long been aware that populations of *Mus*, *Peromyscus*, and other rodents are subject to disturbances of random mating relationships through subdivision and sampling fluctuations in gene frequencies. For both genera there exists a considerable literature on movements and home ranges, but, unfortunately, this type of information can tell us little about gene flow

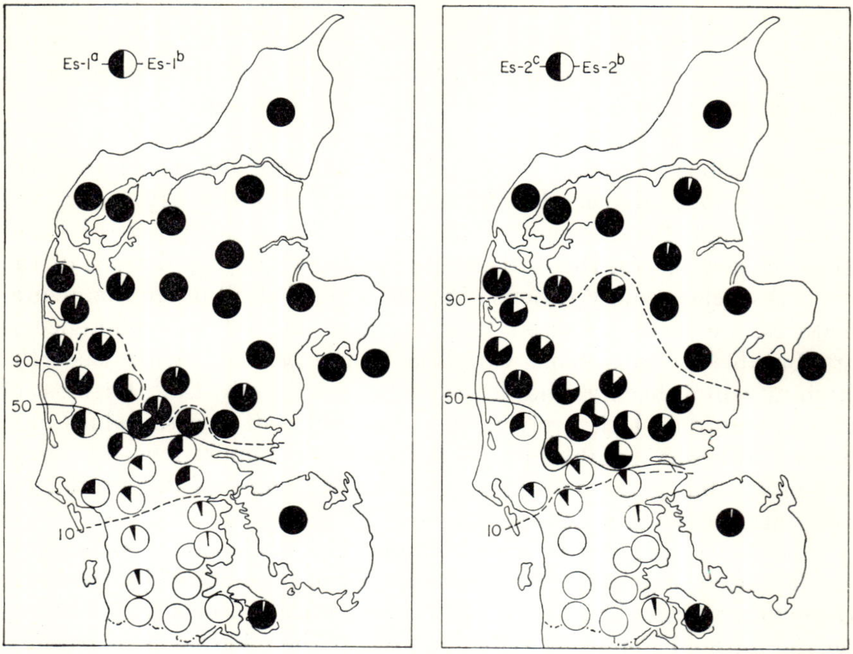

Fig. 1. Introgression of alleles at two genetic loci in hybridizing subspecies of *M. musculus* in Denmark. Filled areas of circles are proportional to allele frequencies. 10%, 50%, and 90% isofrequency lines are shown.

within or among populations, since physical migration cannot be equated with genetic migration.

From theoretical considerations developed in studies of the *t*-alleles in *Mus*, Lewontin and Dunn (1960) predicted that wild populations are subdivided into units (tribes or family groups) with an effective breeding size below that at which random fluctuations and inbreeding play an important role in determining gene frequencies. This interpretation has now been abundantly supported by experimental studies of mice in population cages (Reimer and Petras, 1967) and by field investigations (Crowcroft and Rowe, 1963; Anderson and Hill, 1965; Reimer and Petras, 1968; Rowe and Redfern, 1969). Populations of *Mus* are subdivided into tribes composed of a dominant male, several females, and several subordinate males. Close inbreeding occurs within tribes, and inter-tribal migration is rare, owing to territoriality of tribe members. The genetically effective size of the tribal units is perhaps less than ten.

The genetic consequences of the tribal, territorial social system of *Mus* have been demonstrated by our studies of polymorphic variation

TABLE III

Allele frequencies at three loci in samples from Nieschwietz Farm, Ensinal, Texas

			Es-2			*Es*-3		*Hbb*	
Barn	Number of mice	*Es*-2^a "Silent"	*Es*-2^b Fast	*Es*-2^c Med.	*Es*-2^d Slow	*Es*-3^b Med.	*Es*-3^c Slow	*Hbb*^{d} Diffuse	*Hbb*^{s} Single
1	49		0·77	0·03	0·20	0·34	0·66	0·17	0·83
2	86		0·84	0·01	0·15	0·37	0·63	0·23	0·77
3	101	0·06	0·78	0·01	0·15	0·45	0·55	0·17	0·83
4	25		0·68	0·10	0·22	0·38	0·62	0·22	0·78
5	50	0·13	0·63	0·08	0·16	0·32	0·68	0·10	0·90
7	106		0·78	0·04	0·18	0·43	0·57	0·17	0·83
8	90	0·09	0·76	0·03	0·12	0·52	0·48	0·11	0·89
9	31		0·73	0·11	0·16	0·66	0·34	0·13	0·87
10	47		0·89	0·01	0·10	0·28	0·72	0·22	0·78
11	31		0·86	0·03	0·11	0·42	0·58	0·11	0·89
Heterogeneity χ_2			119·2**	(27 d.f.)		40·1**	(9 d.f.)	19·9*	(9 d.f.)

(Selander, 1970; Selander and Yang, 1970). Populations of mice inhabiting different barns on a single farm often are heterogeneous in allele frequencies (see example in Table III). Since it is hardly possible to attribute this heterogeneity to inter-barn variation in selection intensity or direction, especially where the barns are identical in structure and appearance, we must conclude that barn populations are, in general, established by a few founders, which rapidly increase in numbers to "saturate" the available space, thereby reducing and, eventually, eliminating inter-barn migration. Populations of mice occupying adjacent barns or corn ricks may be as effectively isolated as if they were separated by a physical barrier, and they may maintain their genetic differences for several years at least.

Estimates of average degrees of inter-barn variation in allele frequencies within farms have been computed for the *Es*-3 and *Hbb* loci for 26 farms in Texas for which we have samples of mice from two or more barns (Selander, 1970). Expressed as arc–sin transformations (in degrees) of frequencies, these variances are 52 and 44, respectively. Comparable estimates of inter-farm variances within 18 small geographic regions in Texas are 71 and 73, respectively. The two loci are equally variable, and the inter-barn variances are approximately three-quarters as large as the inter-farm variances.

As predicted from ecological and behavioural evidence of territorial tribe formation, subsampling within barns has revealed spatial heterogeneity in allele frequencies. Moreover, by massive trapping of mice in a grid pattern in two large barns at Hildreth Farm, near Dripping Springs, Texas, we were able to demonstrate a spatial clustering of similar genotypes among adult mice at several loci, thereby providing direct evidence of population subdivision and inbreeding. Allele frequencies within these barns varied in complex mosaic patterns, with "peaks" and "valleys" of frequencies presumably corresponding to tribes or groups of genetically related tribes occupying neighbouring territories.

The degree of genetic subdivision of populations of *Peromyscus* has not been precisely determined. Because *P. polionotus* has a monogamous mating system and the young do not remain long in the parental burrows, it is probable that there is considerably less inbreeding in populations of this species than in those of *Mus*. However, some measurable degree of subdivision exists, since there are significant differences in allele frequencies among samples taken only a few miles apart, and mutant alleles may occasionally drift to moderate frequencies in local populations (Selander *et al.*, 1970*a*).

Several attempts have been made to determine the extent of subdivision of rodent populations through the use of an inbreeding coefficient based on observed deficiencies of heterozygotes in samples (the Wahlund effect) (Rasmussen, 1964; Petras, 1967). However, this approach is theoretically questionable (Workman, 1969), and, at least for *Mus*, simply unproductive (Selander, 1970). Heterosis, directional selection, sexual variation in allele frequencies, and other factors may mask the heterozygote deficiencies produced by subdivision.

INTERPOPULATION VARIATION IN ALLELE FREQUENCIES

The main features of interpopulation variation in allele frequencies in *Peromyscus polionotus* and in the North American populations of *Mus musculus* may be summarized as follows. Allele frequencies at all polymorphic loci are geographically variable; the variation is, for the most part, clinal; there is little concordance among patterns of various loci; and the patterns are difficult to relate to geographical gradients of environmental factors such as temperature or humidity.

Regional mean frequencies of alleles at the *Es*-2 locus in *Mus* in the United States are shown in Fig. 2. The indicated frequencies are the unweighted means for all barns in a given region represented by samples of ten or more individuals. Populations in the Midwest are segregating for only two alleles (*Es*-2^a and *Es*-2^b), whereas three or four alleles are represented in those in the south-western states. Parenthetically, it may be noted that our recent analysis of geographical

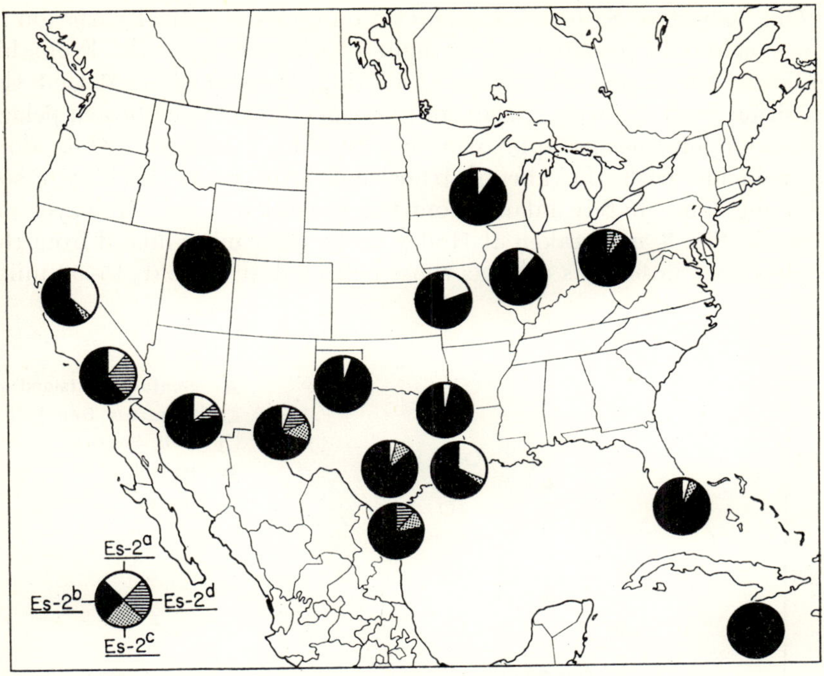

FIG. 2. Regional mean frequencies of alleles at the esterase-2 locus in *M. musculus*. (From Selander, Yang and Hunt, 1969*b*).

variation at the *Es*-1, *Es*-2, *Es*-3, *Es*-5, and *Hbb* loci in 8000 mice collected in 185 barns and fields failed to support an earlier claim (Schwarz and Schwarz, 1943) that North American populations can be divided into southern and northern subspecific entities which meet and hybridize in the central United States.

We have recently completed an analysis of variation at 17 polymorphic loci in samples from 30 populations of *Peromyscus polionotus* (Selander *et al.*, 1970*a*). One of the objectives of this study was to relate patterns of geographical variation in allele frequencies to the subspecific divisions of the species defined by more conventional systematic criteria.

In a series of papers appearing 40 years ago, Sumner (1930) analysed the genetic basis of morphological characters in *P. polionotus*, concentrating primarily on pelage colour, which ranges from dark brown in *P.p. polionotus* of the interior regions of Alabama, Georgia, South Carolina, and Florida, to white in *P.p. leucocephalus* inhabiting white sand dunes on Santa Rosa Island and other barrier islands off the Gulf coast of Florida.

One aspect of Sumner's interpretation of geographical variation in pelage colour is puzzling in the light of our recent work. From his studies in the panhandle region of Florida, Sumner claimed that the transition from the white pelage of the beach mice to the brown pelage of interior mice occurs approximately 40 miles inland and, thus, does not coincide with the abrupt transition from white to brown soil occurring immediately inland from the coastal beaches. However, as shown in Fig. 3, mice taken at Holley, only a few miles inland from the beaches, are, in fact, as dark as those collected at Harold, 15–50 miles

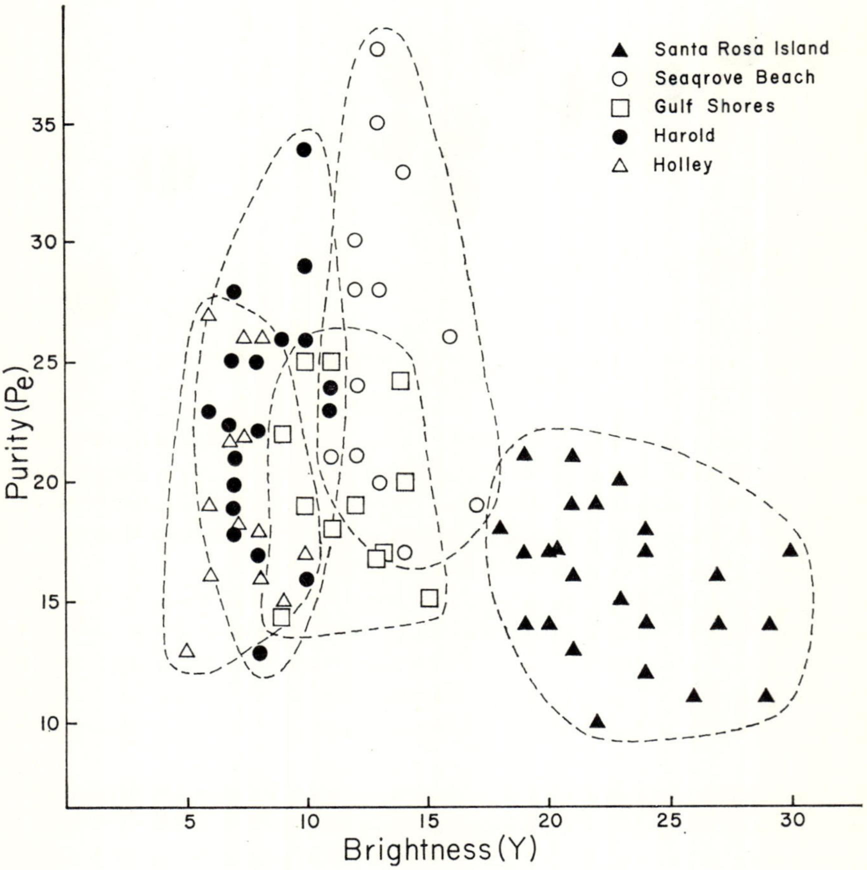

FIG. 3. Dorsal pelage colour in adult individuals of *P. polionotus*, as indicated by values of brightness and purity obtained from diffuse spectral reflectance curves. The Seagrove Beach and Gulf Shores populations, occurring on coastal beaches, are intermediate in colour between the white mice on Santa Rosa Island and the dark brown mice occurring on dark soils of interior mainland Florida (Harold and Holley).

inland. The pattern of geographical variation in pelage colour is actually much simpler than reported by Sumner. Mice living on white sands of islands or peninsulas narrowly connected to the mainland have a white pelage. Mice living on mainland beaches have a buff-coloured pelage, despite the fact that they inhabit white sandy soil similar to that on the islands. We presume that periodic gene flow from populations of brown-pelaged mice inhabiting fields adjacent to the coastal beaches prevents the beach mice from attaining a white pelage.

Two patterns of geographical variation in allele frequencies at protein loci in *Peromyscus* are particularly interesting because they illustrate differences in genetic constitution between the beach and interior populations. At the phosphoglucose isomerase locus (*Pgi*-1), and at several other loci, all insular or peninsular beach populations in the panhandle region of Florida are monomorphic, while most mainland populations are polymorphic (Fig. 4). The insular beach populations are fixed for the allele occurring in highest frequency on the mainland. Presumably, this reduction in variability (mean hetero-

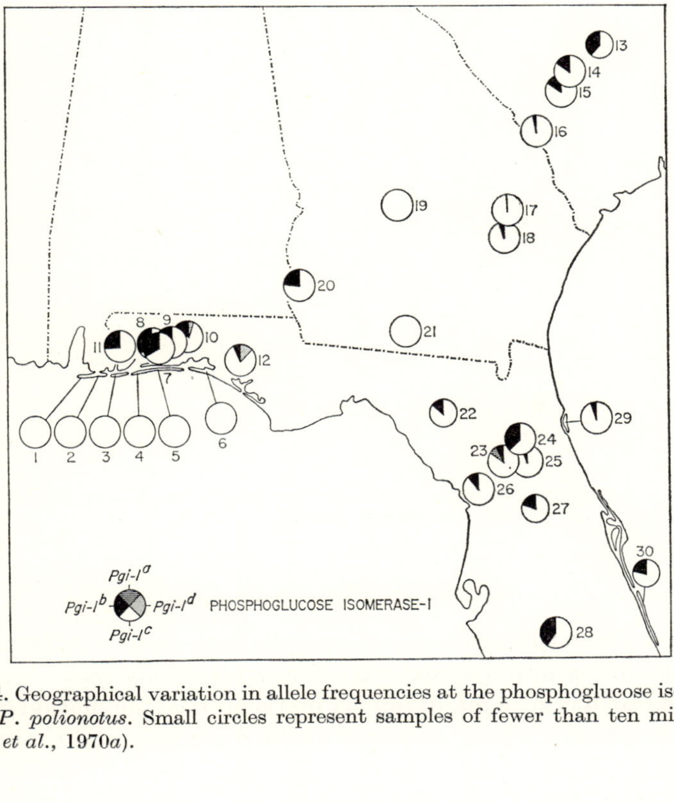

FIG. 4. Geographical variation in allele frequencies at the phosphoglucose isomerase-1 locus in *P. polionotus*. Small circles represent samples of fewer than ten mice. (From Selander *et al.*, 1970*a*).

zygosity is only 2·8%) reflects a tendency for genetic drift in the relatively small insular populations, which, owing to their isolation from one another and from populations on the mainland, cannot readily regain variability through gene flow. In view of this tendency toward monomorphism in the insular populations, it is especially interesting that the situation occasionally is reversed (Fig. 5). Thus, insular populations are polymorphic at the B lactate dehydrogenase (*Ldh*-1) and phosphoglucomutase-2 (*Pgm*-2) loci, while mainland populations are almost uniformly monomorphic. This pattern suggests that there is a strong selective advantage of polymorphism at the *Ldh*-1 and *Pgm*-2 loci in the insular environments.

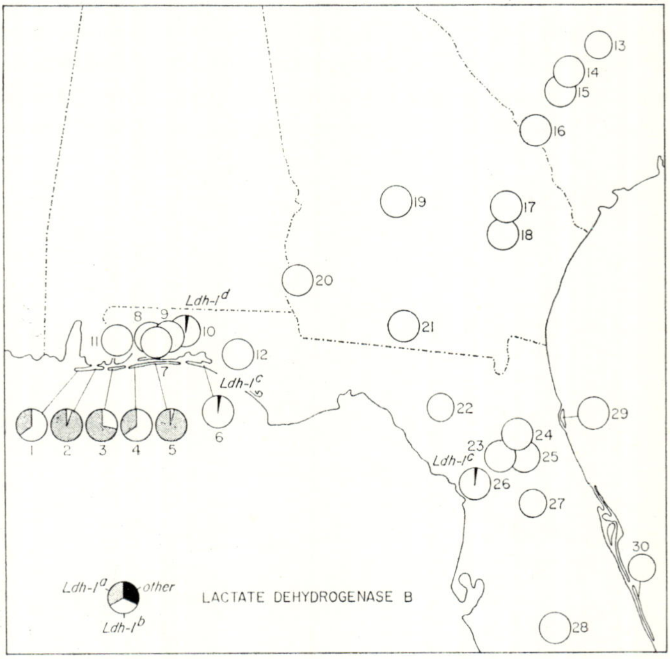

FIG. 5. Geographical variation in allele frequencies at the B lactate dehydrogenase locus in *P. polionotus*. (From Selander *et al.*, 1970*a*).

DISCUSSION AND CONCLUSIONS

Recent demonstrations of surprisingly high degrees of genic polymorphism in natural populations of rodents, *Drosophila*, and other organisms are having a major impact on theoretical population genetics. Numerous questions have arisen as to the functional and evolutionary significance of this variation and the bases for its maintenance in populations. The "isoallelic" hypothesis, which proposes that much if not most of the observed genic variation is physiologically irrelevant (and, hence, selectively neutral), has been supported by several geneticists (Kimura, 1968; Crow, 1968; Mukai, 1968), largely on theoretical grounds, in an effort to avoid the dilemma, first raised by the work of Lewontin and Hubby (1966), that the segregational component of the genetic load would be intolerable if all or any large part of the variation were maintained by selection processes involving overdominance. Even if the dilemma should prove to be more apparent than real (Wallace, 1968), it will be difficult to exclude the possibility that some significant part of the observed polymorphism in populations involves selectively neutral isoalleles. However, we are impressed with the fact that, in every study of protein polymorphism in natural or experimental populations providing data relevant to the question of neutrality or non-neutrality of alleles, direct or circumstantial evidence of the action of selection has been forthcoming.

As noted by Robertson (1968), if a high proportion of mutations at a locus were neutral in their effect on fitness, we would expect to find either of two situations: many alleles would be segregating in large panmictic populations, or few alleles would be segregating in subdivisions of populations, but different sets would occur in each subdivision. Neither situation obtains in *Mus* or *Peromyscus.* At most polymorphic loci, two alleles are more or less ubiquitous, and patterns of geographical variation tend to be clinal. The evidence of differential introgression of alleles at different loci in hybridizing subspecies of *Mus* in Denmark provides a particularly strong argument against the isoallelic hypothesis. Although some significant proportion of the observed allozymic variation in rodents may involve selectively neutral alleles, the available evidence supports the thesis that most of the polymorphisms are maintained by forms of balancing selection.

In concluding this discussion, I should like to consider the important problem of the evolutionary significance of population subdivision. Results of recent investigations of rodent population genetics support Wright's (1931) contention that random drift in small, local populations must be considered, along with selection and other determinant factors, in a comprehensive theory of the evolutionary process. For an individual or small population, stochastic processes may play the dominant role in determining genotype (individual) or gene frequency (population) at a given locus.

With fine-scale subdivision and consequent genetic drift, regional equilibrium gene frequencies become "no longer a single composition to which the population tends, but a statistical ensemble of compositions with an equilibrium set of associated probabilities" (Lewontin, 1967*a*). The implications of this situation for sampling procedures in population genetics, ecology, systematics, and other areas of population biology are apparent.

Elsewhere (Selander and Yang, 1970), we have suggested that the polygamous, territorial, tribal social system of the House mouse is adaptive in terms of efficient exploitation of environmental resources, particularly food, and that the resulting genetic heterogeneity is an inevitable consequence of subdivision, without adaptive significance *per se*. But whatever the ultimate cause of subdivision may be, the population structure of this species clearly corresponds to that described by Wright (1931) as potentially favourable for evolutionary trial and error among local populations. Although evolutionary biologists have in recent years tended to minimize the importance of random fluctuations in the evolutionary process, a balanced assessment would require much additional information from empirical studies. In the case of the House mouse, we should like to know the average "life span" of the tribal units under various environmental conditions in order to judge how long groups may be withdrawn from the common gene pool (Simpson, 1953). For populations of this species living in agricultural areas, the duration of periods of withdrawal of tribes may be very short. Mouse populations may experience repeated cycles involving the finding of a food supply by a few mice, establishment of one or more founding tribes, rapid proliferation of tribes from the founding stock, inbreeding and drift within tribes, and the eventual depletion of the food supply, leading to dispersal of tribe members and the initiation of a new cycle. Also to be considered is the very real possibility that an active, long-distance dispersal of young may be occurring while the parental populations remain statically subdivided. For this reason, the demonstration of subdivision cannot be interpreted as evidence of an absence or low level of gene flow. It is possible that gene flow within and between populations is only slightly impeded by subdivision.

In the long run, drift in individual tribes may have little significance in evolution, other than providing an opportunity for variant alleles to increase in frequency locally, if tribes do not remain intact and isolated long enough for pervasive genetic changes to occur. However, if large populations are founded by a few individuals invading a barn (or field), the founder effect (Mayr, 1954) may play a role in the evolution of this species, with the total variance in gene frequencies among populations significantly augmented by shifts in genome structure arising from the fact of populations being founded by non-random samples of the gene pool.

At present we cannot properly assess the evolutionary significance of genetic drift and other stochastic processes, for this area of evolutionary biology, and especially that aspect concerning the relationships of social and genetic structures of populations, remains largely unexplored. However, we can look forward to rapid progress in this area in the immediate future, now that techniques for determining the genetic structure of populations are available. These same techniques also hold the promise of producing revolutionary advances in the areas of systematics and speciation.

Acknowledgements

Research discussed in this paper was supported by NIH grant GM-15769 and NSF grants GB-6662, GB-15664, and GB-17849. R. S. Ralin assisted in preparation of the manuscript. The coefficient of genetic similarity in Table II was developed by S. Stewart.

References

Anderson, P. K. and Hill, J. L. (1965). *Mus musculus*: experimental induction of territory formation. *Science, N.Y.*, **148**, 1753–1755.

Brown, J. H. and Welser, C. F. (1968). Serum albumin polymorphism in natural and laboratory populations of *Peromyscus*. *J. Mammal.* **49**, 420–426.

Cavalli-Sforza, L. L. and Edwards, A. W. F. (1967). Phylogenetic analysis: models and estimation procedures. *Evolution, Lancaster, Pa.*, **21**, 550–570.

Crow, J. F. (1968). The cost of evolution and genetic loads. In *Haldane and modern biology*: 165–178. Dronamraju, K. R. (ed.). Baltimore: Johns Hopkins Press.

Crowcroft, P. and Rowe, F. P. (1963). Social organization and territorial behaviour in the wild house mouse (*Mus musculus* L.). *Proc. zool. Soc. Lond.* **140**, 517–531.

Dessauer, H. C. and Nevo, E. (1969). Geographic variation of blood and liver proteins in cricket frogs. *Biochem. Genet.* **3**, 171–188.

Harris, H. (1969). Enzyme and protein polymorphism in human populations. *Br. med. Bull.* **25**, 5–13.

Hubby, J. L. and Throckmorton, L. H. (1968). Protein differences in *Drosophila*. IV. A study of sibling species. *Am. Nat.* **102**, 193–205.

Kimura, M. (1968). Genetic variability maintained in a finite population due to mutational production of neutral or nearly neutral isoalleles. *Genet. Res.* **11**, 247–269.

Lewontin, R. C. (1967*a*). Population genetics. *Ann. Rev. Genet.* **1**, 37–70.

Lewontin, R. C. (1967*b*). An estimate of average heterozygosity in man. *Am. J. hum. Genet.* **19**, 681–685.

Lewontin, R. C. and Dunn, L. C. (1960). The evolutionary dynamics of a polymorphism in the house mouse. *Genetics*, **45**, 705–722.

Lewontin, R. C. and Hubby, J. L. (1966). A molecular approach to the study of genic heterozygosity in natural populations. II. Amount of variation and degree of heterozygosity in natural populations of *Drosophila pseudoobscura*. *Genetics*, **54**, 595–609.

Mayr, E. (1954). Change of genetic environment and evolution. In *Evolution as a Process*: 157–180. Huxley, J. *et al.* (eds.). London: Allen and Unwin.

Mukai, T. (1968). Experimental studies on the mechanism involved in the maintenance of genetic variability in *Drosophila* populations. *Jap. J. Genet.* **43**, 399–413.

O'Brien, S. J. and MacIntyre, R. J. (1969). An analysis of gene-enzyme variability in natural populations of *Drosophila melanogaster* and *D. simulans* *Am. Nat.* **103**, 97–113.

Petras, M. L. (1967). Studies of natural populations of *Mus*. I. Biochemical polymorphisms and their bearing on breeding structure. *Evolution, Lancaster, Pa.*, **21**, 259–274.

Prakash, S. (1969). Genic variation in a natural population of *Drosophila persimilis*. *Proc. natn. Acad. Sci., U.S.A.* **62**, 778–784.

Prakash, S., Lewontin, R. C. and Hubby, J. L. (1969). A molecular approach to the study of genic heterozygosity in natural populations. IV. Patterns of genic variation in central, marginal and isolated populations of *Drosophila pseudoobscura*. *Genetics*, **61**, 841–858.

Rasmussen, D. I. (1964). Blood group polymorphism and inbreeding in natural populations of the deer mouse *Peromyscus maniculatus*. *Evolution, Lancaster, Pa.*, **18**, 219–229.

Reimer, J. D. and Petras, M. L. (1967). Breeding structure of the house mouse, *Mus musculus*, in a population cage. *J. Mammal.* **48**, 88–99

Reimer, J. D. and Petras, M. L. (1968). Some aspects of commensal populations of *Mus musculus* in southwestern Ontario. *Can. Fld. Nat.* **82**, 32–42.

Robertson, A. (1968). The spectrum of genetic variation. In *Population biology and evolution*: 5–16. Lewontin, R. C. (ed.). New York: Syracuse Univ. Press.

Rowe, F. P. and Redfern, R. (1969). Aggressive behaviour in related and unrelated wild house mice (*Mus musculus* L.). *Ann. appl. Biol.* **64**, 425–431.

Schwarz, E. and Schwarz, H. K. (1943). The wild and commensal stocks of the house mouse, *Mus musculus* Linnaeus. *J. Mammal.* **24**, 59–72.

Selander, R. K. (1970). Behavior and genetic variation in natural populations. *Am. Zool.* **10**, 53–66.

Selander, R. K., Hunt, W. G. and Yang, S. Y. (1969*a*). Protein polymorphism and genic heterozygosity in two European subspecies of the house mouse. *Evolution, Lancaster, Pa.*, **23**, 379–390.

Selander, R. K., Smith, M. H., Yang, S. Y. and Johnson, W. E. (1970*a*). Biochemical polymorphism and systematics in the genus *Peromyscus*. I. Variation in the old-field mouse (*Peromyscus polionotus*). *Syst. Zool.* (in press).

Selander, R. K. and Yang, S. Y. (1969). Protein polymorphism and genic heterozygosity in a wild population of the house mouse (*Mus musculus*). *Genetics*, **63**, 653–667.

Selander, R. K. and Yang, S. Y. (1970). Biochemical genetics and behavior in wild house mouse populations. In *Contributions to behavior-genetic analysis: The mouse as a prototype*. (In press). Lindzey, G. and Theissen, D. D. (eds). New York: Appleton-Century-Crofts.

Selander, R. K., Yang, S. Y. and Hunt, W. G. (1969*b*). Polymorphism in esterases and hemoglobin in wild populations of the house mouse (*Mus musculus*). *Studies in Genetics V. Univ. Tex. Publ.* No. 6918, 271–338.

Selander, R. K., Yang, S. Y., Lewontin, R. C. and Johnson, W. E. (1970*b*). Genetic variation in the horseshoe crab (*Limulus polyphemus*), a phylogenetic "relic". *Evolution, Lancaster, Pa.*, **24**, 402–414.

Simpson, G. G. (1953). *The major features of evolution.* New York: Columbia Univ. Press.

Sumner, F. B. (1930). Genetic and distributional studies of three subspecies of *Peromyscus. J. Genet.* **23**, 257–376.

Ursin, E. (1952). Occurrence of voles, mice, and rats (*Muridae*) in Denmark, with a special note on a zone of intergradation between two subspecies of the house mouse (*Mus musculus* L.). *Vidensk. Medd. dansk naturh. Foren.* **114**, 217–244.

Wallace, B. (1968). Polymorphism, population size, and genetic load. In *Population biology and evolution*: 87–108. Lewontin, R. C. (ed.). New York: Syracuse Univ. Press.

Workman, P. L. (1969). The analysis of simple genetic polymorphisms. *Hum. Biol.* **41**, 97–114.

Wright, S. (1931). Evolution in Mendelian populations. *Genetics,* **16**, 97–159.

Symp. zool. Soc. Lond. (1970) No. 26, 93–101.

HUMAN MOLECULAR VARIATION

W. H. P. LEWIS

Department of Pathology, St. Helier Hospital, Carshalton, Surrey.

SYNOPSIS

Variants of human peptidases are presented to illustrate the results of the application of starch gel electrophoresis to the study of enzymes in human populations. Polymorphisms of two enzymes are described and some special features of a number of rare variants are discussed.

INTRODUCTION

A significant number of human enzymes have now been studied by electrophoresis and in spite of the fact that this method can reveal only charge or size differences between proteins, none the less, about one in three has been shown to be polymorphic. Among the enzymes which have been shown to occur in different structural forms in human populations are red cell acid phosphatase (Hopkinson, Spencer and Harris, 1963); phosphoglucomutase (Spencer, Hopkinson and Harris, 1964); phosphogluconate dehydrogenase (Fildes and Parr, 1963); adenylate kinase (Fildes and Harris, 1966); adenosine deaminase (Spencer, Hopkinson and Harris, 1968) and peptidases (Lewis and Harris, 1967, 1969). Since so many enzymes have now been studied in man it would be impossible to do justice to them in a paper of this nature. It is proposed therefore to use the group of enzymes which I have been studying to illustrate the type of information one may obtain by the application of starch gel electrophoresis to the study of human enzymes in different populations and families.

The group of enzymes with which I have been concerned are grouped under the trivial name of peptidases. They are widely distributed in all mammalian tissues which have so far been examined, and are characterized by their ability to catalyse the hydrolysis of peptide bonds of small peptides consisting of 2–4 amino acids. The source of the enzyme for human population studies has been mainly red blood cells but placentae have also been used.

CHARACTERIZATION OF PEPTIDASES

After electrophoresis in starch gel at pH 7·5 of crude haemolysates or tissue extracts, the peptidase activity was detected by utilizing the reaction sequence shown in Fig. 1. The free amino acid released by the peptidase is deaminated by the snake venom amino acid oxidase. The hydrogen peroxide which is one of the products of this reaction is reduced by the catalytic action of peroxidase with the concomitant

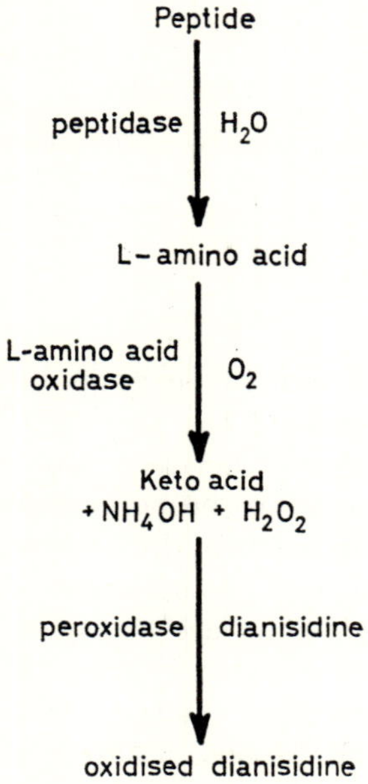

FIG. 1. The reaction sequence employed for the detection of peptidase activity after starch gel electrophoresis.

production of oxidized dianisidine, resulting in a brown precipitate at the site of enzyme activity. The amino acid oxidase shows significant activity against only 6 amino acids, leucine, iso-leucine, phenylalanine, tyrosine, tryptophan and methionine. However, it is possible with the use of peptides from which at least one of these amino acids can be released, to characterize the substrate specificity of these enzymes quite precisely. The results are shown in Table I. In human red blood cells at least 6 distinct peptidases are demonstrable and their relative electrophoretic mobilities are illustrated in Fig. 2. These enzymes may also be distinguished on the basis of their reactivity with –SH reagents, suggesting that peptidases C, D and E contain at least one reactive –SH group per enzyme molecule. Peptidases A, B, and F do not apparently react with reagents of this type.

TABLE I

Relative activities of the human red blood cell peptidases against various amino acid peptides and leucyl-β-naphthylamide (Leu-β-NA)

	Peptidase					
	A	B	C	D	E	F
Val-Leu	+ + +	−	−	−	−	−
Gly-Leu	+ + +	−	+	−	−	−
Leu-Gly	+ + +	−	+	−	−	−
Gly-Phe	+ +	−	+	−	−	−
Leu-Ala	+ +	−	+ +	−	−	−
Leu-Leu	+ +	−	+ +	−	+	−
Gly-Trp	+ +	−	+ +	−	−	−
Lys-Leu	+	+	+ + +	−	+	−
Lys-Tyr	+	+	+ + +	−	+	−
Pro-Phe	+ +	+	+ + +	−	−	−
Pro-Leu	+ +	+	+ + +	−	−	−
Phe-Leu	+ +	+ +	+ + +	−	+	−
Ala-Tyr	+ +	+ + +	+	−	±	−
Phe-Tyr	+ +	+ +	+ + +	−	+	−
Leu-Tyr	+ + +	+ +	+ + +	−	+	−
Leu-Gly-Gly	−	+ + +	−	−	±	−
Leu-Gly-Phe	−	+	−	−	±	−
Leu-Leu-Leu	−	+ +	−	−	+	+
Tyr-Tyr-Tyr	−	+	−	−	+	+
Phe-Gly-Phe-Gly	−	−	−	−	+ +	−
Leu-β-NA	−	−	−	−	+	−
Leu-Pro	−	−	−	+ +	−	−
Phe-Pro	−	−	−	+ +	−	−

So far, only peptidases A, B, C and D have been studied in large population groups. In all some 5000 individuals of European, African, Indian and Australasian origin have been studied.

PEPTIDASE A

The various phenotypes of peptidase A are shown in Fig. 2. With the exception of the common phenotype, Pep A1, only Pep A2-1 occurs at an appreciable frequency in any of the population groups, and has so far been found, with the exception of a single individual, only among individuals of African origin, where it occurs in about one individual in seven. The frequencies of the various phenotypes of peptidase A in different population groups are shown in Table II.

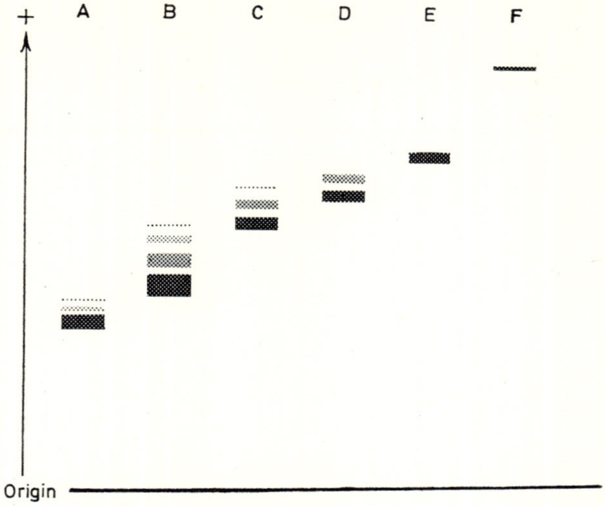

Fig. 2. Relative electrophoretic mobilities in starch gel at pH 7·5 of human red blood cell peptidases.

TABLE II

Frequencies of peptidase A phenotypes in various population groups

		Peptidase A							
	Total	1	2–1	2	3–1	4–1	5–1	6–1	7–1
Europeans	3129	3123	—	—	1	3	1	1	1
British Resident Negroes	636	554	78	3	1	—	—	—	—
"Bantu" (Southern Africa)	100	87	11	2	—	—	—	—	—
Yoruba (Nigeria)	155	127	25	3	—	—	—	—	—
Asiatic Indians	615	613	—	—	1	—	1	—	—

Peptidase A2-1 appears on the basis of family studies to occur in individuals heterozygous for a pair of alleles (*Pep* A^1 and *Pep* A^2) which in homozygotes give rise to the Pep A1 and Pep A2 phenotypes. Pep A3-1, Pep A4-1, Pep A5-1, Pep A6-1 and Pep A7-1 are all rare, and family studies suggest that they occur in individuals heterozygous for the common allele *Pep* A^1 and one of a series of rare alleles, *Pep* A^3, *Pep* A^4 etc.

This particular set of rare variants illustrates a number of interesting features of enzyme variation. It can be seen from Fig. 2 that the electrophoretic patterns of the presumed heterozygotes all show three main bands of activity. This type of electrophoretic pattern in heterozygotes is thought to occur when the enzyme in homozygotes consists of at least two identical sub-units (Shaw, 1965). If the sub-unit coded for by the common allele *Pep* A^1 is represented by α^1 and that coded for by the variant allele by α^*, and if two different sub-units are synthesized in the heterozygote and combine at random, then there will be three possible enzyme structures: $\alpha^1\alpha^1$, $\alpha^1\alpha^*$ and $\alpha^*\alpha^*$.

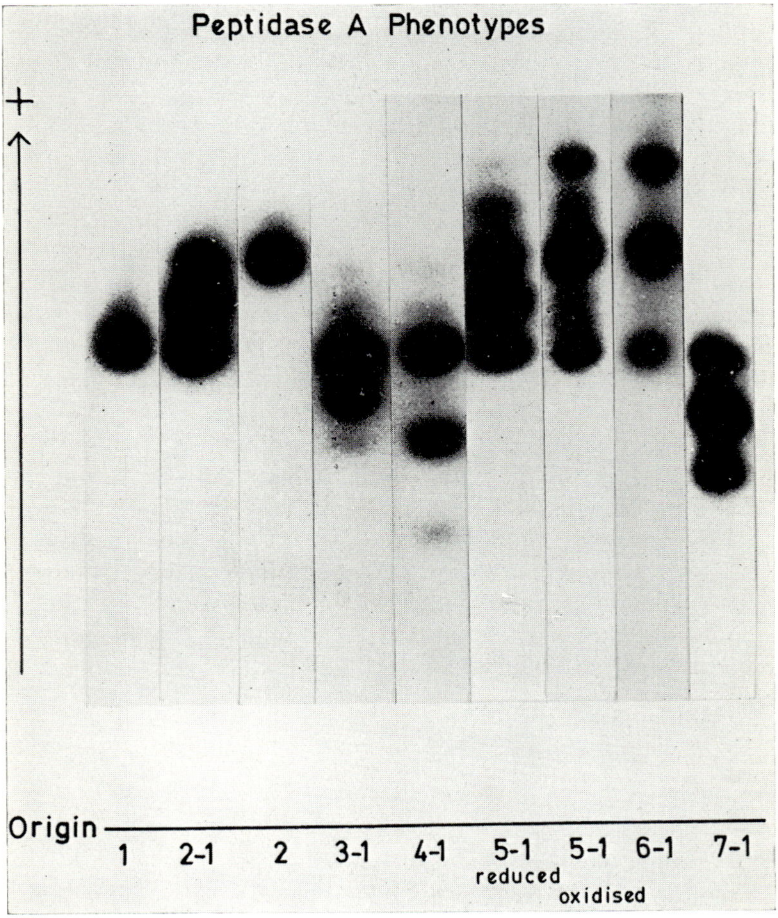

FIG. 3. Electrophoretic patterns of some of the phenotypes of peptidase A in man.

Another interesting point is the asymmetric pattern in some of the phenotypes, particularly Pep A3-1 and Pep A4-1. This phenomenon, in which the components associated with the possession of the *Pep* A^3 and *Pep* A^4 alleles are very much weaker than the other main band of activity, might result from these sub-units being catalytically less active, synthesized at a slower rate or less stable. Both Pep A3-1 and Pep A4-1 have been observed in placentae, and in each case the electrophoretic pattern was symmetrical with the three bands showing activity in order of mobility approximately in the ratio of 1 : 2 : 1. It would appear therefore that the sub-units coded for by *Pep* A^3 and *Pep* A^4 are less stable and that this is seen in the red blood cell because synthesis of protein has ceased. It is clear that the choice of tissue in the study of enzyme variation can play an important part in the determination of whether or not subtle differences of this sort are revealed.

Pep A5-1 is unique in that the electrophoretic pattern seen in this phenotype can occur in two forms (Fig. 3). If a fresh sample is subjected to electrophoresis then the so-called reduced pattern is seen, but if the sample is stored at 4°C for a few days the pattern labelled oxidized is seen. If the stored sample is treated with mercaptoethanol at a concentration of 25 mM immediately before electrophoresis then the reduced pattern is again found. On the other hand if a fresh haemolysate is treated with oxidized glutathione at a concentration of 25 mM about 15 min before electrophoresis, the pattern observed is identical with that of the stored sample. It seems likely that the sub-unit coded for by the variant allele *Pep* A^5 differs from that coded for by the common allele *Pep* A^1 in that it contains a reactive –SH group which undergoes an exchange reaction with oxidized glutathione forming a disulphide bond with a half molecule, while the other half molecule is converted to reduced glutathione. The addition of the acidic peptide glutathione to the enzyme molecule could explain the increases in electrophoretic mobility which follow storage (for a more detailed discussion of this particular variant see Lewis, Corney and Harris, 1968).

PEPTIDASE B

Peptidase B shows no common variants in the European, African or Indian groups, but a number of rare variants, some of which are illustrated in Fig. 4, have been found. Pep B2-1, Pep B3-1 and Pep B4-1 have all been found among the European group, Pep B5-1 was found in a pair of Nigerian (Yoruba) twins, and Pep B6-1, Pep B6 and Pep B7-1 have so far been found only among Australian aboriginees.

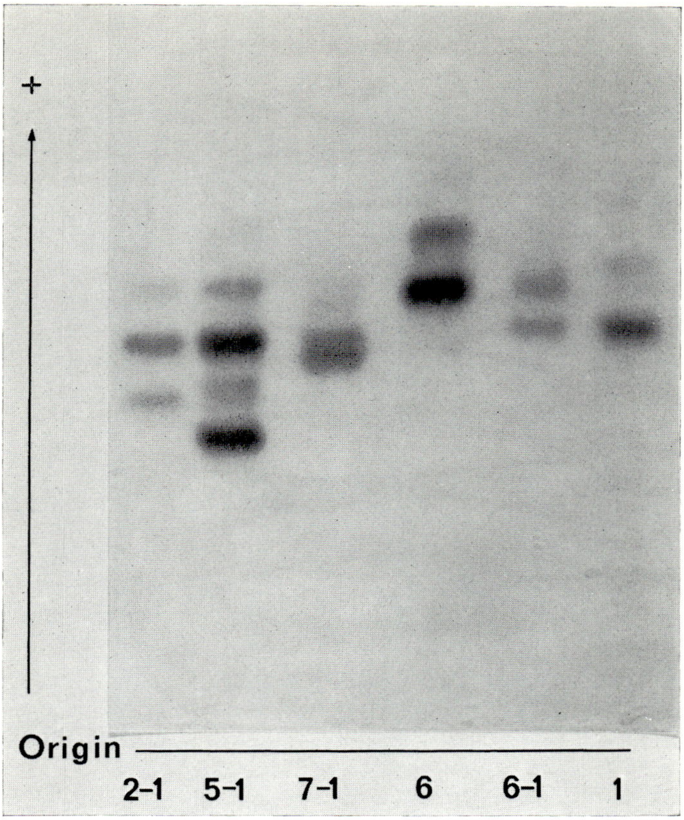

Fig. 4. Electrophoretic patterns of some of the phenotypes of peptidase B in man.

These last 3 variants are quite common in some tribes and are absent in others, suggesting that they may be quite recent mutations. Family studies suggest that the rare variants occur in individuals who are heterozygous for a common allele, *Pep B^1*, and a series of rare alleles *Pep B^2*, *Pep B^3*, etc., which occur at an autosomal locus (Lewis and Harris, 1967; Blake, Kirk, Lewis and Harris, 1969).

As shown in Fig. 4, the electrophoretic pattern of Pep B1 consists of a series of bands which have diminishing activity towards the anode. In Pep B6 the electrophoretic pattern is similar to Pep B1 except that all the components migrate further towards the anode by a constant amount for each isozyme. Thus it appears that all the isozymes of peptidase B are controlled by a single gene locus. This phenomenon is seen in a number of other human enzymes including adenylate kinase

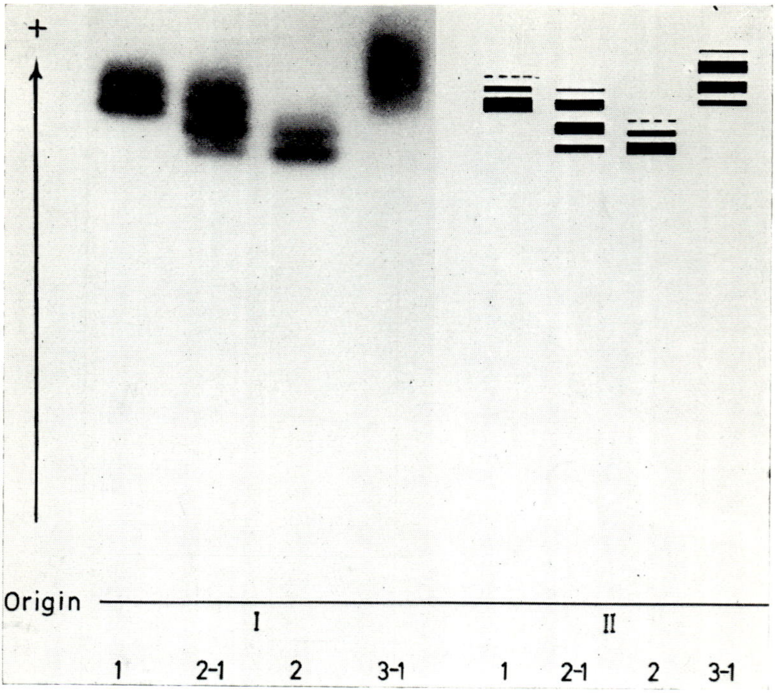

FIG. 5. Electrophoretic patterns of the phenotypes of peptidase D in man (From Lewis and Harris, 1969, by permission).

(Fildes and Harris, 1966), phosphoglucomutase (Spencer *et al.*, 1964) and adenosine deaminase (Spencer *et al.*, 1968).

PEPTIDASE D

Peptidase D (prolidase) shows a different order of variation from the other peptidases. The electrophoretic patterns of the phenotypes which have so far been found in the three main population groups are shown in Fig. 5. The incidence of Pep D2-1 in the European and Indian groups is around 2% and in the African group this phenotype is about twice as common. Pep D3-1, on the other hand, occurs at an appreciable frequency among the African group, where it is about as common as Pep D2-1 (Lewis and Harris, 1969). One example of the Pep D3-1 phenotype was found in the Indian group where it may presumably be regarded as a rare variant. The relative similarity between the gene frequencies in the European and Indian groups, and the relative dissimilarity of the African group is a common feature of human population studies.

So far, only a limited amount of information is available about the linkage relationships of the peptidases. However, studies on a family in which variants of peptidases A and B were segregating and another in which variants of peptidases A and D were segregating, suggest that neither the loci controlling the structures of peptidases A and B nor the loci controlling peptidases A and D are closely linked. These pairs of loci are either on different chromosomes or well separated on the same chromosome.

References

Blake, N. M., Kirk, R. L., Lewis, W. H. P. and Harris, H. (1969). Some further peptidase B phenotypes. *Ann. Hum. Genet.* **33**, 301–305.

Fildes, R. A. and Harris, H. (1966). Genetically determined variation of adenylate kinase in man. *Nature, Lond.* **209**, 261–263.

Fildes, R. A. and Parr, C. W. (1963). Human red cell phosphogluconate dehydrogenases. *Nature, Lond.* **200**, 890–891.

Hopkinson, D. A., Spencer, N. and Harris, H. (1963). Red cell acid phosphatase variants, a new human polymorphism. *Nature, Lond.* **199**, 969–971.

Lewis, W. H. P., Corney, G. and Harris, H. (1968). Pep A5–1 and Pep A6–1: two new variants of peptidase A with features of special interest. *Ann. Hum. Genet.* **32**, 35–42.

Lewis, W. H. P. and Harris, H. (1967). Human red cell peptidases. *Nature, Lond.* **215**, 351–355.

Lewis, W. H. P. and Harris, H. (1969). Peptidase D (prolidase) variants in man. *Ann. Hum. Genet.* **32**, 317–322.

Shaw, C. R. (1965). Electrophoretic variation in enzymes. *Science, N.Y.*, **149**, 936–943.

Spencer, N., Hopkinson, D. A. and Harris, H. (1964). Phosphoglucomutase polymorphism in man. *Nature, Lond.* **204**, 742–745.

Spencer, N., Hopkinson, D. A. and Harris H. (1968). Adenosine deaminase polymorphism in man. *Ann. Hum. Genet.* **32**, 9–14.

TAXONOMY AND PHYLOGENY

Symp. zool. Soc. Lond. (1970) No. 26, 105–116.

PATTERNS OF SUBSPECIFIC VARIATION

G. B. CORBET

Mammal Section, British Museum (Natural History), London, England

SYNOPSIS

Geographical variation within a species may be random, clinal or discontinuous, or a combination of these may be found in one species. The chief obstacle to determining the pattern of variation in a species is the availability of samples that are sufficiently large, and sufficiently random with respect to locality, season, age, etc., to allow the true pattern of variation in nature to be accurately inferred from the pattern of variation seen in the collected sample. Accurate mapping of distribution combined with study of habitat selection assists in determining the degree of continuity or discontinuity of range and therefore, probably, of variation. The European hedgehog, *Erinaceus europaeus*, shows how provisional is our knowledge of variation even in a relatively well known species. The elephant-shrews, Macroscelididae, show how variation tends to be continuous in steppe species, whereas in forest species it is discrete where the forest is fragmented but clinal within large blocks of forest or along the linear habitat of riverine forest.

Arguments against the formal recognition and naming of subspecies are convincing when applied to species in which the whole pattern of variation has been well demonstrated and shown to lack discretely recognizable entities, but there is a strong case for retaining the use of named subspecies in lesser known species and especially where the limits of the species are still uncertain. The concept of subspecies is meaningless unless it is restricted to discrete segments of a species. A distinction should be drawn between "definitive subspecies" that have been convincingly shown to be discrete entities in nature and "provisional subspecies" that are based on discretely definable samples that are too small or non-random to indicate with a high degree of probability whether the populations from which they were drawn are or are not discrete entities.

INTRODUCTION

A frequently employed definition of the species is that of Mayr (1963): "groups of actually or potentially interbreeding natural populations, which are reproductively isolated from other such groups". In spite of some difficulty in application this is a useful concept of the relationship between different species, but it also carries an implication about the nature of variation within the species. By describing a species as a group of "natural populations" some pattern of intraspecific variation involving recognizable lower categories seems to be implied. Likewise, looking at the subject of variability from the other end, the geneticist recognizes aggregates of individuals within the species that he variously calls "demes" or "populations". Too often, both in taxonomic and genetical contexts, the existence of a hierarchy of discrete groups within the species is taken for granted. The object of this paper is to consider the reality and the nature of such patterns within the species, and especially the nature of the primary subdivisions or major trends of variation within the species as a whole rather than the more local levels of variation that chiefly concern the population geneticist.

THE NATURE OF THE SPECIES

It will first be necessary to elaborate a little on the nature of the species itself. There has been much argument about whether species should be conceived as groups of animals defined in morphological terms or in terms of reproductive isolation, as if these were incompatible alternatives. In fact they are usually thoroughly compatible since it is the inability of two species to interbreed that allows morphological divergence to take place and to persist. The morphological differences between reproductively isolated groups may be very small and difficult to detect, but it is not the magnitude of a specific difference that makes species objective realities but the existence of a discontinuity of variation that is caused by the inability of the two groups to interbreed and thereby to exchange characteristics.

The first task of the taxonomist is not to create, nor to describe nor even to define species, but to detect them, by recognizing, amongst the often bewildering complex of variation correlated with age, sex, locality or individual genotype, those differences between individuals that are correlated with the inability to interbreed. In order that two species should retain their distinctive characters, it is not, of course, necessary for interbreeding to be absolutely impossible, only that it should be sufficiently rare, or that the offspring should be at a sufficient selective disadvantage, not to swamp the differences that already exist between the species.

The principal difficulty in applying this concept of the species is in determining the relationship between completely isolated, allopatric groups that show a small, but consistent, amount of phenotypic difference. In view of the existence of every degree of difference between allopatric populations from virtual identity to a level of distinctness normally found between sympatric species, no absolute criterion of a specific difference can be laid down. But as a guide towards stabilizing the concept of a species the definition based on reproductive isolation is valuable. Furthermore in practice it is often no more difficult to apply than one based on morphology, in view of the sampling difficulties involved and the variable degree of phenotypic difference between discrete, sympatric species.

THE NATURE OF SUBSPECIFIC VARIATION

Many of the divergences of opinion as to the nature of subspecific variation and how it should be described stem from the difference between the approach of the academic student of speciation, who can afford to choose species in which the variation is well known or easily observed, and that of the practising taxonomist who, in revising a group, must deal with species ranging from the well known to the unknown. But even in well known species the limitations imposed by the exceedingly small samples available have rarely been appreciated. Considering a species as a geographical unit, variability amongst its members may theoretically be considered to have (1) no geographical component (i.e. to be spatially random), (2) to show continuous geographical components (i.e. to be clinal) or (3) to be discontinuous; or each of these situations may apply in different parts of the range. The nature of this variation can be deduced only from the study of samples. Compared with the size of the "populations" from which these samples have been drawn, even the best museum collections represent a minute sample, and even where, as in the case of the larger mammals, these can be supplemented by field observations, the available data still represent, in most cases, an exceedingly small sample. The pattern of variation, random, continuous or discontinuous, seen in the collected (or observed) sample can be taken to represent the pattern present in nature only if the sample is adequate in size and is random with respect to space and time, a condition rarely realized. Because the samples available for study are always spatially discontinuous, the degree to which the variation amongst the living animals is continuous or discontinuous must be deduced from our knowledge of the species' distribution and in particular of its habitat selection.

The European hedgehog—a relatively well known species

The European hedgehog, *Erinaceus europaeus,* can be used to illustrate some of these problems. Being rather easily recorded, its distribution is known with fair accuracy, and from what is known of its habitat requirements it is likely that it is distributed with a considerable degree of continuity throughout the main parts of its range. On the other hand the samples that have been available to students of variation have usually been extremely small and most of the taxonomic results must be considered as "sample taxonomy", i.e. it has been carried out on samples that are inadequate to reveal the actual patterns found in nature. Subspecies have been described and subspecific boundaries drawn with little regard to the probability of either the range or the variation being continuous or discontinuous.

The overall range (Fig. 1) comprises two major, widely isolated segments, in Europe and western Asia, and in eastern Asia. The samples that have been examined have probably been sufficient to establish that the morphological differences between these major

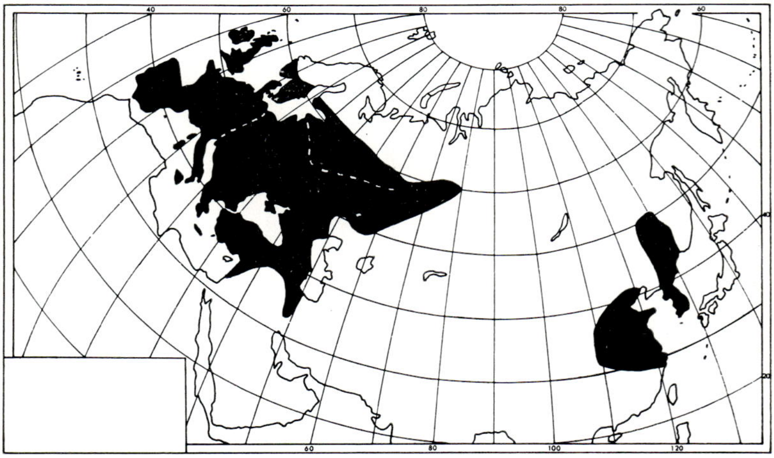

FIG. 1. The distribution of the *Erinaceus europaeus* group. The white broken lines indicate the position of possible species boundaries.

segments are fairly constant, but recent workers have disagreed on the rank to be accorded to the difference: species according to Jones and Johnson (1960), subspecies according to Stroganov (1957). Within the western segment of the range, several lesser isolates are found, e.g. in Ireland, Britain, Scandinavia and the Mediterranean islands. One of these, on Crete, is distinguishable from the mainland population (but only on the basis of small size); others, e.g. on Rhodes, Sicily and Britain, have been named as subspecies on the basis of wholly inadequate samples.

Variation within the continental population is of considerable interest. Twelve names have been proposed, as subspecies or species, but the only boundary that has been demonstrated with any conviction is that in eastern Europe between the western form *europaeus* with dark breast and short maxillae, and the eastern form *concolor* (=*roumanicus*) with white breast and long maxillae. Assessment of this difference has alternated between the specific and subspecific level. The boundary, at least in Czechoslovakia, has only recently been shown to be objective, and additional differences in the skull and in the chromosome complement tend to support the two-species hypothesis (Kratochvil, 1966; Kral, 1967). However, the situation is complicated by the fact that northern Russian animals appear to belong to the western form although an equally objective boundary between the two forms in Russia has not yet been demonstrated.

Other reputed continental subspecies are considerably more hypothetical. Southern Spanish animals have been named as *E.e. hispanicus* on the basis of pale colour. They are indeed pale and the proportion of pure white spines is greater, but the limited material available suggests that this is smoothly clinal and any boundary, such as that shown on a map by Herter (1963), would be quite arbitrary.

The Elephant-shrews—a less well known group

The above example illustrates how little we know about the pattern of variation in a relatively well known species. The situation in less well known species can be illustrated from the Elephant-shrews of Africa, family Macroscelididae. Elephant-shrews are found in much of Africa in habitats ranging from rain forest to semi-desert. Within this family about 45 forms have been described as species and an additional 45 as subspecies. In a recent revision (Corbet and Hanks, 1968) these were reduced to 14 species, although subsequent work would increase this by one (Corbet, in press). The number of specimens available for a study of variation is not greatly different from the situation with regard to hedgehogs, but we lack the detailed knowledge of distribution and habitat that would help us in extrapolating from the pattern of variation in our collected samples to the actual pattern in nature.

The object of the revision mentioned above was to delimit the species rather than to study subspecific variation in detail, but it became clear that the nature of the subspecific variation itself varied from species to species.

All Elephant-shrews are ground-dwelling, predominantly diurnal insectivores. The East African species, *Elephantulus rufescens*, is a typical plains species, found in the short-grass plains of East Africa where it shelters in clumps of thorn-bush from which it makes short forays into the surrounding open ground (Fig. 2). It was considered that 13 previously named forms probably belonged to this species. These had mostly been named on the basis of colour, which varies dorsally from rufous brown to pale yellowish brown or dark greyish brown whilst the hairs of the ventral pelage are either pure white or have more or less prominent grey bases. The major part of the available collection, consisting of series from discrete, widely separated localities,

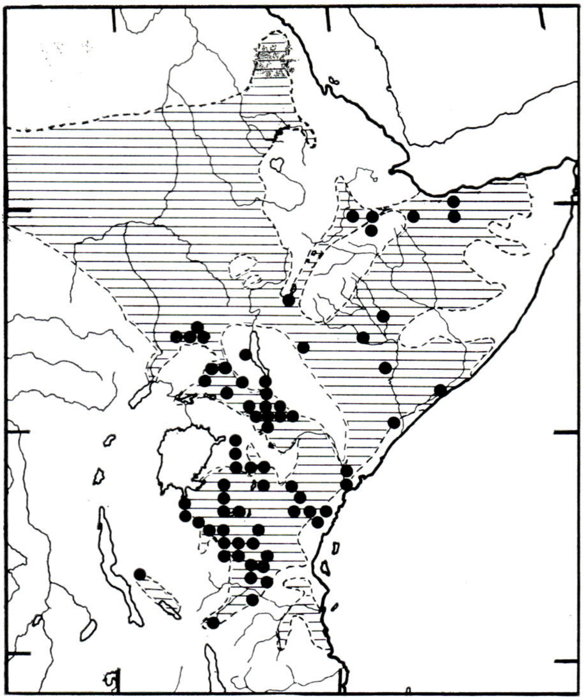

FIG. 2. The known distribution of *Elephantulus rufescens* (spots) related to the distribution of the short-grass steppe/dry savanna vegetation zone (horizontal shading). (Adapted from Corbet and Hanks (1968) and Keay (1959) respectively.)

does indeed give the impression of discontinuous variation, but the more specimens that are collected from intermediate areas, the more difficult it becomes to postulate any discrete boundaries between these forms. In view of the continuity of apparently suitable habitat through most of the range, it seems likely that most of the variation is continuous, with the possible exception of certain marginal populations.

In addition we find that there seems to be a strong correlation between colour and habitat, and it is probable that similar pelages have been independently developed in different parts of the range with similar soils or vegetation. The very rufous form (*rufescens* s.s.) found at Voi in Kenya, where red soils predominate, is very similar to the form *boranus* described from southern Ethiopia. Likewise the sandy coloured form with pure white ventral pelage found in the drier parts of Tanzania, e.g. *ocularis* from Dodoma, closely resembles that in the dry parts of northern Kenya, e.g. *dundasi* from Lake

Baringo; and the greyish form found in the Serengeti area of Tanzania closely resembles one from south-eastern Ethiopia (*peasei*). Subspecific variation in this species has yet to be studied in detail, but it seems likely that there is a complex mosaic showing correlation with local habitat, without discrete boundaries, and that no pattern of major discrete subspecific units can be recognized.

The opposite extreme with regard to habitat is represented by members of the genus *Rhynchocyon*, the giant Elephant-shrews of the forest (Fig. 3). Two species, *R. chrysopygus* and *R. petersi*, have very limited ranges in the coastal forests of Kenya and north-eastern

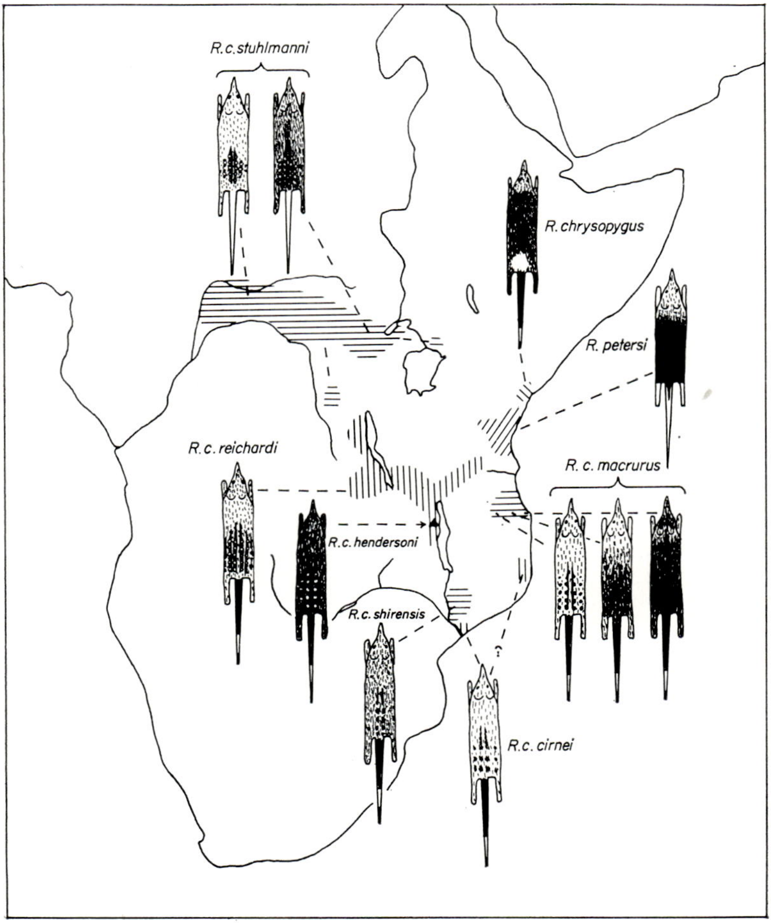

FIG. 3. The range and variation of members of the genus *Rhynchocyon*, showing both discrete and clinal variation in *R. cirnei*.

Tanzania. The remaining members of the genus have been tentatively treated as a single species, *R. cirnei*, and have a wide range from the Congo forest to Uganda and south to the Zambezi. This species occurs in forest with closed canopy including rain forest, the drier forests of eastern Africa (apparently including some riverine thicket), and the lower montane forests of the Tanganyika–Nyasa Rift and south-western Tanzania. Much of the range is therefore fragmented and this is reflected in the extreme variation of colour and pattern (illustrated in colour by Corbet and Hanks, 1968: plate I).

The animals in the Congo Forest have one rather constant character, the white tail, that separates them from all the more southern forms, but the dorsal colour is variable, showing an increase in intensity from west to east with the darkest animals in the Ituri Forest. The eastern members of this group are isolated in several relict forests in Uganda. The few specimens available suggest that these isolates are not strongly differentiated but the samples are inadequate. It is possible that this northern group should be given specific rank, *R. stuhlmanni*, but information is not available to show how this form interacts with the adjacent form *R.c. reichardi* that extends in montane and relict forest from the northern end of Lake Tanganyika to near the southern end of Lake Nyasa with remarkably little variation considering the apparent fragmentation of the forest. This form is characterized by clear white spots on the back and a very dark tail.

Within the range of *R.c. reichardi*, a single very dark individual from near Livingstonia in northern Malawi was named as *R. hendersoni* and subsequently treated as a race of *R. cirnei*. It seemed likely that this was no more than a melanic individual of the adjacent race, until two further identical specimens came to light from the same place. These very distinctive animals are therefore known only from one small patch of forest within a few miles of other forests occupied by the widespread and uniform *R.c. reichardi*.

In southern Malawi *R. cirnei* is represented by a subspecies *R.c. shirensis* quite discretely different from *R.c. reichardi*, lacking white spots, and showing very little variation within its own range. In contrast, the form found in the forests of southern Tanzania, *R.c. macrurus*, shows great variation. At the coast, e.g. around Lindi, the pattern of dark lines and spots in the dorsal pelage, characteristic of the whole species, is almost obliterated by an overall suffusion of rufous colour approaching black on the rump. This gives these animals a superficial resemblance to the adjacent coastal species *R. petersi*, but there are differences in the head and tail, and the relationship with *R. cirnei* is shown by the fact that away from the coast the red colour is reduced, apparently in a gradual sequence, until at Liwale, about 180 km inland, it is scarcely perceptible. This cline between two superficially very different forms could not easily have been predicted. Before the intermediate stages of the cline came to light the coastal form was in fact treated as a subspecies, *melanurus*, of the adjacent species *Rhynchocyon petersi*.

Rhynchocyon cirnei therefore shows a number of very discrete allopatric segments which must be considered on the borderline of speciation. Some of these are characterized by great uniformity, suggesting that any fragmentation of the forest within the range is recent, e.g. *R.c. reichardi* and *R.c. shirensis*. Others show clinal variation in what is probably still a continuous habitat, whilst *R.c. hendersoni* probably represents a recent, but rather extreme, single-character difference that has arisen since the local fragmentation of forest in the Nyika area of Malawi has taken place.

DISCUSSION

The very provisional state of our knowledge of the nature of the regional variation in Palaearctic *Erinaceus*, in spite of the long history of acquaintance with the group, can be attributed very largely to the "political" fragmentation of its range, resulting in fragmentary and uncoordinated work published in a great diversity of languages and based on collections widely scattered in a multitude of museums. No one has attempted the task of analysing even the existing data on the continental scale necessary to reveal the overall pattern. This is the situation that applies to most Palaearctic species, and it is consequently dangerous to use the current subspecific taxonomy for any generalization with regard to variation or zoogeography without appreciating the very provisional nature of the data.

These differences are less serious with regard to Nearctic species and several fairly detailed studies of regional variation have been undertaken. Hagmeier (1958) for example studied the American marten, *Martes americana*, and showed that the previously described subspecies had little relevance to the true nature of the variation. He demonstrated that the variation was clinal without concordance between the clines in different characters, and concluded that no discrete subspecies could be recognized. This is likely to be the case with most of the larger, highly mobile mammals with large and apparently continuous ranges, but one must always guard against the possibility that discrete boundaries, such as has been found in *Erinaceus* in eastern Europe, might be interposed even in a continuous range.

The situation with regard to tropical mammals is in one sense less satisfactory in that very few genera or families have been satisfactorily revised in recent years even at the species level. On the other hand the relevant material, although less abundant than in the case of most temperate groups, is generally available in a few major museums and the literature is much less scattered and in fewer languages. Of the two species of Macroscelididae described above, *Elephantulus rufescens* is more typical of the situation generally found in small mammals. The numerous forms already described are often based on subtle differences of colour and size, but sufficient examples of very similar sibling species have been demonstrated, amongst rodents, shrews and bats, that extreme caution must be exercised in "lumping" even slightly differentiated forms.

The kind of gross geographical variation of pattern and colour seen in *Rhynchocyon* is more characteristic of birds than of mammals, but it is also found in some squirrels and monkeys. However, there are few cases where the variation has been analysed in sufficient detail to distinguish between discrete differences across barriers, and gradual transitions which can often seem discrete by the haphazard application of subspecific nomenclature (Corbet, 1966, p. 7). A classical example of this kind of variation is the squirrel, *Callosciurus prevosti* in Borneo where rivers appear to separate discrete subspecies (Banks, 1931). A more recently studied example is the tamarin, *Saguinus fuscicollis*, in the Amazon Basin, which also shows discretely delimited subspecies, differing in colour and pattern and separated by rivers (Hershkovitz, 1968).

THE NOMENCLATURE OF INFRASPECIFIC CATEGORIES

The major difficulty that has confounded most attempts at the description of subspecific variation in any comprehensive way is the difficulty of judging whether an apparently discrete difference between two observed samples represents an equally discrete difference within the species from which the samples were taken, or is due to non-random sampling of a population showing continuous variation. Because of this kind of difficulty the formally named subspecies of taxonomic workers is almost meaningless as indicating any particular level or kind of subspecific unit. The very large majority of subspecies, as listed in checklists, have not been shown to have any objective reality as discrete units within a species. For these reasons some authors, e.g. Wilson and Brown (1953), have advocated the complete abandonment of formally applied trinomials for subspecies. Burt (1954) argued along the same lines with special reference to mammals. The subspecies has also had its defenders, e.g. Inger (1961) dealing with vertebrates in general and Anderson (1966) dealing with the gophers of the genus *Thomomys*. A compromise solution is that of Pimental (1959) who discussed the possible patterns of subspecific variation and concluded that formal naming of subspecies should be confined to its highest level, i.e. physically isolated entities differing in several correlated characters at the 84% level (i.e. 84% of each subspecies showing non-overlap of variation within each of these characters). Pimental recognized the gulf between the studied sample and the living population and maintained that "most proposals of taxa below the genus are little more than hypothesis". But he appeared to take the view that the study of subspecific variation was concerned only with the academic study of evolution and advocated a perfectionist policy of choosing a suitable species and collecting optimum samples from optimally spaced localities before rushing into print.

Undoubtedly a great deal of taxonomic effort is now wasted on studies of variation that are neither sufficiently detailed to contribute to any general theory, nor sufficiently well documented to provide a contribution to classification. But there is a great need for a general purpose taxonomy that represents our present knowledge of the diversity of mammals, however incomplete that knowledge is. It is tempting to say that formal subspecific names should be applied only to segments of a species that have been shown to be discrete and recognizably different, e.g. complete isolates or contiguous populations where a very narrow zone of intergradation between two otherwise discrete forms has been adequately demonstrated. But this is unrealistic in that the necessary information is rarely available. The only solution, to make subspecific names meaningful, seems to be to reject all names based on average differences or that have been shown to represent points on a cline; to treat as "provisional subspecies" groups that can be discretely diagnosed on the basis of presently available data but cannot yet be confidently considered to represent discrete groups in nature; and as "definitive subspecies" groups whose presence as discrete entities in nature has been shown by adequate sampling. It is probable that the majority of existing subspecific names will finally be rejected on this criterion; but at the opposite extreme others will undoubtedly prove to represent distinct species—it would be a pity to throw out the baby with the bathwater.

References

Anderson, S. (1966). Taxonomy of gophers, especially *Thomomys*, in Chihuahua, Mexico. *Syst. Zool.* **15**, 189–198.

Banks, E. (1931). The forms of Prevost's squirrel found in Sarawak. *Proc. zool. Soc. Lond.* 1931, 1335–1348, pl. 1.

Burt, W. H. (1954). The subspecies category in mammals. *Syst. Zool.* **3**, 99–104.

Corbet, G. B. (1966). *The terrestrial mammals of western Europe.* London: Foulis.

Corbet, G. B. (in press). Order Macroscelidea. In *The mammals of Africa: an identification manual.* Meester, J. (ed.). Washington: Smithsonian Institution.

Corbet, G. B. and Hanks, J. (1968). A revision of the elephant-shrews, family *Macroscelididae. Bull. Br. Mus. nat. Hist.* (*Zool.*) **16**, 47–111.

Hagmeier, E. M. (1958). Inapplicability of the subspecies concept to North American marten. *Syst. Zool.* **7**, 1–7.

Hershkovitz, P. (1968). Metachromism or the principle of evolutionary change in mammalian tegumentary colors. *Evolution, Lancaster, Pa.* **22**, 556–575.

Herter, K. (1963). *Igel.* Ziemsen: Wittenberg Lutherstadt. (English edition: 1965, *Hedgehogs.* London: Phoenix House.)

Inger, R. F. (1961). Problems in the application of the subspecies in vertebrate taxonomy. In *Vertebrate speciation.* (Blair, W. F. ed.). 262–285. Austin: University of Texas Press.

Jones, J. K. and Johnson, D. H. (1960). Review of the insectivores of Korea. *Univ. Kans. Publs. Mus. nat. Hist.* **9**, 549–578.

Keay, R. W. J. (1959). *Vegetation map of Africa.* Oxford: University Press.

Kral, B. (1967). Karyological analysis of two European species of the genus *Erinaceus. Zool. Listy* **16**, 239–252.

Kratochvil, J. (1966). Zur Frage der Verbreitung des Igels (*Erinaceus*) in der CSSR. *Zool. Listy* **15**, 291–304.

Mayr, E. (1963). *Animal species and evolution.* Cambridge, Mass.: Harvard Univ. Press.

Pimental, R. A. (1959). Mendelian infraspecific divergence levels and their analysis. *Syst. Zool.* **8**, 139–159.

Stroganov, S. U. (1957). *Insectivores of Siberia.* Moscow.

Wilson, E. O. and Brown, W. L. (1953). The subspecies concept. *Syst. Zool.* **2**, 97–111.

Symp. zool. Soc. Lond. (1970) No. 26, 117–125.

OVERLAP OF VARIATION IN BRITISH AND EUROPEAN MAMMAL POPULATIONS

MICHAEL N. DADD

The Zoological Society of London, London, England

SYNOPSIS

Twenty-three of the 38 British species of mammals have had endemic subspecies described, 15 of which are still in current use. The validity of these has not hitherto been examined in the light of modern taxonomic theory, or of additional material obtained since their description. Their designation does not, therefore, imply any constant level of differentiation from the continental forms, although it is often taken to.

This paper sets out to investigate the validity of some of these cases in which the British form of a species has at any time been regarded as distinct from the continental—regardless of existing views. Irish forms are also considered, but not those endemic to the smaller islands. The level of difference required for two populations to be considered distinct is that 90% of one population should be clearly distinguishable from the same proportion of the other.

Of the 6 subspecies examined here, the validity of one is supported, one previously considered invalid is now considered valid, and the remainder are rejected.

INTRODUCTION

There are, at present, 38 indigenous species of mammals to be found on the mainland of the British Isles. Twenty-three of these have been considered sufficiently distinct from the continental forms to warrant separate status. Of the 26 available subspecies, Miller (1912) accepted 18 as valid. Thirteen of these, together with two described more recently, are commonly accepted today. In the past, any local population whose members could be distinguished from the nominate form by some character were regarded as distinct species or subspecies. This was possibly a valid approach in the nineteenth century when collections usually consisted of a few specimens from widely scattered localities. With the increased amount and range of material now available, and in the light of modern ideas on variation and taxonomic theory, the validity of many of these early descriptions must be in doubt. At present therefore, the designation of the British forms as "endemic subspecies" should not be taken to imply, although it often is, a constant level of differentiation from the continental forms.

The present study set out to investigate the validity of those cases in which the British form of a species has at any time been regarded as distinct from the continental—regardless of any existing views as to their status. Cases of subspecific variation within the British Isles are also being considered, except for those endemic to the smaller islands. The aim so far has been to test the validity of the original descriptions—that is, of those characters claimed by the authors to show differences sufficiently consistent to warrant separate status. In most cases no attempt has been made to search for other characters which might demonstrate subspecific differences where previous diagnoses are considered invalid, although note has been made of any relevant work by other authors.

The current concept of a species is more or less precisely definable in terms of reproductive isolation. This provides a standard to which other categories may be compared, which is not likely to vary at the whim of an individual—as sometimes happened in the past when the species was regarded more as a unit of convenience. The subspecies must obviously be more arbitrarily defined. However, it appears reasonable that formal description and terminology should be used only where they can be related to definable discontinuities in the range of variation within a species. The level of difference required for 2 populations to be considered subspecifically distinct in this study, is that 90% of one population should be clearly distinguishable from the same proportion of the other. This criterion should apply to samples from any points in their ranges, other than in the immediate area of contact where intermediate forms might occur. Since in the case of the British Isles there is no such area of contact, any subspecies which do exist should be clearly defined. In the few cases of variation within the mainland it is to be expected that the definition will be less exact.

The present paper is restricted to the results of an investigation of the status of 6 of the British subspecies, all of which are members of the Order Carnivora.

ORDER CARNIVORA

Mustela erminea *Linnaeus*, 1758

Three subspecies of the stoat have been described from within the British Isles. One of these, *Mustela erminea ricinae* Miller, 1907 will not be considered here since it is restricted to the islands of Islay and Jura off the west coast of Scotland. The remaining subspecies are *Mustela erminea hibernica* from Ireland and *Mustela erminea stabilis* from the mainland.

Mustela erminea hibernica

The Irish form of the stoat was first given separate status by Thomas and Barrett-Hamilton (1895*a*) as a distinct species, *Putorius hibernicus*.

This paper was a preliminary note and was followed almost immediately by a fuller description (Thomas and Barrett-Hamilton, 1895*b*). The description was based on specimens from Eniskillen in County Fermanagh. The characters used were the smaller size compared with the mainland and continental form (*Putorius ermineus*), and also the colour. The lighter colour of the ventral surface was described as being contracted to a narrow line on the chest and belly, while stretching the full width of the animal between the limbs. In addition the upper lip was said to be dark compared with the white colour in *Putorius ermineus.* Miller (1912), while relegating the form to a subspecies of *Mustela erminea*, commented that the ear also lacked a white margin found in the other forms, that the colour pattern was slightly more variable than suggested by the original description, and that the skull could not be distinguished from the continental form, *Mustela erminea aestiva* Kerr, 1792, though it was undoubtedly smaller than the English *Mustela erminea stabilis.*

Figure 1 shows the results of an examination of the condylo-basal length of the skull. There is considerable overlap between the 3 forms and this cannot therefore be considered a good character for taxonomic purposes. Similar results were obtained with other measurements of the skull. As far as colour is concerned, all of the Irish skins examined showed clearly the dark upper lip and ear margins, together with the typical ventral pattern described above. All of the English and continental forms showed the white upper lip and ear margin and the lighter colour extending right across the ventral surface. The Irish form of the stoat, *Mustela erminea hibernica*, may therefore be considered to constitute a distinct subspecies.

Mustela erminea stabilis

The English stoat was first described by Barrett-Hamilton (1904) as *Putorious ermineus stabilis*, based on a series of specimens from Blandford in Dorset. The description consists principally of the colour pattern, this being slightly darker above and less yellow below than in the continental form. Average dimensions were also given for the skull, though no direct comparison was made with the other forms. Miller (1912) added that the size of the English form was slightly larger than in *Mustela erminea aestiva* and that the teeth, especially the carnassials, demonstrated this point. He also stated that the British forms lacked the lighter undersurface of the tail found in continental specimens.

On examining the available material, it was evident that the original description was of little assistance in separating the populations. There was a considerable overlap between the populations in the depth

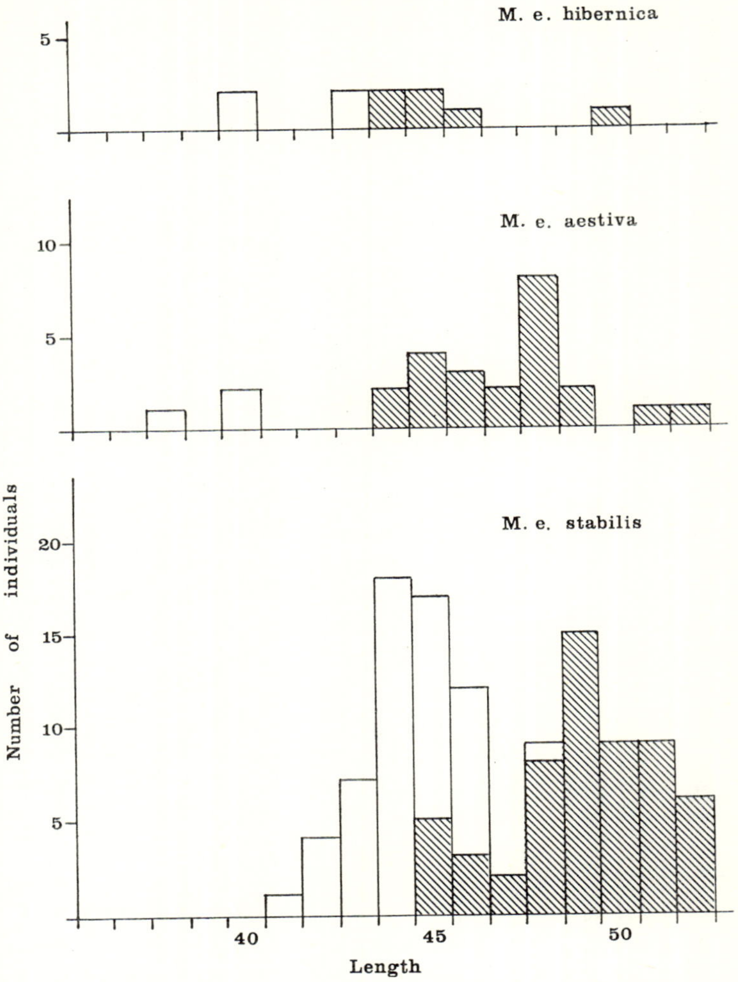

FIG. 1. The variation in condylo-basal length of the skull of the British forms of *Mustela erminea*. (Males shown diagonally shaded, females open).

of colour. The light undersurface of the tail described by Miller (1912) occurs in both. Measurements of the skull (Fig. 1) and of the teeth also show an overlap. On the basis of the original diagnosis therefore, the English form of the stoat, *Mustela erminea stabilis*, cannot be considered a valid subspecies and must be relegated to the synonymy of the continental *Mustela erminea aestiva*.

Mustela putorius *Linnaeus*, 1758

Two forms of polecat have been described from the British Isles, *Mustela putorius anglius* for England and Wales and *Mustela putorius caledoniae* for Scotland.

Mustela putorius anglius

This was described by Pocock (1936) from Llanganmarch, Breconshire, as having in the winter coat a frontal band posterior and median to the eyes which was darker than, and separate from, the greyish-white cheek patch. In contrast, in *Mustela putorius putorius* there was a pale frontal band, continuous with and of the same colour as, the cheek patch. In addition the cheek patch was described as continuous with the pale colour of the throat in *Mustela putorius putorius* but separate in *Mustela putorius anglius*. Pocock also stated that the English form had darker fur and shorter guard hairs.

The status of this subspecies has recently been revised by Poole (1964), who states that neither the colour of the frontal band nor the continuity of the cheek patch are reliable characters. The two forms do not even appear to be distinct varieties—captive specimens were found sometimes to show the two patterns alternately in successive years. Measurement of the guard hairs showed that these were of equal length in both populations. There does appear to be some variation between the English/Welsh and continental forms in colour, but this is not enough to justify separation. Poole therefore states that the English polecat must be considered part of the nominate race, *Mustela putorius putorius*.

Mustela putorius caledoniae

The Scottish polecat was described by Tetley (1939) as having a marked post-orbital constriction in the skull, distinguishing it from the straight post-orbital region in the English/Welsh form. He also commented on the large size of the skulls and that the skins did not differ from those of the English/Welsh.

The diagnostic feature of this form is also the only constant diagnostic feature of the ferret, *Mustela putorius furo* Linnaeus, 1758, as demonstrated by Ashton (1955). The two show great similarities and differ to a considerable extent from *Mustela putorius putorius*. Since only four specimens of *Mustela putorius caledoniae* are known, all from the same area in Sutherland and all caught within two years, they might represent escaped ferrets, polecat-ferrets, or hybrids between escaped ferrets and polecats.

Against this theory must be placed Tetley's statement that the skull of *Mustela putorius caledoniae* is large. This is at variance with the smaller size of the ferret skull compared with the polecat. However, Pocock (1932) reported some polecat-ferrets from Mull which on the evidence of the skull alone are large enough to be considered true polecats. Possibly similar specimens could have been found in Sutherland. Ashton (1955) states that the skins of *Mustela putorius caledoniae* agree with the original diagnosis in that they are very similar to *Mustela putorius putorius* (including *M.p. anglius*).

The basic need to solve this problem is a further supply of specimens. Unfortunately there appears to be no such material available in any collection at present (Ashton, 1955). This, together with the general opinion that the polecat has been virtually extinct on the mainland of Scotland for the last 50 years (Walton, 1964), suggests that in all probability *Mustela putorius caledoniae* is not a valid subspecies.

Meles meles *Linnaeus*, 1758

A single subspecies of the badger has been described from the British Isles, *Meles meles britannicus* Satunin, 1906. Satunin's description was based solely on the larger size indicated by the cranial measurements of 6 specimens given by Barrett-Hamilton (1899). Miller (1912) relegated it to the nominate subspecies, where it has remained since. Re-examination of the available material, together with a large series of measurements of Scandinavian specimens given by Hysing-Dahl (1954) supports this view.

Lutra lutra *Linnaeus*, 1758

The Irish form of the otter was first considered to be a distinct taxonomic entity by Ogilby (1834). Basing his remarks on a single specimen from near Newton Lemavaddy in Londonderry, he described it as a distinct species to which he gave the name *Lutra roensis*. The most important character separating the Irish and English forms was said to be the much darker colouration of the former, which approached very nearly to black on both dorsal and ventral surfaces. Ogilby also stated that the pale colour beneath the throat was smaller in area in the Irish form, and that there were some differences in the size of the ears and in the proportion of other parts. He did not however give details of these differences. He also considered that the habit of the Irish otter of spending much of its time in the sea, rather than the rivers, was of considerable importance.

MacGillivray (1838) associated the Irish form with otters from the north of Scotland, which were also semi-marine in their habits, but considered that there were not sufficient reasons to admit the existence of two British species. Miller (1912) relegated the Irish otter to the synonymy of *Lutra lutra*, regarding this as the only European species. However Hinton (1920), while describing some new specimens which the British Museum (Natural History) had just received from Galway, noted that they were very much darker than any English specimens previously examined. He therefore gave the Irish form subspecific status as *Lutra lutra roensis*. He also commented that Ogilby's type specimen had faded from its original near black to a deep reddish-brown, apparently because it had been on public exhibition for some time.

Figure 2 shows the result of an examination of the dorsal colour of the Irish and English forms, using a spectrophotometer. It was neces-

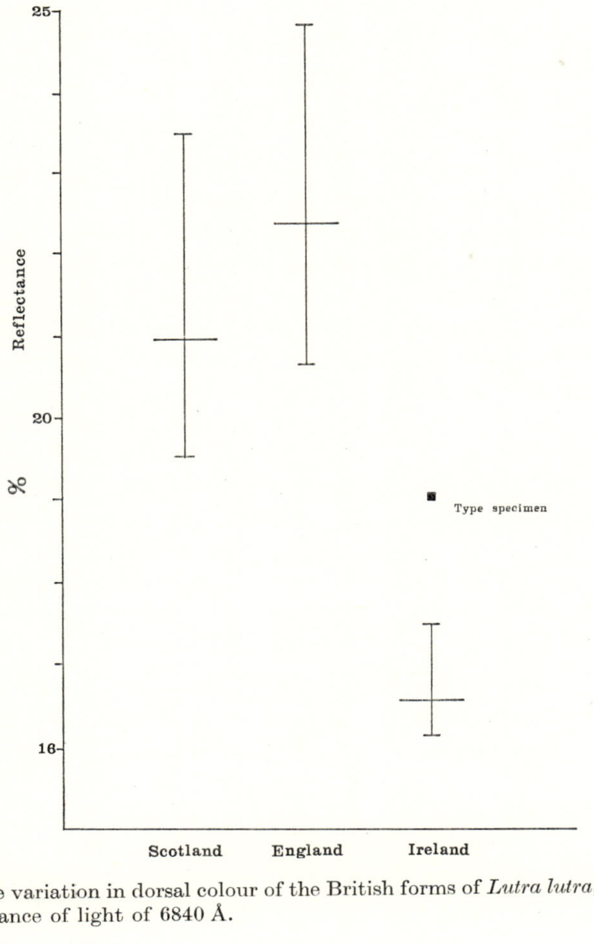

FIG. 2. The variation in dorsal colour of the British forms of *Lutra lutra*, as measured by the reflectance of light of 6840 Å.

sary to make use of this instrument since the differences, although evident to the naked eye, were relatively slight and difficult to define objectively in words. (On the figure, Ogilby's type specimen is shown separately since it is known to have faded.) It is evident that there is no overlap between the populations. The pale colour beneath the throat is also much smaller in area, sometimes absent altogether, in the Irish form while quite prominent in the English. There is however, considerable overlap in the ventral colour, although the English specimens are lighter on average. In addition, it has been realized that semi-marine habits are a common characteristic of otters living in coastal areas.

It would appear therefore that, based on the specimens available at present, the Irish otter must be granted subspecific status. However, since most of the English specimens examined came from the eastern part of the country, it may well be that a further series from the western part may show an intermediate range of colouration.

CONCLUSION

The judgements made above as to the subspecific status of some of the British mammals are based on the original diagnoses of the authors, who were not usually in possession of a great many specimens and were not always able, therefore, to take account of the range of variation to be found within a species. Even today, in the majority of cases there are too few specimens and the range from which they come is far too restricted for the full pattern of variation to be seen. All of the British subspecies must therefore be considered provisional until such time as a much greater quantity of material is available for examination. Meanwhile the use of the subspecific names, if felt necessary as a matter of convenience, should not be taken to imply any constant level of difference between the continental and British mammals.

Acknowledgement

I should like to thank Dr. G. B. Corbet of the British Museum (Natural History), for allowing me access to the collections of the Museum, and also for much valuable advice.

References

Ashton, E. H. (1955). Some characters of the skulls of the European polecat, the Asiatic polecat and the domestic ferret. *Proc. zool. Soc. Lond.* **125**, 807–809.

Barrett-Hamilton, G. E. H. (1899). Note on the beechmarten and badger of Crete. *Ann. Mag. nat. Hist.* (7) **4**, 383–384.

Barrett-Hamilton, G. E. H. (1904). Notes and descriptions of some new species and subspecies of Mustelidae. *Ann. Mag. nat. Hist.* (7) **13**, 388–394.

Hinton, M. A. C. (1920). The Irish otter. *Ann. Mag. nat. Hist.* (9) **5**, 464.

Hysing-Dahl, C. (1954). Den norske grevling. *Årbok Univ. Bergen* (*Mat.-nat.*) 1954 (16), 1–55.

Kerr, R. (1792). *The animal kingdom.* London.

Linnaeus, C. (1758). *Systeme de la nature.* 10th edition. Uppsala.

MacGillivray, W. (1838). *British quadrupeds.* Edinburgh.

Miller, G. S. (1907). Some new European Insectivora and Carnivora. *Ann. Mag. nat. Hist.* (7) **20**, 389–398.

Miller, G. S. (1912). *Catalogue of the mammals of Western Europe* (*Europe exclusive of Russia*) *in the collections of the British Museum.* Trustees of the British Museum, London.

Ogilby, W. (1834). Notice of a new species of otter from the north of Ireland. *Proc. zool. Soc. Lond.* **1834**, 110–111.

Pocock, R. I. (1932). Ferrets and polecats. *Scott. Nat.* **1932**, 97–108.

Pocock, R. I. (1936). The polecats of the genera *Putorius* and *Vormela* in the British Museum. *Proc. zool. Soc. Lond.* **1936**, 691–723.

Poole, T. (1964). Observations on the facial pattern of the polecat (*Putorius putorius* Linn.). *Proc. zool. Soc. Lond.* **143**, 350–352.

Satunin, K. A. (1906). Die Säugetiere des Talyschgebiets und der Mugansteppe. *Izv. kavkaz. Muz.* **2**, 87–394.

Tetley, H. (1939). On the British polecats. *Proc. zool. Soc. Lond.* **109B**, 37–39.

Thomas, O. and Barrett-Hamilton, G. E. H. (1895*a*). The Irish stoat distinct from the British. *Ann. Mag. nat. Hist.* (6) **15**, 374–375.

Thomas, O. and Barrett-Hamilton, G. E. H. (1895*b*). The Irish stoat distinct from the British species. *Zoologist* (3) **19**, 124–129.

Walton, K. C. (1964). The distribution of the polecat (*Putorius putorius*) in England, Wales and Scotland, 1959–62. *Proc. zool. Soc. Lond.* **143**, 333–336.

Symp. zool. Soc. Lond. (1970) No. 26, 127–134.

TAXONOMIC AND INDIVIDUAL VARIATION IN GIBBONS

COLIN P. GROVES

Duckworth Laboratory of Physical Anthropology, Cambridge University, Cambridge, England

SYNOPSIS

Varying between different species and subspecies, gibbons may be uniform, polymorphic or dimorphic in colour. Those forms which are dimorphic include representatives of four out of the six species; colour type may be dependent on, or independent of, sex. Incidence of the two colour types, dark and light, varies geographically as do the precise shades of colour; family data deduced from wild populations seem consistent with the hypothesis that the light colour type is due to a simple autosomal recessive gene influenced by modifiers.

It has long been appreciated that there is considerable individual variation in colour in many forms of gibbon (genus *Hylobates*), but the precise nature of the variation or polymorphism is still unclear. As late as 1927, Pocock could state that *Hylobates lar lar* ranged in colour, "irrespective of sex, from black through various shades of brown to buff or almost cream", and that *Hylobates hoolock*'s colour phases were also irrespective of sex, or only approximately correlated.

The fact that the nature of colour polymorphism differs from species to species, even from subspecies to subspecies, has become clear from my own study of gibbons. Only two species, the Siamangs (*H. syndactylus* and *H. klossii*), are completely monochromatic; in both, individuals are black throughout life. In two other species, the Hoolock (*H. hoolock*) and the Concolor or Crested gibbon (*H. concolor*), which are not otherwise closely related, colour is related to both age and sex: both sexes are born grey-white, turning black within a few months; after 6–7 years, the female becomes light-coloured, the male remaining black. Within both species, males show subspecific variation in the amount and patterning of white on the face; the females show the respective patterns in their black juvenile coat.

The remaining gibbons were placed by Pocock (1927) in a single species, *H. lar*, but divided by Kloss (1929) into 3 species—*lar*, *agilis* and *cinereus* (subsequently changed to *moloch* in accordance with the rules of priority). Because of the close relations between them, and the mosaic of characters differentiating them, I prefer not to separate them specifically until there is field evidence one way or the other. However, field observations have shown that one member of the complex, *pileatus*—referred to as *H. lar* even by Kloss—is in fact a good species, as it is sympatric with another representative in part of Thailand (G. Berkson, personal communication). This leaves the following 7 subspecies of *Hylobates lar*:

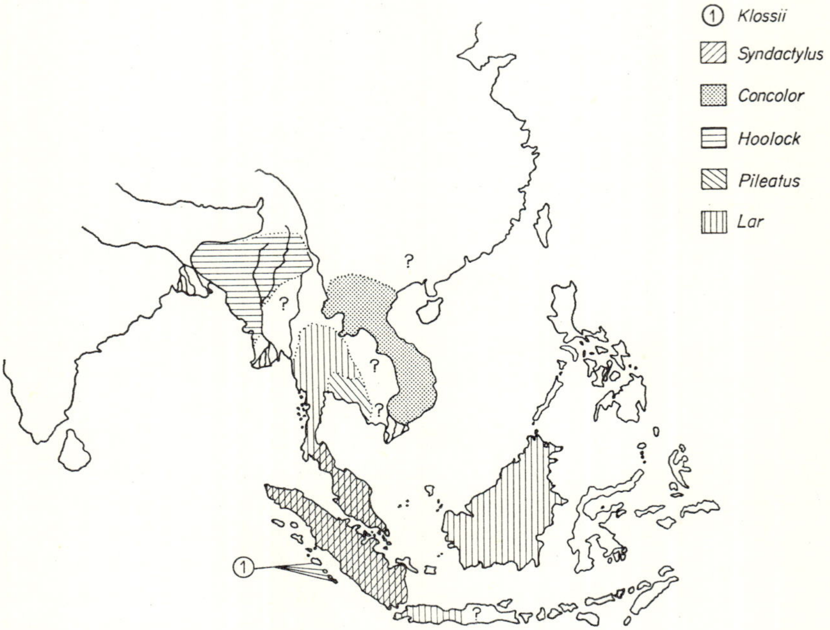

FIG. 1. Distribution of the 6 species of gibbon (*Hylobates*). Note that the only 2 confirmed cases of sympatry are:
a. *syndactylus/lar* in Malaya and Sumatra,
b. *pileatus/lar* at Khao Yai, Saraburi province, Thailand.

(1) White-handed group
- (a) *carpenteri* Groves, 1968. N. Thailand.
- (b) *entelloides* I. Geoffrey St. Hilaire, 1842. S. Thailand, Tenasserim.
- (c) *lar* Linnaeus, 1771. Malaya (except north-west).
- (d) *albimanus* Vigors and Horsfield, 1828. N. Sumatra.

(2) Without white hands
- (e) *agilis* F. Cuvier, 1823. S. Sumatra, N.W. Malaya.
- (f) *moloch* Audebert, 1800. Java.
- (g) *muelleri* Martin, 1940. Borneo.

Within these two species (*H. lar* and *H. pileatus*) all the types of gibbon colour polymorphism are found:

(1) Monochromatism. *H.l. moloch* of Java is consistently silvery grey, with little variation;

H.l. albimanus (N. Sumatra) is buffy-brown.

(2) Polychromatism. *H.l. muelleri* of Borneo shows a variety of colour types at any one locality; the average hue varies from place to place in Borneo, but the same types occur nearly everywhere. No variation due to sex is discoverable, and there is only a slight blackening due to age.

(3) Dichromatism. The remaining 4 subspecies, and *H. pileatus* show sharply defined colour-phases, a dark and a light, with no intermediate but very rare blotched or banded individuals which may be intergrades of a special type.

(a) *H. pileatus* has sexual dichromatism: adult males are black (with white on face, hands and feet, and pubic tuft), adult females creamy-buff with a black "cap" and breast-patch. This is similar to *H. hoolock* and *H. concolor*, but comes about in quite a different way: both sexes are born creamy-buff and gradually develop black spots on the crown and breast. In the male these spread rapidly producing the black colour by the time maturity is reached; in the female they spread only gradually, never doing more than blackening the throat and the inner surfaces of the upper arms and thighs.

(b) *H.l. lar*, *entelloides*, *carpenteri* and *agilis* have non-sexual dichromatism: either sex may be either dark or light after the initial white and neonate coat.

Table I shows how the colour phases "shift" from one to another of these 4 races. The lightest in both phases is *lar*; in *entelloides* both become one step darker, so that the light phase goes from whitish to honey, and the dark phase from medium to dark brown. This simple relationship is complicated by consideration of *carpenteri*, which appears superficially to have a dark phase like *entelloides* but a light phase like *lar*. However, examination of the individual hairs puts a different complexion on the matter: the three fit into the scheme proposed by Hershkovitz (1968), forming a quasi-evolutionary series in the sequence *carpenteri-entelloides-lar*.

In the dark phase, all 3 are unsaturated, i.e. they have hairs with at least 2 colours, a basal and a terminal. The basal portion is always much lighter than the terminal. The terminal portion lightens from black in *carpenteri via* dark brown in *entelloides* to light brown in *lar*; the basal portion, from dark brown *via* grey-brown to light grey-brown.

In the light phase, which is more advanced along the metachromatic paths (as being more bleached) under Hershkovitz's scheme, only

TABLE I

Colour Phases in Hylobates lar.

Subspecies	No.	Dark phase Hair colour (basal: terminal)	% frequency of dark phase nor-thern	sou-thern	Light phase Hair colour (basal: terminal)
carpenteri	(154)	silvery brown: blackish	75·4	37·5	light grey: creamy
entelloides	(43)	greyish brown: dark brown	67·7	25·0	honey
lar	(100)	light grey brown: yellow-brown	93·8	55·5	whitish
agilis	(58)	black	92·9	59·1	buffy

carpenteri remains unsaturated; the other two are saturated, with solid-coloured hairs. Thus again *carpenteri* is least advanced; *lar* is lighter than *entelloides* and hence more evolved.

The fourth dichromatic race, *H.l. agilis*, is fully saturated in both dark and light phases. The dark phase is solid black; the light phase buffy, not dissimilar from some *entelloides*, but differs from the other three in having the hands and feet of the same colour as the body; and the white face-ring may be either reduced to a browband or expanded to form copious cheek-whiskers. It is interesting to note that *H.l. albimanus*, which occurs further north in Sumatra than *agilis*, is coloured like a light phase *agilis* or *entelloides*; it differs from the former in its white hands and feet, and from both in its smaller size.

All the mainland-Sumatran group of races of *H. lar* occupy extended ranges within which there are no essential colour variations, and intergrade over only a narrow area. Intermediates between *lar* and *entelloides* occur at Trang in peninsular Thailand; between *entelloides* and *carpenteri* two possibly intermediate specimens have been taken at Lampha and Paungdaw, on the Burma-Thailand border (see below). I have seen no intermediates between *agilis* and either *lar* (in Malaya) or *albimanus* (in Sumatra); whether they will be found, or whether *agilis* will turn out to be a distinct species sympatric with its neighbours, cannot at the moment be predicted. At any rate the trinomial is both useful and meaningful when applied to *Hylobates lar*; it seems as certain as it can be without actual fieldwork that the recognized subspecies are "discrete entities in nature", not merely "provisional . . . based on discretely-definable samples" in the words of Corbet (1970).

Within each subspecies, as can be seen from Table I and Fig. 2, there are different frequencies of light and dark phases in different portions of the range. Moreover, these follow a definite pattern within each subspecies, the dark phase having a higher frequency in the northern part of the range of each, a lower frequency in the southern. *H.l. entelloides* has lower frequencies of the dark phases than the other races; the similarity in colour between the North-Sumatra *albimanus* and the light phase of *entelloides* has already been noted, and if the former is regarded as essentially a small-sized southern extension of the latter, then it can be seen as continuing the trend towards reducing the frequency of the dark phase in which, as far as can be told from 13 skins, the end point is zero.

In parenthesis, the holotype skin of *albimanus* (in the British Museum) is in the dark phase and exactly resembles *entelloides*. Either there is a mistake in the type locality (given merely as "Sumatra") or occasional dark specimens also do occur in North Sumatra, giving a dark-phase frequency there of 1 in 14, i.e. 7%.

Little has been recorded about the inheritance of colour in *H. lar*, especially since zoos tend to cage the colour phases separately; but Carpenter (1940) records family data for gibbons in the Chiangmai area (these gibbons now form the type series of *carpenteri*, in the Museum of Comparative Zoology at Harvard). These data must be taken cautiously: gibbons live in mated pairs with up to 4 immatures per group, so that although the probability is that the immatures are the offspring of the pair in most cases, the possibility that a gibbon may mate again after the death of its previous consort makes all genetic conclusions tentative.

In Carpenter's data, there are 11 dark × dark matings, with 23 black, 7 buff youngsters attached to these pairs; 3 light × light matings with 2 black, 4 buff youngsters; and 5 dark × light matings with 12 black, 2 buff youngsters. The absence of intermediates suggests a simple Mendelian relationship. As these figures stand, they cannot be interpreted as simple parent-offspring relationships, although if one assumes little re-pairing of the adults, the weight of evidence is that light colouration is probably recessive.

J. O. Ellefson (personal communication) recorded family data for gibbons in various semi-isolated areas in Malaya with different frequencies of dark and light phases. Broken down by locality, the records are as follows:

Geranggau, Trengganu:	11 dark × dark, with 17 dark youngsters.
	2 dark × light, with 2 dark youngsters.

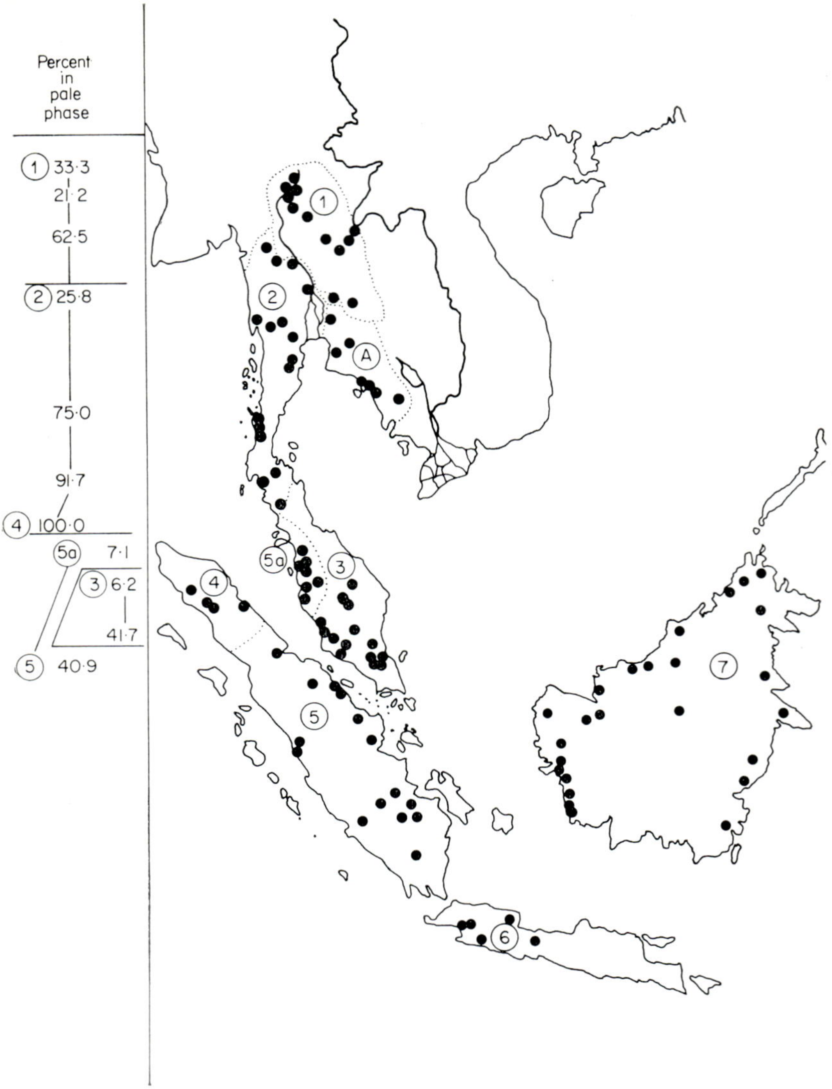

FIG. 2. Distribution of *Hylobates pileatus* and of the subspecies of *H. lar*, showing gradients in colour frequency.

A—*H. pileatus*; 1—*H.l. carpenteri*, 2—*H.l. entelloides*, 3—*H.l. lar*, 4—*H.l. albimanus*, 5 and 5a—*H.l. agilis*, 6—*H.l. moloch*, 7—*H.l. muelleri*.

Filled circles indicate localities where the presence of the taxon concerned has been substantiated, either through museum specimens or through field observation. The column headed "percent in pale phase" demonstrates the frequency gradients for this colour phase in the four racial groups (arranged geographically):

1—*carpenteri*; 2 and 4—*entelloides/albimanus*; 3—*lar*; 5a and 5—*agilis*.

In each of the 4 groups, the frequency of the pale phase increases from north to south.

Endau, Johore:	3 dark × light, with 3 dark youngsters.
	2 light youngsters.
Gombak valley:	1 light × light, with 2 light youngsters.
Kg. Londah, N.S.:	1 dark × dark, with 1 dark youngster.
Ranjan Panjang:	1 dark × dark, with 1 dark youngster.
Lobak Paku, Pahang:	1 dark × dark, with 2 dark youngsters, plus one group of 3 dark, 1 light of uncertain relationships.

These data do not greatly add to the picture except circumstantially. In the Trengganu sample, where only 4·4% (46 specimens) are in the light phase, the frequency of the pale gene would be only 0·2, and the proportion of dark × dark matings in which both partners were heterozygous would be sufficiently infrequent—14% of all such pairings—for the lack of light offspring in all dark matings to be no cause for surprise. There is no evidence of re-pairing in Ellefson's data.

In the National Zoological Park at Washington, D.C., a dark male of *H.l. carpenteri* has mated with a light female of *H.l. entelloides*, and the two offspring are still in the zoo, caged with their parents. Both, when last seen by me a year ago, were a peculiar blotchy or brindled colour with the dark colour mainly on the ventral surface, extremely reminiscent of the two specimens from the Tailand-Burma border, at Lampha (Tenasserim) and "the forest south-east of Paungdaw". Both these localities are on or near the *carpenteri-entelloides* boundary. The most likely explanation is that the precise expression of colour within the broad range of "light" and "dark"—or, in Hershkovitzean terms, the precise position on the metachromatic scale—is determined by modifier genes, and that when different phase members of two subspecies hybridize the simple Mendelian relationship determining colour within each of them breaks down, producing a particoloured hybrid. The concentration of dark colour on the ventral surface is reminiscent of a stage in the development of *H. pileatus*. There is also a brindled skin from Johore in the United States National Museum; in this case only a mutation can be adduced in explanation, as this locality is well within the range of *H.l. lar* and far from any hybrid zone.

Acknowledgements

Many thanks are due to the curators of collections, in which specimens of *H. lar* were studied, for access to the collections in their charge, and for assistance in studying them: Dr. G. B. Corbet in the British Museum (Natural History), London; Dr. B. Lawrence in the M.C.Z.,

Harvard; Drs C. O. Hardley and J. R. Napier in the U.S.N.M., Washington; and Dr. R. G. Van Gelder in the Amer. Mus. N.H., New York. Thanks are also due to Dr. J. O. Ellefson, Dr. G. Berkson and Mr. D. J. Chivers for access to information collected by them.

References

Carpenter, C. R. (1940). A field study in Siam of the behaviour and social relations of the gibbon (*Hylobates lar*). *Comp. Psychol. Monogr.* **16** (5), 1–212.

Corbet, G. B. (1970). Patterns of subspecific variation. *Symp. zool. Soc. Lond.* No. 26, 105–116.

Hershkovitz, P. (1968). Metachromism or the principle of evolutionary change in mammalian tegumentary colour. *Evolution, Lancaster, Pa,* **22**, 556–575.

Kloss, C. B. (1929). Some remarks on the gibbons with the description of a new subspecies. *Proc. zool. Soc. Lond.* **1929**, 113–127.

Pocock, R. I. (1927). The gibbons of the genus *Hylobates*. *Proc. zool. Soc. Lond.* **1927**, 719–741.

Symp. zool. Soc. Lond. (1970) No. 26, 135–147.

THE EXTENT OF VARIATION IN FOSSIL *HIPPOPOTAMUS* FROM AFRICA

S. C. CORYNDON

Department of Palaeontology, British Museum (Natural History) London, England

SYNOPSIS

The history of the family Hippopotamidae can be traced back about 10 million years (Early Pliocene). The earliest known members of the family, though more primitive than the two extant species, are already well specialized suines. Various suggestions have been made connecting the Hippopotamidae with either pigs or anthracotheres but there are no very clear affinities within the sub-order.

Specimens representing the common hippopotamus, *Hippopotamus amphibius* Linn., are seldom found in any quantity in museums, due mainly to their large size and weight. This presents difficulties in determining the normal variation within the species. A series of mandibles from a small area in Uganda was studied to see what variation, if any, could be found in an isolated community, and if there were any differences from the generally accepted normal characteristics of the species as a whole. Only minor variations were found, but some were consistent within the population and may be significant.

Two series of fossil hippopotamus from East Africa were also studied, and these show not only characters representative of the particular species complex, but where deposits are shown to cover a long time sequence, evolutionary trends are also seen. It is hoped that with further investigation of East African deposits new information relevant to the phylogeny of the family will emerge.

INTRODUCTION

The Hippopotamidae, exclusively an old world family, made their appearance towards the end of the Tertiary Era. Until recently, the earliest known were from the Pinjor stage of the Siwalik Hills in Asia, and from late Pliocene deposits in East Africa. Recent discoveries in East Africa have uncovered hippopotamus remains in even earlier deposits, but so far these consist mainly of isolated teeth, which indicate only the presence of a primitive hippopotamus (Fig. 1).

Hippopotami can be conveniently divided into two main groups—tetraprotodont with 4 incisors in each jaw, and hexaprotodont, with 6 incisors. This division is rather misleading, and other characters, particularly in the facial region of the skull and the form of the canine teeth should also be taken into consideration. As the names indicate, the main criteria for separation are based on the numbers of incisor teeth (Fig. 2). When applied to the two living forms, *Hippopotamus amphibius* and *Choeropsis liberiensis* Morton, the former demonstrates

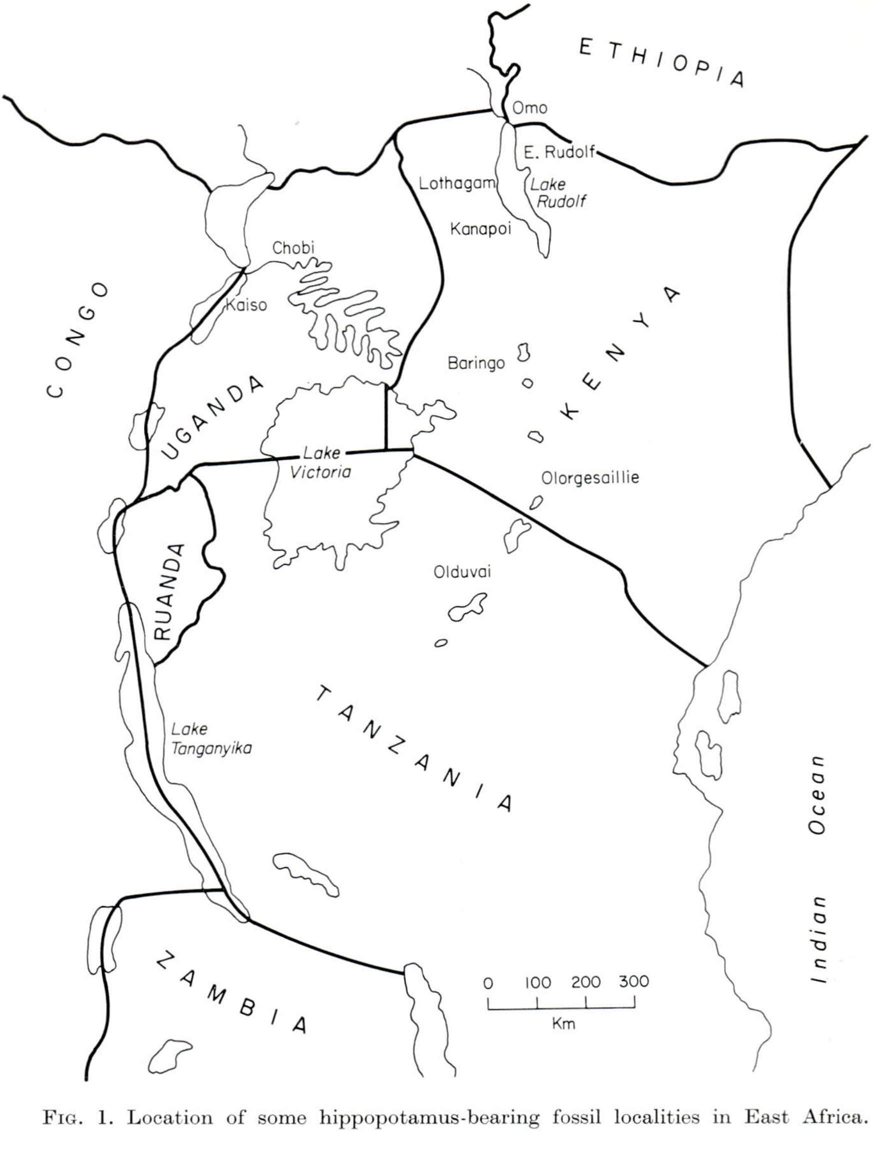

FIG. 1. Location of some hippopotamus-bearing fossil localities in East Africa.

typical tetraprotodont characters, but with the latter difficulties arise; *Choeropsis* has 4 incisors in the upper jaw, but only 2 in the lower, and for this reason has been called diprotodont; this species however is most probably an aberrant hexaprotodont. Hexaprotodonts, known only in fossil form, are the more primitive group and were first recognized in the nineteenth century among specimens from the Siwalik Hills. Recently a large hexaprotodont complex has been recognized in East Africa.

In studies of fossil hippopotami, the question of placing individual specimens into one or other category presents certain difficulties, due to the fact that so many remains consist of molar or canine teeth only, with just the occasional more complete specimen. Discoveries in East Africa over the past few years have brought to light a hippopotamus fauna extending in time over nearly 10 million years, with a considerable

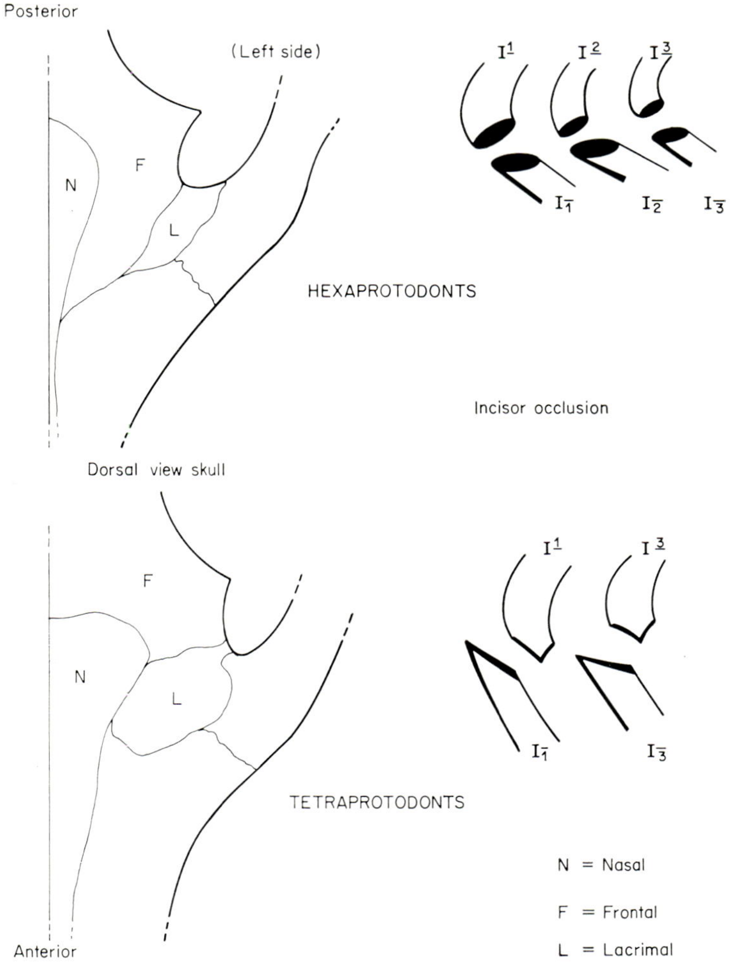

FIG. 2. Lacrimal area and incisor occlusion in hexaprotodont and tetraprotodont hippopotamus.

concentration from deposits of around 4 million years through to the present. These new specimens, representing both tetraprotodont and hexaprotodont forms have thrown completely new light on affinities

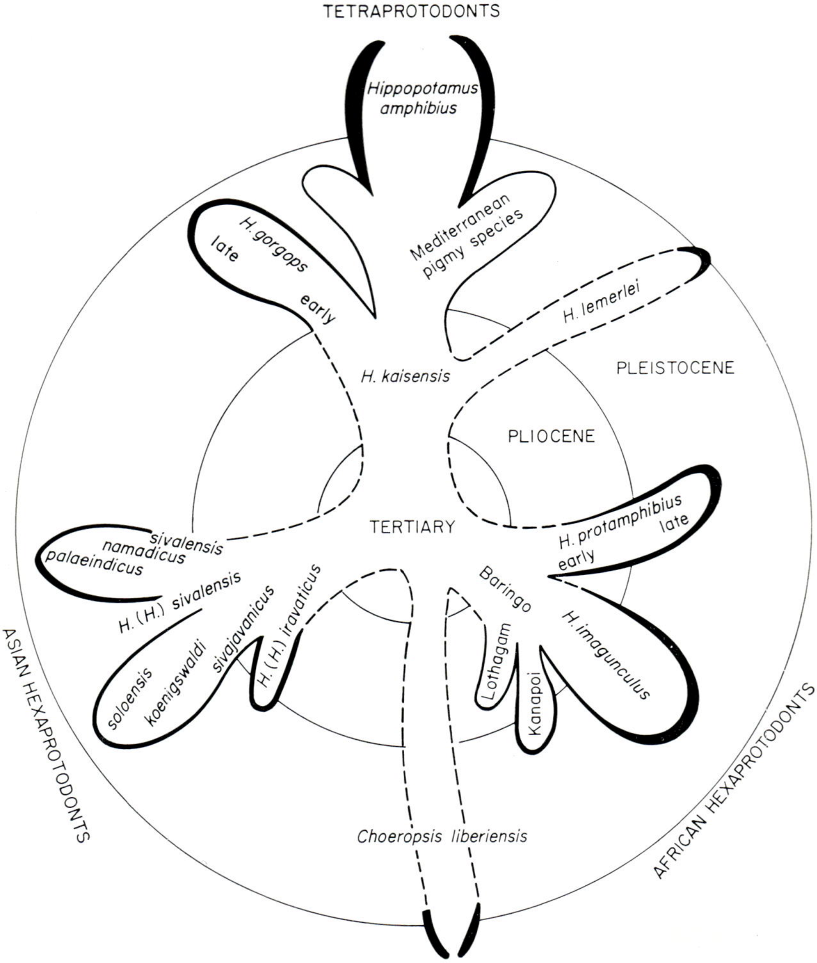

Fig. 3. Phylogeny of the Hippopotamidae.

within the family, and have made it possible to separate the tetraprotodonts and hexaprotodonts on characters other than just the number of incisor teeth, and to separate the East African hexaprotodonts from those of Asia (Fig. 3).

EXTANT HIPPOPOTAMI—*Hippopotamus amphibius* AND *Choeropsis liberiensis*

In order to determine the basic characters relevant to the study of fossil Hippopotamidae, it was necessary first to try and determine the important features in the living forms. Although *Choeropsis* is a small animal with a skull of convenient size to handle, it is also a very rare animal, and remains only occasionally find their way into museums; it is an animal that seems to thrive and breed in captivity, but unfortunately the zoo specimens preserved in museums invariably show anomalies and are not very reliable for basic taxonomic study.

Choeropsis liberiensis has recently been the subject of an investigation by Corbet (1969). He recognized consistent differences in specimens from Nigeria compared with other areas to the west. This is probably the result of isolation of one group from the other due to changes in environment. Unfortunately, because of its rarity, very little is known about its habits and ecology.

A study of *Hippopotamus amphibius*, though a common beast in the right environment, presents practical difficulties in study due to its great size—one skull takes up a great deal of space on a work bench, even if you are strong enough to lift it there in the first place. Although the British Museum (Nat. Hist.) has several specimens, they originate from various parts of Africa and no two specimens come from the same place (except those from London Zoo). With this situation, any variation seen between one specimen and another may just indicate normal variation in a species. It has been stated several times in the literature that *H. amphibius* "is very variable" (Hooijer, 1950).

Hippopotamus amphibius FROM UGANDA

In 1966, I had the opportunity to visit the Game Management camp at Chobi, in the Murchison Falls Park in Uganda. This camp was sited on the banks of the Victoria Nile, where a very large hippopotamus population was being systematically culled. Most parts of the animals killed, and of the elephants which were likewise being controlled, were converted into saleable goods for the tourist trade, meat for the local market, or bone meal for fertilizer. The only parts of the animals that were not used were the intestines, which were buried. Each animal was weighed, measured and sexed when shot, and all the mandibles of the hippopotami and elephant were retained. Unfortunately the skulls, which would have been far more valuable for study, were converted to bone meal after the incisors and canines had been removed. However, some 700 hippopotamus' mandibles had been saved (again with the canines and incisors removed), and it was hoped that these, even in a limited way, might give some indication of the extent of variation within this homogeneous population. Many of the specimens were damaged, with most or all of the teeth missing; 300 mandibles were in sufficiently good condition to be studied in general, of which 68 were investigated more thoroughly and detailed measurements taken.

In the collections at Chobi, only one out of the 700 mandibles showed any dental abnormality, and in this case had a diprotodont anterior dentition, with absolutely no sign of there ever having been a lateral incisor on either side. Other than this, the specimen was perfectly normal.

Nearly 50% of the specimens had either an alveolus for a P_1 or a rugosity in that position, but rarely were they found on both sides in the same mandible. This tooth, when present, is always a single rooted tooth in *H. amphibius* and is probably the first milk molar retained in the adult. I have not come across any specimen, extant or fossil, in which I can see evidence of the first premolar in the adult jaw being a true permanent tooth.

In fossil Hippopotamidae, the lower second and third premolars are very similar teeth both in size and morphology, with a rather complicated enamel pattern. In the living animal however, the premolars are more simple and the second premolar is smaller than the third, the latter being closer to the fourth premolar in shape and form. On the other hand, the third premolar in the Chobi specimens was unusual in having a strong accessory cusp on the lingual side of the tooth, a feature only rarely seen in *H. amphibius*. This character however, occurs amongst virtually all the early fossil hippopotami, hexaprotodont or tetraprotodont. The fourth premolar in the Chobi animals also had this unusual accessory cusp, higher and more robust than in the P_3. There is a strong tendency in both fossil and living hippopotami for the P_4 and P^4 to be misaligned, and set in the jaw at an angle to the long axis rather than parallel to it. It is interesting to note that, although this tooth is so commonly out of alignment, it is always twisted in the same way with the normal posterior lobe on the buccal side, and the anterior lobe on the lingual (Fig. 4).

As would be expected, the sockets for the canines and incisors of the male hippopotami from Chobi were significantly larger than those of the females, and this was presumably reflected in the teeth themselves. Clearly, a study of mandibles alone has very severe limitations, but the unexpected form of the premolar teeth, so consistent in the Chobi specimens, may possibly be a character indicating a degree of isolation

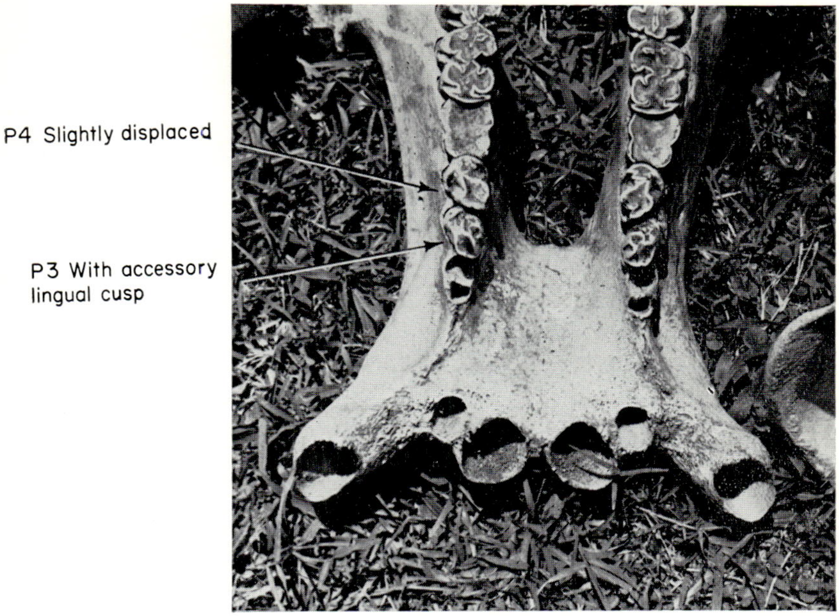

FIG. 4. Mandible of *Hippopotamus amphibius* from Chobi, Murchison Falls Park, Uganda. Occlusal view of dentition (lacking incisors, canines, P1, P2). Photograph by S. C. Coryndon.

of the Victoria Nile hippopotamus over a fair period of time. The reproductive rate of the hippopotamus, with a gestation period of about 240 days, must make for a slow evolutionary rate, and thus any consistent variation seen in a population probably reflects a fair length of time of isolation. If we could have examined 60 skulls from Chobi as well as the mandibles it might have proved a much more useful exercise. Indeed, if one could investigate similar groups of living hippopotami and get some idea of possible variations in other compact communities, it would be very valuable to an understanding of evolutionary rates and normal variation in the African hippopotamus.

FOSSIL HIPPOPOTAMI

The fossil members of the family, often common elements in a fossil fauna, are becoming more widely known due to extensive excavations in deposits of Plio/Pleistocene age in East Africa. The tetraprotodont *amphibius* complex is known from Pleistocene deposits throughout Africa and Europe, and as with so many mammals in the late Pleistocene throughout the world, some forms attained an exceptionally large size. Others, particularly in the Mediterranean islands, developed pygmy forms, presumably due to their isolation in a restricted habitat. Although the smallest of these island forms are about the size of the living West African Pygmy hippopotamus, morphologically they stand much closer to *amphibius*. *H. amphibius* was extensive in Europe and Africa during the Pleistocene and numerous specimens have been recovered from the Upper Villafranchian deposits of the Val d'Arno, the middle Pleistocene deposits of the Cromer Forest bed in East Anglia and elsewhere, but it is most common in Late Pleistocene deposits of the last Interglacial. These specimens appear to differ little from the extant *H. amphibius*, and could well be developments from an earlier form trending towards the modern species.

EAST AFRICAN FOSSIL HIPPOPOTAMI

In Africa, the tetraprotodont *Hippopotamus amphibius* lineage can be traced back to the late Pliocene form *H. kaisensis* Hopwood, from the Kaiso Formation of Uganda (Hopwood, 1926). Although there are many gaps in the record, this species could well be an ancestral form from which the later true tetraprotodonts have developed. Unfortunately little is known of the skull of *H. kaisensis*, and until the morphology of the skull bones is known, particularly in the lacrimal area, the true position of this species cannot be confirmed. Several mandibles of *kaisensis* are preserved however, which clearly show 4 well developed incisor teeth, the lateral incisors being more than half the size of the central; in the extant *amphibius* the lateral incisors are a great deal smaller than the central. The enamel of the lower canines in *kaisensis* is also of the tetraprotodont *amphibius* type with marked ridging and rugose texture, unlike that in the hexaprotodonts where the enamel is not ridged, and has a fine texture (Cooke and Coryndon, 1970).

Hippopotamus gorgops DIETRICH

The only tetraprotodont fossil group which can be traced over a reasonable time span in an enclosed area is *Hippopotamus gorgops* from deposits at Olduvai Gorge in Tanzania, where deposits span the better part of 2 million years from Early to Late Pleistocene (Dietrich, 1928). *H. gorgops* is known from isolated specimens at 3 other sites, but at Olduvai it is not only virtually the only species of hippopotamus, but is among the most common of all mammals in the deposits from Bed I upwards (Leakey, 1965).

The hippopotamus from the base of the Olduvai levels, Bed I, in many ways is very similar to a primitive *amphibius*, but exhibits clearly some of the characters found only in *gorgops*. In Bed II, of mid-Pleistocene age, the *gorgops* characters are more precise—elevated orbits, long tooth row, shallow palate, diastema between P2 and P3, and a characteristic area between the glenoid and the parietal. The uppermost level, Bed IV of Late Pleistocene age, has a hippopotamus in which the *gorgops* characters are shown in their ultimate state. Thus, at Olduvai, *H. gorgops* can be seen evolving through some one and a half million years (Fig. 5), from an early unspecialized *amphibius*-like form to the very specialized true *gorgops*.

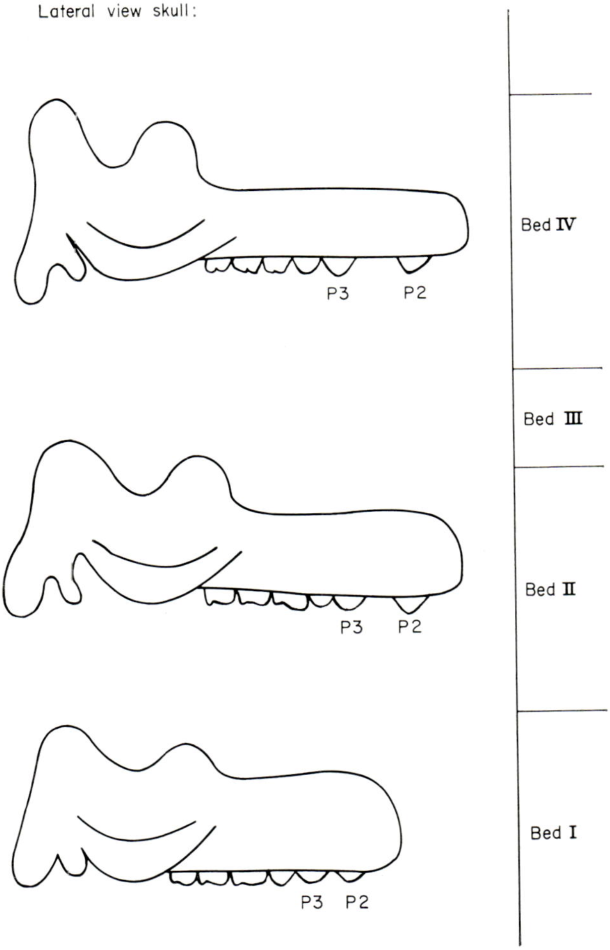

Fig. 5. Evolution of *Hippopotamus gorgops* through Olduvai.

Hippopotamus protamphibius ARAMBOURG

A similar type of evolutionary sequence is seen in a group of hexaprotodonts found in the deposits of the Omo basin in Ethiopia, which range from 4 million years through to the type Omo deposits of upper Villafranchian age (equivalent to the upper part of Bed I Olduvai) (Arambourg, Chavaillon and Coppens, 1967; Howell, 1968). In these Omo deposits the hippopotamus is again among the most common of all mammals preserved, and is found throughout the series. By far the most common hippopotamus is that species described by Arambourg (1944, 1947) as *Hippopotamus protamphibius*. This species, in spite of its name, is not a forerunner of *amphibius*, but is in reality the end of a hexaprotodont line. The story is confused by the fact that typical *protamphibius* has only 4 incisors in upper and lower jaws, whilst retaining all the other characters of a hexaprotodont—the position of the lacrimal bone in relation to the nasals, the shape of the mandible and the gracile limb bones. That it is a hexaprotodont *sensu lato* can be seen in specimens, both from the type locality and from lower levels, in which a hexaprotodont dentition is seen.

In the lowest of the Omo levels, at Yellow Sands, the hippopotamus is of medium size with low orbits and a very small lacrimal bone which is completely separated from the nasal by the frontal. It is also hexaprotodont. In the slightly higher levels of Brown and White Sands, the hippopotamus is still hexaprotodont. From the lowest tuffs in the type area, the hippopotamus is similar, but rather more advanced with a larger lacrimal bone, though still separated from the nasal, and is still hexaprotodont. In the highest levels, both hexaprotodont and tetraprotodont forms are found, in fact there are specimens in which both types of dentition are found in the same specimen. The lacrimal bone in these typical specimens just makes contact with the nasal (Fig. 5).

The Omo hippopotami form a gradually evolving group, with a gradual transition from the primitive hexaprotodont in the lowest levels (Yellow Sands) to the highest (Shungura) from which the species *protamphibius* was described. In the case of the Omo hippopotami, it seems that the tendency is for the hexaprotodont condition to be reduced and the lacrimal area to be enlarged; this latter character may possibly be linked with an increasingly amphibious habitat.

The nomenclature of these specimens in which a progression of characters is seen presents problems. At the moment, the terms "early form" and "late form" are used, not only with the hippopotamus but with fossil pigs and elephants from these same Plio/Pleistocene East African deposits as well, where similar transitions exist (Fig. 7).

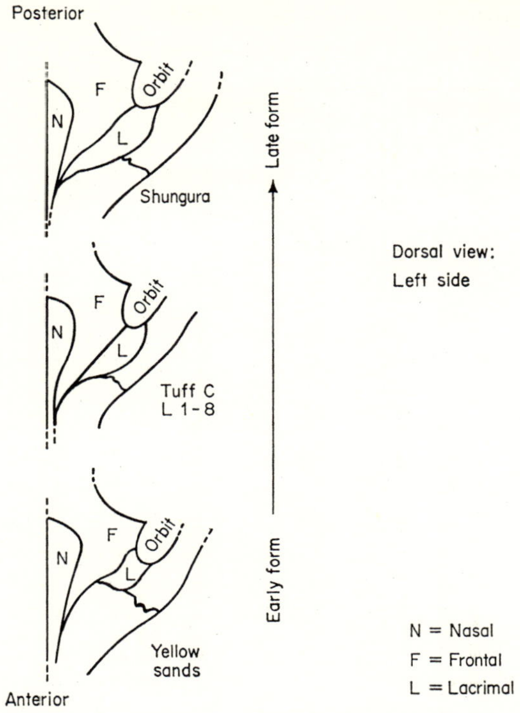

FIG. 6. Lacrimal area of skull in *Hippopotamus protamphibius* from Omo.

CONCLUSION

Of the 2 fossil series quoted, at Olduvai and Omo, in the former area only one other species is present, and that a very rare pygmy tetraprotodont from one site in Upper Bed II. At Omo, as well as *protamphibius* there are at least 2 other species, both in the upper levels, but found in fair quantity. One of these may be similar to *H. gorgops*, but more specimens are needed to confirm the true designation. The other species is a small hexaprotodont form, very similar to, if not identical with, *H. imagunculus* from Kaiso in Uganda.

Many other sites in East Africa have yielded fossil hippopotamus, and in the Late Pliocene/Early Pleistocene sites they are exclusively hexaprotodont, except for *H. kaisensis* which has so far been recognized only in the Kaiso Formation. Sites in the Eastern Rift of Kenya such as Lothagam and Kanapoi in Turkana, the Baringo area, and the East side of Lake Rudolf have all produced exciting new hippopotamus faunas in recent years. The greatest difficulty at the moment is in

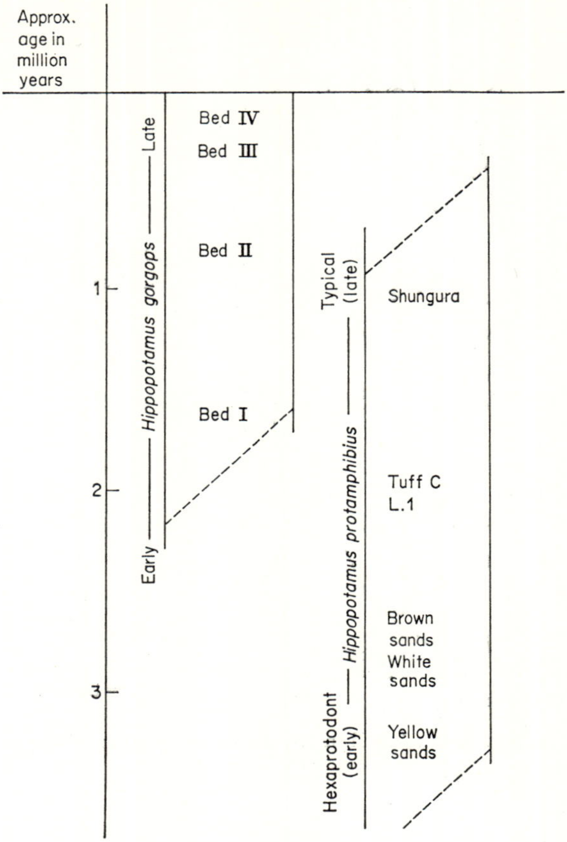

FIG. 7. Hippopotamus evolution at Olduvai and Omo.

deciding where each fits into the general scheme of hippopotamus taxonomy. Each new collection from any one site seems to contain a new group of hippopotami, unlike any other. It is unlikely that there are a multitude of different species, but it is possible that throughout the Pliocene and Pleistocene there has been isolation of communities, tending towards special character formation, similar to those seen in the hippopotamus of Chobi.

ACKNOWLEDGEMENTS

Financial help in these studies from the Wenner-Gren Fund for Anthropological Research is gratefully acknowledged.

I would like also to express my sincere thanks to many people who have helped me in so many ways, Dr. L. S. B. Leakey, Dr. A. J. Sutcliffe, Professor F. Clark Howell, Mr. Christopher Buckland-Wright and most of all Dr. H. B. S. Cooke, without whose constant help and encouragement none of this work would have been achieved.

References

Arambourg, C. (1944). Les Hippopotames fossiles de l'Afrique. *C. r. hebd. Séanc. Acad. Sci., Paris* **218**, 602–604.

Arambourg, C. (1947). Contribution à l'étude géologique et paléontologique du bassin du Lac Rodolphe et de la basse vallée de l'Omo. *Mission scient. Omo* **1**, 232–562.

Arambourg, C., Chavaillon, J. and Coppens, Y. (1967). Premiers résultats de la nouvelle mission de l'Omo (1967). *C. r. hebd. Séanc. Acad. Sci., Paris* **265**, 1891–1896.

Cooke, H. B. S. and Coryndon, S. C. (1970). Pleistocene mammals from the Kaiso Formation and other related deposits in Uganda. In *Fossil vertebrates of Africa* (L. S. B. Leakey and R. J. G. Savage, eds). **2**, 107–224, London and New York: Academic Press.

Corbet, G. B. (1969). The taxonomic status of the Pygmy hippopotamus, *Choeropsis liberiensis*, from the Niger Delta. *J. Zool. Lond.* **158**, 387–394.

Dietrich, W. O. (1928). Pleistozäne deutsch-östafrikanische Hippopotamus-Reste *Wiss. Ergebn. Oldoway-Exped.* **1913**, 1–40.

Hooijer, D. A. (1950). Fossil Hippotamidae of Asia. *Zool. Verh., Leiden* **8**, 1–124.

Hopwood, A. T. (1926). Fossil Mammalia. *Occ. Pap. geol. Surv. Uganda* **2**, 13–36.

Howell, F. C. (1968). Omo Research Expedition. *Nature, Lond.* **219**, 567–572.

Leakey, L. S. B. (1965). *Olduvai Gorge 1951–1961.* Vol. I. Cambridge: University Press.

Symp. zool. Soc. Lond. (1970) No. 26, 149–161.

VARIATION AND SPECIATION IN FOSSIL VOLES

KAZIMIERZ KOWALSKI

Institute of Systematic and Experimental Zoology, Polish Academy of Sciences Kraków, Poland

SYNOPSIS

Voles are a rapidly evolving, geologically "young" group of mammals with great importance for the stratigraphy of continental deposits. In fossil material, voles and lemmings are represented almost exclusively by molars; all the following remarks are therefore limited to the morphology of these teeth.

The fact that tooth-pattern varies so widely in populations of voles has not been taken sufficiently into account by palaeontologists. Some species have been described on the grounds of single teeth, which were in all probability individual variants in populations of already known forms. Examples are given of individual deviations in the molar-pattern of recent populations and also examples of populations with unusually high variability. The reasons for this augmentation in the variability of particular species are not clear. The genetic polymorphism in the tooth-pattern known in some recent species is to be found also in fossil voles and lemmings (e.g. *Dicrostonyx*).

Another problem in the study of fossil materials is the care which must be taken to identify changes in teeth during ontogenetic development, especially in the forms with rooted molars.

Differences between populations of voles living in different regions or biotopes may involve the dimensions of teeth. Thus, in fossil localities from the Quaternary the consecutive layers may contain populations of one and the same species but with different dimensions, reflecting climatic changes during the sedimentation of these strata.

Some aspects of the evolution of vole molars are discussed in this paper. This evolution may lead to complication, or, in some cases (in fossil species), to simplification of teeth. Examples of unusually high variability of voles in different stages of their geological history are given. Two alternatives are discussed: does this unusual variability precede speciation, or is it the result of it? Another unsettled question concerns the population variability of rapidly evolving forms: is it diminished, as would be expected from the theoretical point of view, or augmented, as observations of fossil voles suggest?

INTRODUCTION

The fact of great tooth-pattern variability in vole populations has not been taken sufficiently into account by palaeontologists. This paper gives examples of individual deviations in the molar-pattern of recent populations and also examples of populations with unusually high variability. The genetical polymorphism in the tooth-pattern known in some recent species is to be found also in fossil voles and lemmings.

Differences between populations of voles living in different regions or biotopes may involve the size of teeth. Thus, in fossil localities from the Quaternary, consecutive layers may contain teeth from populations of one and the same species but of different sizes, presumably reflecting climatic changes during the sedimentation of these strata.

Examples of unusually high variability of voles in different stages of their geological history will be described. They imply that the variability of rapidly evolving forms may be greater than normal.

Voles and lemmings form one of the youngest subfamilies of mammals. They appeared in the Pliocene as a branch of the cricetids specialized for feeding on cellulose-rich green parts of plants, and underwent a rapid differentiation, populating the temperate and cool zones of the Northern Hemisphere.

Owing to their quick evolution and abundant occurrence in the continental deposits of the Pliocene and Quaternary, voles are important index fossils which make it possible to determine and correlate the age of sediments (Kowalski, 1966). It is the teeth of these mammals that are chiefly met with in fossil material, and of these the molars, especially the highly differentiated first lower and third upper molars (M_1 and M^3), are mainly used for the purposes of systematics. Other elements of the skeleton are rather scarce in the fossil record and, however interesting the studies dealing with them may be (*e.g.* Repenning, 1968, on the structure of the mandible), they are as yet of no great importance to stratigraphy and systematics.

Most papers on the systematics of contemporary voles contain references to the variability of their molars. However, the existence of this great variation does not interfere with the correct recognition of species, since neontologists have at their disposal other systematic characters, such as the morphology of the skull and soft parts of the body, the number of chromosomes, and data concerning ethology and geographical distribution. This is not so for palaeontologists: having at our disposal only a few characters of the teeth, we must give special attention to variation in order to distinguish correctly the species and separate evolutionary from intrapopulation variation.

THE RANGE OF VARIATION

Variation in the structure of molars may be remarkable in contemporary populations of voles. Without going into its causes (*i.e.* differences in the circumstances of development, genetic differences, pathological changes), it is necessary to note that in some populations there occasionally appear specimens with a structure deviating strikingly from the typical one. An extreme example is the so far unpublished specimen of *Lagurus lagurus* from Siberia, in which one row of upper molars is normal, whereas the other contains 4 teeth (Fig. 1). The occurrence of individual deviations of this sort demands extreme caution in describing new forms of fossil voles on the basis of single specimens of teeth, although this is not infrequently done, *e.g.* in the papers by Heller (*Pliomys proavus* Heller, 1958) and others.

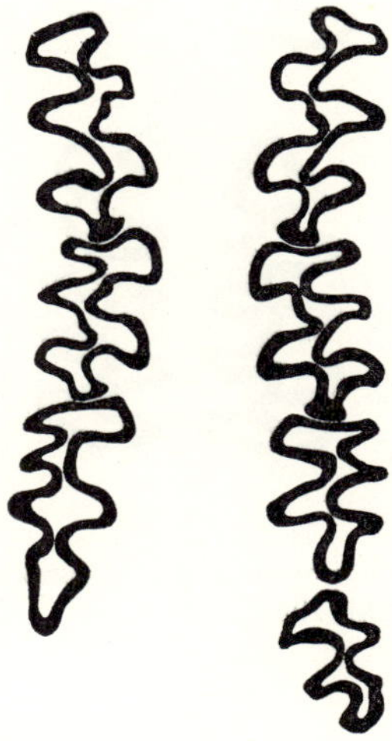

FIG. 1. The right and left upper molars of a specimen of *Lagurus lagurus* from Siberia. The right row consists of four molars, the left one is normal.

In voles the range of variation in the structure of molars differs from species to species. Some characters are very consistent, *e.g.* the presence of the fifth loop of enamel on M^2, which, as a rule, allows the unequivocal determination of *Microtus agrestis*. Nevertheless, even in this species the degree of development of the fifth loop is variable and there occur rare specimens in which this loop is completely lacking (Reichstein and Reise, 1965).

Unusual variation in the structure of molar crowns may also be found. This is true, for example, of the population of Snow voles *Microtus nivalis mirhanreini* from the Tatra Mountains, in which M_1 is marked by exceptional structural variation (Fig. 2) (Kowalski, 1957). An uncommonly high degree of variation of M^3 is characteristic also of the alpine species from the Altai Mountains, *Alticola macrotis* (Fig. 3) (Geptner and Rossolimo, 1968). In both these examples we are concerned with alpine species having limited ranges. This may reflect the rule that a species inhabiting very diversified environments

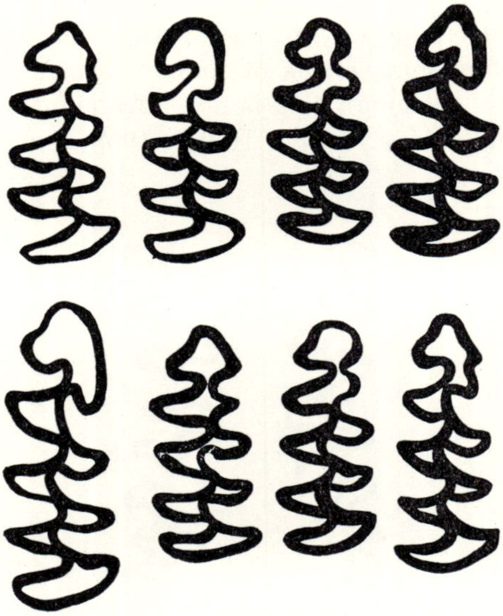

FIG. 2. Variation of M_1 in the population of *Microtus nivalis mirhanreini* from the Tatra Mountains. (After Kowalski, 1957.)

situated in a small area (and thus in the situation which makes the genetical isolation of their particular populations possible) shows increased individual variation.

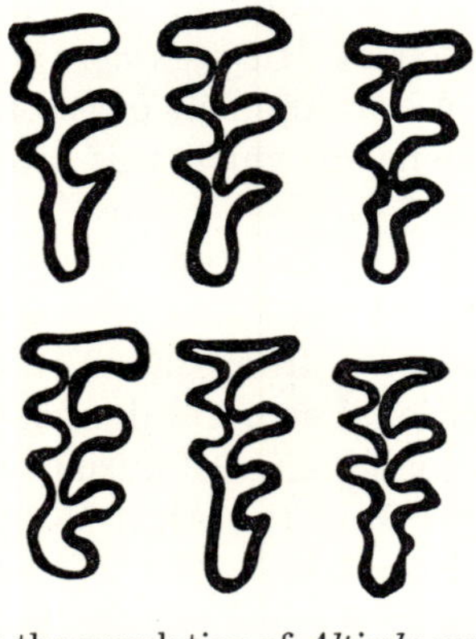

FIG. 3. Variation of M^3 in the population of *Alticola macrotis* from the Altai Mountains. (After Geptner and Rossolimo, 1968.)

INTRA-POPULATION VARIATION

There may occasionally be two or more types of tooth structure in one population (genetical polymorphism). Such a situation exists, *e.g.* in *Microtus arvalis* which shows several structural types of M^3, which can be classified into two forms, a normal and a "*simplex*". The *simplex* form is genetically controlled as a recessive character and occurs abundantly in populations near the northern limit of the species' range (Fig. 4).

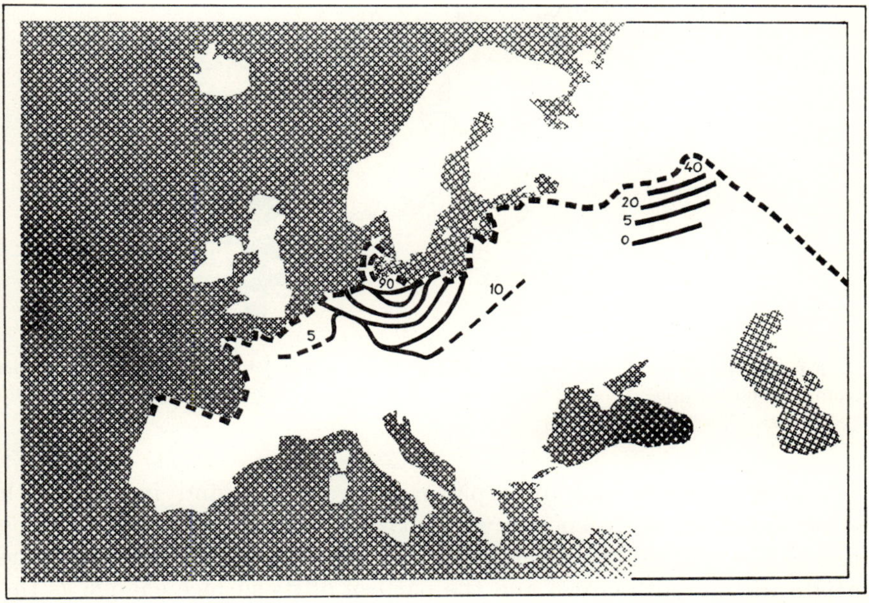

FIG. 4. Frequency of the *simplex* form in *Microtus arvalis*. The figures indicate the percentage of specimens with this type of dentition. (After Zimmermann, 1952.)

The frequency of the *simplex* form fluctuates between 0% and 90%. It is still a matter for dispute whether the form, which is undoubtedly linked with other characters, is of positive selective value (Stein, 1958; Zimmermann, 1952, 1958; Farbiszewska and Makarzec, 1959). Two types of the structure of M^3 (*complex* and *simplex*) occur in varying proportions also in the Bank vole *Clethrionomys glareolus*. In populations examined by Zejda (1960) in Czechoslovakia the abundance of the *simplex* form changed with geographic situation from 3% to 62% (Fig. 5). Similar variation is found in different populations from the British Isles (Corbet, 1964).

Two species of the fossil Collared lemming distinguished by Hinton (1926), *Dicrostonyx gulielmi* Sanford and *D. henseli* Hinton, are probably morphs of one and the same species. Both these forms occur together in fossil materials, though in various proportions; intermediate forms are also encountered.

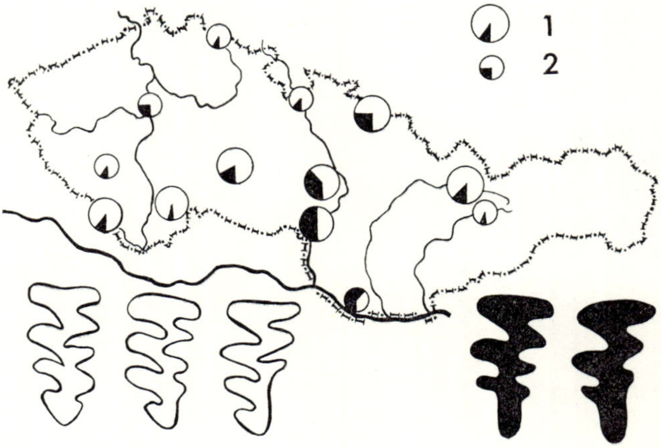

FIG. 5. The frequency of *simplex* molars (black sectors) in *Clethrionomys glareolus* from Czechoslovakia. 1—samples of more than 60 specimens; 2—samples of 40–60 specimens. (After Zejda, 1960.)

No signs of sexual dimorphism have been found in the tooth structure of voles. In species with rooted teeth including most of the fossil voles and, of the recent ones (*e.g.*) the genus *Clethrionomys* the height of the tooth-crown and the pattern of its surface change as the teeth become worn (Fig. 6). Zejda (1960) proved by experimentally grinding down the teeth of *Clethrionomys glareolus* that M^3 passes from

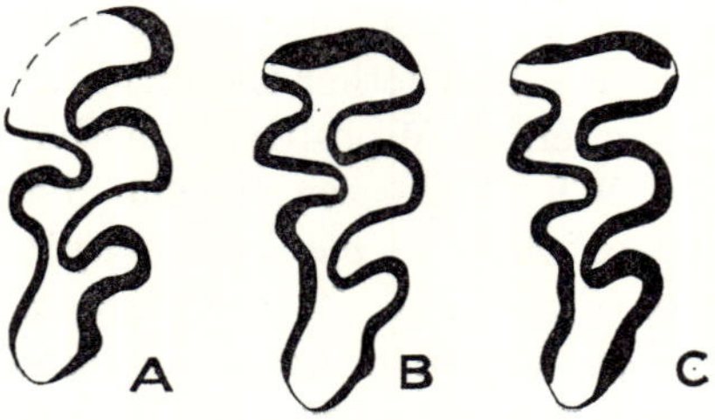

FIG. 6. A change in the pattern of right M^3 in *Clethrionomys glareolus* owing to wear. The drawings A, B and C show the same tooth at successive stages of experimental grinding. (After Zejda, 1960.)

the normal (*complex*) form into a simplified (*simplex*) one. This means that the description of the *simplex* condition must be carried out on teeth showing the same degree of wear. There is no doubt that some species of fossil voles have been erected on the basis of descriptions of single teeth derived from rather old specimens.

INTER-POPULATION VARIATION

In addition to differences in the form of teeth within a population, there are also differences between populations of the same species, living under different geographical conditions. Special importance is ascribed to the clinal type of variation, particularly to that in the size of teeth. For example, as we move from south-west to north-east, the length of the tooth-row of *Clethrionomys glareolus* increases (Table I). This phenomenon is connected with a change in food, which in the case of northern populations consists for the most part of green parts of plants.

TABLE I

Variation in the mandibular tooth-row length in Clethrionomys glareolus *(after Voronzov, 1961)*

Subspecies	Distribution	Tooth-row length Range (mm)	Mean (mm)
C.g. istericus Mill.	Parkland of the European part of the USSR and Central Europe.	4·6–5·3	5·0
C.g. glareolus Schr.	Deciduous and mixed forests of Europe.	4·7–5·6	5·0
C.g. suecicus Mill.	Zone of taiga in Europe.	4·7–5·6	5·1
C.g. devius Strog.	Riverside meadows of the Lower Pechora.	5·7–6·2	5·9

In Goyet Dave in Belgium Sickenberg (1939) found two different forms of the Bank vole *Clethrionomys glareolus*, a small form in the layers with a forest fauna and a large one in the layers with an Arctic fauna.

In contemporary subspecies of *Microtus gregalis*, some of which inhabit the tundra and some the steppes of Asia, the tundra subspecies have been found to be larger (Ognev, 1950). This species occurs in abundance also in the Pleistocene deposits of Europe. In the Nietoperzowa Cave in Poland I found (Kowalski, 1961) that the length of M_1 in specimens of this species from the layer corresponding to the climatic minimum of the last glaciation was 2·8 mm, whereas it was 2·6 mm on average in the strata immediately above and below, laid down in milder climates. Jánossy (1955) has reported a similar result from the Istállóskö Cave in Hungary, which shows that this species has reflected the influence of external conditions on morphological change. Kurtén (1960) found similar fluctuations in cricetids in the Early Pleistocene. It is of minor importance whether these changes are defined as the replacement of different subspecies of the same species in a stratigraphic series or as modifications of morphological characters caused by external conditions. The important point about these changes is the fact that they are reversible, and, as such, do not result from a unidirectional process of evolution but are adaptations of a population of the species to the external conditions acting on them directly.

VARIATION AND SPECIATION

As has already been mentioned, voles and lemmings were subject to considerable evolutionary changes in the course of their short geological history. As regards the molars, these changes are expressed by an increase in hypsodonty, leading to the formation—in most modern species—of open-rooted molars. The thickness of the enamel layer changes, too; this layer as a rule, becomes thinner and thinner, its thickness becomes irregular and cement appears in the enamel folds. These changes are paralleled in many evolutionary lines, giving rise to a lot of difficulties for systematists. Some genera of fossil microtines, *e.g. Mimomys*, represent grades rather than clades *sensu* Huxley (see Wood, 1965). It is open to discussion whether the evolution of voles occurred through the simplification of the tooth structure, as accepted by Hinton (1926) and by Zimmermann (1955), or through the complication of the original pattern (Guthrie, 1965). Both processes seem to be encountered in microtines, in which the tubercular structure of teeth (typical of their ancestors, the cricetids) underwent various transformations connected with different methods of grinding food. This resulted in a complicated structure such as can be seen in *Dicrostonyx* on the one hand, and, on the other, in a very simple form such as exists in *Ellobius*. The process of tooth evolution in the Microtinae was so rapid that we sometimes find remains representing various stages mixed together in one layer, although the species represented usually occur in different fossil localities, especially those of the

Pliocene and Early Pleistocene in somewhat different stages of evolution (this particularly concerns the height of the crown).

The process of evolution does not fall within the scope of this review, although how we treat successive morphological stages of an evolutionary line is relevant. It is much more difficult to find examples of the splitting of one evolutionary line into two descendent ones in fossil material than of continuous evolution within one line.

Specimens of voles which show intermediate characters between the species *Dolomys hungaricus* and *Mimomys stehlini* are met with in the faunas from the transition between the Pliocene and Villafranchian. Alongside them in the same localities occur specimens with a structure typical of both the parental species (Fig. 7). Such assemblages of specimens have been found at Węże in Poland (Kowalski, 1960), at Sète in France (Thaler, 1966), and also in some localities in Czechoslovakia. The presence of transitional forms may be interpreted in

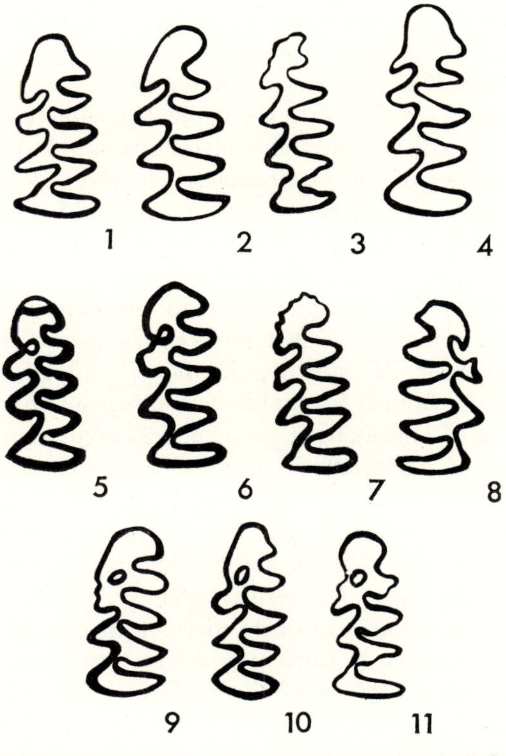

FIG. 7. M_1 of fossil voles from the Upper Pliocene of Węże in Poland. 1–4—*Dolomys hungaricus*; 5–8—transitional forms; 9–11—*Mimomys stehlini*. (After Kowalski, 1960.)

3 ways. It is possible that we are concerned with a single very variable species; secondly and extremely unlikely, that specimens of different geological ages occur mixed together in each of these localities, representing transitions from one species to another; finally, we may be dealing here with two distinct species, whose morphology is so variable that it is impossible to tell them apart on the basis of the characters which can be examined in fossil material. According to Thaler (1962), this variation is a sign of original polymorphism (*polymorphisme originel*—the presence of a greatly variable homogeneous population). The variation next becomes limited in the species descending from the original form. The fact is that in later strata the genera *Dolomys* (still existing as a relict form in the mountains of the Balkan Peninsula) and *Mimomys* (which has evolved into the modern genus *Arvicola*) are morphologically well divided.

Chaline (1966) described a similar phenomenon from the Middle Pleistocene of France. At the Mas Rambault site he found a very variable population of *Allophaiomys pliocaenicus*. In the somewhat later Valerats site (from the time of the Mindel glaciation) there was a population of the same species with the crown-pattern resembling that in the species described earlier as *Microtus nivalinus*, *Pitymys hintoni* and *P. gregaloides*. Together with the typical specimens of *Allophaiomys pliocaenicus* they still formed a continuous spectrum of variation. Four different types of the Microtinae, belonging to the genera *Allophaiomys*, *Microtus* and *Pitymys*, have been found in the more recent locality at Bourgade. Later, *Allophaiomys* became extinct and the other forms persisted as quite distinct genera.

The fauna from the Riss glaciation in the British Isles provides a similar example. A medley of specimens of the genus *Microtus*, belonging to three species, *M. nivalis*, *M. malei* and *M. oeconomus*, are encountered in the faunas from this period, *e.g.* in the fauna from the Glutton Stratum of Tornewton Cave. They form an almost uninterrupted series of graded variation. In particular, the specimens referred to as *M. malei* Hinton always occur with *M. nivalis*, and one can easily find all intermediate stages in the tooth structure (most specimens in the collection of Hinton, who described this species, have been labelled as intermediate between *M. nivalis* and *M. malei*). Nowadays, *M. nivalis* and *M. oeconomus* are quite distinct species, of which the first inhabits the mountains of Europe and Western Asia and the second lives in the northern regions of the Holarctic. In the modern fauna, however, there are also specimens with intermediate characters in the structure of skull and teeth (Gaffrey, 1943) and, consequently, the identification of a single skull is sometimes impossible.

It is an open question whether in these cases the attainment of genetic isolation, which eventually initiates the process of independent evolution and differentiation of species, took place within one area (sympatrically) or after the two populations had been isolated geographically. Chaline (1966) assumed that the differentiation of the variable population described by him occurred on the spot. However, the alternating glaciations and interglacials of Europe seem to have caused repeated discontinuities in the originally uniform ranges of the species of voles. In the glacial periods this was a discontinuity between particular refuges of thermophilous forms in the south of Europe, and in the interglacials a separation between the northern and alpine ranges of cold-adapted species. The effect of the Quaternary discontinuities on the formation of species has been found many times in the European fauna. In the case of species which show a relatively slow evolution, the Pleistocene period may, roughly speaking, be treated as a whole. In this way one can explain, for example, the allopatric differentiation of *Desmana* into two species, *Desmana moschata* in south-eastern Europe and *D. pyrenaica* in the south-west. As regards the then fast-evolving voles, each glaciation isolated populations and brought about the differentiation of species at various stages of evolution. The picture is, in addition, complicated by migrations from Asia.

It is open to dispute whether the particularly great variability of some vole populations at different stages of evolution was a phenomenon preceding their speciation (as suggested by Thaler, 1962), or whether it resulted from the speciation and subsequent meeting—in the same area—of forms which were genetically independent but still little changed morphologically. In such a case they would appear to us—on account of the limited number of characters that we can examine in fossil material—as a variable but homogeneous population.

Guthrie (1965) has carried out an analysis of variation of different structural characters in the molars of two species of the genus *Microtus*, the fossil *M. paroperarius* and the recent *M. pennsylvanicus*. He showed that the structure of the anterior part of M_1 and that of the posterior part of M^3 are marked by a higher degree of variability than the structure of the other portions of the molars. In Guthrie's opinion, this indicates that the characters which undergo rapid evolution show greater individual variation. It might be supposed on the basis of general evolutionary theory that the characters which undergo rapid evolution and are, therefore, subject to strong selection, should have reduced individual variation. Guthrie, however, adduces examples both from palaeontological studies on various mammalian species and from experimental studies on *Drosophila*, which indicate that this does not always correspond with the facts and, consequently, the characters which undergo rapid change and are subject to intense selection show increased variation. Although Guthrie's theory that the terminal portions of tooth-rows in *Microtus* actually undergo rapid evolution leading to their extension is questionable, yet his discussion is very pertinent. It may be supposed that the foregoing examples of increased variation during the periods of intense evolution of voles support the inference about increase in variation whilst undergoing intense selection. Further studies carried out in this respect on contemporary and fossil voles should throw more light on this problem.

References

Chaline, D. (1966). Un example d'évolution chez les Arvicolidés (Rodentia): les lignées *Allophaiomys*, *Pitymys* et *Microtus*. *C. r. hebd. Séanc. Acad. Sci. Paris* **263**, 1202–1204.

Corbet, G. B. (1964). Regional variation in the bank-vole *Clethrionomys glareolus* in the British Isles. *Proc. zool. Soc. Lond.* **143**, 191–219.

Farbiszewska, J. and Makarzec, B. (1959). Gebissvariabilität bei *Microtus arvalis* (Pallas, 1779) in Ost-Polen. *Acta theriol.* **3**, 302–307.

Gaffrey, G. (1943). Zur Unterscheidung von *Microtus ratticeps* (Keys et Blas.) und *Microtus nivalis* (Martins). *Zool. Anz.* **143**, 157–164.

Geptner, V. G. and Rossolimo, O. L. (1968). Vidovoi sostav i geografitseskaia izmentzivost aziatskikh gornykh polevok roda *Alticola* Blanford, 1881. *Sb. Trud. zool. Mus.* **10**, 53–93.

Guthrie, R. D. (1965). Variability of characters undergoing rapid evolution, an analysis of *Microtus* molars, *Evolution, Lancaster, Pa,* **19**, 214–233.

Heller, F. (1958). Eine neue altquartäre Wirbeltierfauna von Erpfingen (Schwäbische Alb). *Neues Jb. Geol. Paläont. Abh.* **107**, 1–102.

Hinton, M. A. C. (1926). *Monograph of the voles and lemmings* (*Microtinae*), Vol. I. pp. 1–488. London: British Mus. (N.H.).

Jánossy, D. (1955). Die Vogel- und Säugetierreste der spätpleistozänen Schichten der Höhle von Istállóskö. *Acta. archeol. Acad. Sci. Hung.* **5**, 149–181.

Kowalski, K. (1957). *Microtus nivalis* (Martins, 1842) (Rodentia) in the Carpathians. *Acta theriol.* **1**, 159–182.

Kowalski, K. (1960). Cricetidae and Microtidae (Rodentia) from the Pliocene of Węże (Poland). *Acta zool. cracov.* **5**, 447–488.

Kowalski, K. (1961). Plejstoceńskie gryzonie Jaskini Nietoperzowej w Polsce. *Folia quatern.* **5**, 1–22.

Kowalski, K. (1966). Stratigraphic importance of rodents in the studies on European Quaternary. *Folia quatern.* **22**, 1–16.

Kurtén, B. (1960). Chronology and faunal evolution of the earlier European glaciations. *Commentat. biol.* **21**, 1–62.

Ognev, S. I. (1950). *Zveri SSSR i prilezhaschchikh stran.* **7**, 1–707. Moskva-Leningrad: Akad. Nauk. SSSR.

Reichstein, H. and Reise, D. (1965). Zur Variabilität des Molaren-Schmelzschlingenmusters der Erdmaus, *Microtus agrestis* (L.). *Z. Säugetierk.* **30**, 36–47.

Repenning, C. A. (1968). Mandibular musculature and the origin of the subfamily Arvicolinae (Rodentia). *Acta zool. cracov.* **13**, 29–72.

Sickenberg, O. (1939). Die Insectenfresser, Fledermäuse und Nagetiere der Höhlen von Goyet (Belgien). *Bull. Mus. r. Hist. nat. Belg.* **15**, 1–23.

Stein, G. H. W. (1958). Über den Selektionswert der Simplex-Zahnform bei der Feldmaus, *Microtus arvalis* (Pallas). *Zool. Jb.* (*Syst.*) **86**, 27–34.

Thaler, L. (1962). Campagnols primitifs de l'Ancien et du Nouveau Monde. *Collogues int. Cent. natn. Rech. scient.* **104**, 387–397.

Thaler, L. (1966). Les rongeurs fossiles du Bas-Languedoc dans leurs rapports avec l'histoire des faunes et la stratigraphie du tertiaire d'Europe. *Mém. Mus. natn. Hist. nat., Paris* n.s., C, **17**, 1–295.

Voronzov, N. N. (1961). Ekologicseskie i nekotorye morfologicseskie osobennosti ryzhikh polevok (*Clethrionomys* Tilesius) europeiskogo severo-vostoka. *Trudy zool. Inst., Leningr.* **29**, 101–136.

Wood, A. E. (1965). Grades and clades among rodents. *Evolution, Lancaster, Pa,* **19**, 115–130.

Zejda, J. (1960). The influence of age on the formation of third upper molar in the bank-vole *Clethrionomys glareolus* (Schreber, 1780) (Mammalia: Rodentia) *Zool. Listy* **9**, 159–166.

Zimmermann, K. (1952). Die *simplex*-Zahnform der Feldmaus, *Microtus arvalis* (Pallas). *Zool. Anz.* **17** suppl., 492–498.

Zimmermann, K. (1955). Die Gattung *Arvicola* Lac. im System der Microtinae. *Säugetierk. Mitt.* **3**, 110–112.

Zimmermann, K. (1958). Selektionswert der *simplex*-Zahnform bei der Feldmaus? Eine Entgegnung. *Zool. Jb.* (Syst.) **86**, 35–40.

REPRODUCTIVE AND SEASONAL VARIATION

Symp. zool. Soc. Lond. (1970) No. 26, 165–188.

VARIATION AND ADAPTATION IN REPRODUCTIVE PERFORMANCE

H. G. LLOYD

Ministry of Agriculture, Fisheries and Food
Infestation Control Laboratory, Field Research Station
Worplesdon, Guildford, Surrey, England

SYNOPSIS

Variation in reproductive performance, extending from onset and duration of the breeding season to the productivity of young, is more readily seen in polyoestrous, polytocous species than in species producing single or twin young annually. The ecological adaptations of these variations, in general terms, are understood, but the precise effects of each environmental component upon reproduction in mammals is difficult to elucidate in the wild. Births occur at a season most favourable to the survival of mother and offspring, and food availability to the young at weaning is also of considerable importance. Instability of numbers in cyclic species seems not to be primarily connected with reproductive productivity. Proximate factors such as light fix the general breeding patterns of species, but ultimate factors appear to be responsible for variations in performances in any season. The more important environmental influences upon reproduction are: population density, food availability and quality, age structure of the population, and climate and physical factors. Field investigations on the reproduction of wild rabbits in Britain, mainly on offshore islands, reveal that in spite of lack of evidence on the effects of food upon breeding, the density of population seems to have an over-riding importance upon breeding productivity. In detail, each island population examined showed different breeding characteristics, but the general patterns of relationships were similar.

INTRODUCTION

The reproduction of wild mammals is enormously varied and complex. Some species show considerable variation in reproductive performance, whilst others tend to be more steady; the ecological adaptations of these variations are known in some cases, but the mechanism whereby the environment exerts its influence upon the range of behavioural and physiological stages involved in breeding is far from being understood. Mammalogists studying reproductive performance in different situations are faced with a lack of information on the requirements of the species and with an inability to qualify and quantify the environmental components; in the field, difficulties in the experimental manipulation of the natural components prevent the direct study of even one component. Furthermore, mammals as a group, unlike birds, do not show stereotyped visual displays, and their vocal communications may be difficult to study; many are active at night; some seldom expose themselves outside snow or ground cover; they may be difficult to count and their breeding places may be inaccessible or obscured; the counterpart of clutch sizes are hidden from view and can only be studied by killing or handling a captive animal; the quality and quantity of nourishment that the unweaned young receive are not easily deduced; the nourishment of weaned young can be deduced only by indirect means; and most important, there is a large gap in our understanding of olfactory communication. Small wonder, therefore, that much remains to be elucidated.

Much that is known about the reproduction of mammals and its relationship to the environment is gathered from a mass of often conflicting evidence. It is conflicting because there is no defined standard situation and even the disarmingly simple concept of population density is complex and possibly meaningless unless it is related to behaviour and other environmental variables. How, for example, does one measure and compare the population densities of House mice (*Mus musculus*) on an island such as Skokholm and those in corn ricks? And which of these is the more hostile environment?

Some generalized observations on the adaptive significance of reproductive performances are possible, but exceptions are many and bear out the improbability of reproductive patterns being determined by the same set of factors in all species or even in different communities of the same species. Nor is there any reason to suppose that in seasonal breeders, variations in such aspects as onset, duration and productivity of the breeding season are mediated by environmental components acting in exactly the same way in each case.

SEASON OF BIRTHS AND FOOD RESOURCES

It seems to me that Lack's hypothesis that the reproductive performance has evolved, through natural selection, to yield the greatest number of surviving offspring is applicable to mammals, but until population density in mammals can be related to or qualified by population age-structure, space, food availability, cover, social interaction and so on, it will be difficult to investigate the qualification that the reproductive performance is mediated in a strictly density-dependent manner. Obviously, young must be born at a time when food of the required quality and quantity is available, either for the nursing mother, for the weaned offspring or for both. The nutritional demands upon a pregnant female are not high but become considerable during lactation. In some post-partum ovulatory species the demands of lactation prolong gestation (mice, rats, voles), or inhibit post-partum oestrus according to the size of the litter being nursed (domestic rabbits). The duration of lactation is relatively long and development of young is slow in some species. This occurs in Grey squirrels (*Sciurus carolinensis*) in Britain where under optimum conditions they produce two litters a year, in spring and late summer. Food supplies in spring are less certain than in late summer and autumn; the slow development of the young may be an adaptation to this, and only by ensuring against poor food supplies in the one season of the year could the maximum potential of reproduction for the year consisting of two breeding periods, be obtained. The mean nest litter-sizes in Grey squirrels are 2·50 in the spring and 3·23 in the autumn, the young weigh 13–17 g at birth, and 7 weeks later when solid food is also taken they weigh approximately 150 g (Shorten, 1954). In contrast, the coypu (*Myocastor coypus*), after a long gestation period, produces precocious young which suckle for 6–10 weeks in captivity, but can be weaned at 5 days of age (Newson, 1966). Coypu are non-seasonal breeders and the variable weaning period may be an adaptation to enable solid food to be taken early according to the availability at the time of birth and perhaps also to birth litter sizes.

Mammals of temperate and Arctic regions are subject to pronounced seasonal climatic changes and the optimum periods for production of young are better defined than in tropical or equatorial regions. Desert animals are closely dependent upon climatic changes and the irregularity of the seasons is the dominating factor in their reproduction, whereas species in temperate zones appear to be influenced by a wider variety of environmental factors. In temperate and Arctic species the fundamental breeding pattern of mammals is related to the turnover of the population—the reproductive rate, mortality rate, physiological and ecological longevity. At the two extremes of this pattern are species:

(a) which produce single young annually,

(b) those which may breed continuously. (These may be seasonal breeders in less favourable circumstances.)

Between these two extremes are:

(c) seasonal breeders characteristically showing post-partum oestrus producing many, multiple-young litters in a season,

(d) non-post-partum oestrus breeders producing two or three multiple-young litters,

(e) single littering species producing multiple-young litters.

Mammals showing type (c) breeding, notably myomorphs and lagomorphs, seem to be characterized by the optimistic nature of their breeding performance. If in this group, the onset of breeding in the season (overtly demonstrated by mating in most species) is a rapid but variable response to improvement in food quality, there is some possibility of this increase in supply being still available at lactation, or weaning, since the gestation and lactation periods are short. If, however, total disaster overtakes these offspring it matters little since there are many more litters to follow in that season. Type (a) breeders produce single young annually and are usually characterized by having long gestation periods; nutritional changes at the mating period can have little relationship to the food availability at birth, lactation or weaning, and therefore the period when they mate is subject to less variation. Their young are born at the most favourable period in terms of food resources, and possibly cover or concealment. This must be early enough in the year to allow the young to survive independently before adverse climatic conditions prevail and possibly, to permit the young to become sexually mature in the next mating season. It is also characteristic of these species that lactation is prolonged. Their longevity is relatively great, survival of young is high and in the long term, if conditions are highly favourable, they can show a dramatic increase in numbers in spite of their low reproductive performance. The intermediate types (d) are specialized species, such as tree squirrels, moles and hedgehogs, occupying difficult or unstable niches. Type (e) carnivores, the exploiters of situations, have comparatively short gestation periods, and in temperate regions have restricted but asynchronous mating periods. In Britain the number of ova ovulated by the fox (*Vulpes vulpes*) may have some relationship to the physical condition of the female, but in the tundra, Arctic and Red foxes may respond to the increase in availability of food during the winter and spring resulting from winter breeding of rodents in pre-peak years of the rodent cycle. For scavenging hill foxes in Britain, winter is not by any means the lean period supposed, nor has the availability of carrion diminished at lambing time when foxes may be nursing or rearing litters. The range of birth litter sizes in foxes in Wales and Kent is astonishingly wide, varying from 1–13, with a mean of 4·6 (H. G. Lloyd, unpublished). Both the wide range of weights of embryos in a set nearing term and field observations on the decreasing sizes of unweaned litters of cubs from April to June suggest that unfavourable conditions for lactation may affect only those young unable to obtain as much nourishment as others during the suckling free-for-all which is characteristic of carnivores.

Food requirement, however, is not the only factor determining the period of production of young. For example, rabbits (*Oryctolagus cuniculus*) in pre-myxomatosis populations in good grass-growing areas in Britain, showed peak breeding in April and May, followed by a rapid decline and termination of breeding in July. Present populations breed into August, but in both circumstances food, by subjective estimates, appears to be in greater supply and of better quality in July and August than in January, February or March. Thus the termination of breeding in the season in this species may be unconnected with food availability. Rabbits however may not be the best of examples since their original breeding patterns in Spain may have been and may still be bimodal, in spring and summer and late autumn and winter, as it tends to be in Australia. The termination of breeding at apparently optimum times may be connected with such factors as the health of the female, increased exposure of late litters to disease and predators or to hostile influences of others of its own kind. If cessation of breeding does not occur because of exhaustion of the female, the cause can be found only in the increasing hostility of the environment which may diminish the chances of survival of mother and offspring.

VARIATION OF PRODUCTION IN CYCLIC SPECIES

Instability of numbers, especially in species which show some periodicity in population cycles with occasionally spectacularly high or low numbers at approximately 4 or 10 year intervals, occurs in the Arctic, boreal and temperate regions. The mammals showing these cycles characteristically have high rates of reproduction—voles, lemmings and Snow-shoe hares—but in the same regions the slow breeders do not fluctuate with any periodicity or in such numbers. Elton (1924) considered that climate was the primary cause of these fluctuations.

Cycles of microtine rodents in temperate areas can also be spectacular in some years. Those which have been extensively studied do not seem to show any variation in intensity of breeding during the actual breeding periods though the seasons do vary in duration. Litter sizes appear not to change with the phase of the cycle but surprisingly little is known of the actual production of young per pregnancy in different years of the cycle and most references refer to litter sizes *in utero* so that the extent of pre-natal mortality is not known. One enigmatic feature of these cycles is that breeding continues during the decline, but the length of the breeding season may be slightly curtailed. In the case of animals showing 3-, 4- or 5-year cycles it seems that variation in duration of breeding occurs at different phases of the cycle and the increase in numbers is due to high survival rate and the decline to low survival rate. The reproductive patterns do not appear to be connected with the rise and fall in numbers though they may have some relationship to the periodicity of the cycle. Microtine cycles are, however, by no means regular in occurrence and many populations may remain at low densities for many years (Chitty and Chitty, 1960). In Snow-shoe hares (*Lepus americanus*) and Muskrats (*Ondatra zibethica*) which exhibit 10 year cycles, the reproductive rate is lower than in species showing 4 year cycles, and opportunist carnivores (lynx, fox and mink) show periodic cycles which seem to be determined by the abundance of prey. If cycles with any degree of periodicity do occur in slow breeders, such as the Cervidae, it is likely that they would occur at greater intervals of time than in the rodents and lagomorphs although irregular catastrophes due to weather or food might obscure the effect. It is interesting to learn that Wolf (*Canis lupus*) fur returns showed irregular fluctuations before the extermination of the bison (*Bison bison*) in the late nineteenth century. After this large and constant biomass of herbivores disappeared the wolf appeared to have regular peaks of abundance at approximately 10 years (Hewitt, 1921) but the time interval during which these have been studied is too short to permit definite conclusions.

Species with high reproductive productivity, such as voles and lemmings have few specific measures to counter predation and their high birth rates are associated with low survival rates. Some species such as the Wood mouse (*Apodemus sylvaticus*) (Jewell, 1966) and the House mouse (*Mus musculus*) (Southwick, 1958) show very early sexual maturity so that the first, and possibly the second, generations may under some conditions participate in the same breeding season as the parents. It is also interesting to note that the more "efficient" types of placentae are found in the highly productive breeders—haemoendothelial and haemochorial in lagomorphs and myomorphs (and sciuromorphs, histricomorphs, insectivores and primates), while the groups with long gestation periods—pigs, horses and ruminants—have the least efficient, epitheliochorial and syndesmochorial placentae, and the intermediate species in terms of annual productivity (carnivores) have an intermediate endotheliochorial placenta.

Mammals are committed to breeding at oestrus and, unlike grebes among birds for example, are not influenced by the suitability of the habitat for nest building. In multi-littering species the habitat may not be well suited for survival of the offspring throughout the distribution of the species at all times of the season. For example I have observed that the offspring of early pregnancies of Brown hares (*Lepus europaeus*) are dropped in inadequately concealed forms, or near hedgebanks or road verges which receive much attention from predators, whereas in mid-season young have adequate concealment. This is especially noticeable in extensive areas of plough land. The dual nest building behaviour of wild rabbits—some breeding in stops and others in main warrens—may be a reflection of variation in production of groups of individuals in rabbit communities.

ENVIRONMENTAL INFLUENCES AND BREEDING COMPONENTS SUBJECT TO VARIATION

In temperate zones the proximate factor determining the onset of breeding or oestrus activity seems to be the influence of light, at times of increasing or decreasing daylength according to the species, and, in addition, possibly the effects of temperature. In some species, sheep for example, the onset of reproductive activity recurs at a remarkably constant period each year according to the geographical location, but in other species, European rabbits for example, there is a considerable variation in the time of breeding. Thus in some species, characteristically those producing many litters in a season, other factors, usually referred to as ultimate, repress or delay the effect of the proximate stimulus and it is to these that variations in performance can mainly be attributed.

It is therefore relevant to consider the many aspects of reproduction which are subject to variation. Some of these are:

(a) Onset of breeding season.
(b) Duration of breeding in the season.
(c) Age at onset of reproduction or puberty in individuals.
(d) Proportion of animals participating in breeding—or pregnancy rate.
(e) Extent of out-of-season breeding.
(f) Ovulation rate per pregnancy and mean rate per breeding female per season.
(g) Uterine litter sizes and intra-uterine mortality.
(h) Birth litter sizes.
(i) Proportion of successful to unsuccessful pregnancies.

These seem to show causal relationships to changes in some of the environmental components but since all the external factors affecting reproduction may not be known, and since there is difficulty in measuring the few components that can be recognized and even more in assessing the synergistic or antagonistic effects of different components, it is rarely possible to reveal exclusive cause and effect relationships between one environmental component and reproduction.

Recognizing the limitations imposed by field study techniques, 4 factors of fundamental importance emerge:

(a) "Population density"—closely connected with social intercourse especially in gregarious species.
(b) Food availability and quality.
(c) Age-structure of the breeding population.
(d) Climate and other physical factors.

The physical factors are of fundamental importance in determining breeding patterns and can, in addition, produce variations within any season in some species. For example, a recurrence of cold weather in late winter/early spring inhibited post-partum oestrus in wild rabbits dropping litters at that time (Brambell, 1944). Improvement in grazing conditions following rainfall may boost or cause a resumption of breeding in rabbit populations in mid-summer in drier areas of Australia (Poole, 1960). Where breeding is influenced by precipitation it may be mediated by reinforcement of endogenous gonadotrophins by the oestrogenic activity (Moule, Braden and Lamond, 1963) of new plant growth, or it may simply be due to increased availability of water. An adverse effect of precipitation occurs occasionally in breeds of small sheep when uninterrupted wetting and chilling by heavy rainfall in the last month of gestation often produce a distinct lack or derangement of maternal behaviour at parturition.

The apparent variation in onset of breeding—earlier breeding in short day breeders (Mule deer (*Odocoileus hemionus*): Einarson, 1956) and later breeding in long day breeders (the fox in Sweden: Jan Englund, personal communication) at increasing latitudes—may, if genetical differences are not involved, be entirely due to the difference in the light regimen. Table I shows the variation in amount of daylight and the degree of change of light during the first 3 months of the year. At the equinoxes (17th/18th March and 17/18th September) at all latitudes in each hemisphere, the ratio of light to dark will be the same, but on any other day of the year the ratio will be different. For example, at 65°N as compared with 52°N there will be a delay of 32 days in experiencing 8 h of daylight. If the proximate stimulus in any given species is not mediated by other factors, species mating in the autumn should show earlier mating with increasing latitude. If this is not so, other physical factors or genetical differences may be involved.

In some species there is a strong possibility of genetical differences influencing timing of breeding. Moles (*Talpa europaea*) are monoestrous: in North Wales they breed about one month later than moles in southern England and in populations north and south of the river Trent at

TABLE I

Latitude and day-length between 17 January and 17 March

	Hours of daylight 17 January	Increase (minutes) and % increase in daylight between 17 January and 17 February	Increase (minutes) and % increase in daylight between 17 February and 17 March	8 hours daylight
65°N (Skelleftea, N. Sweden)	5·25	227 m (72%)	191 m (36%)	12 Feb.
58°N (Lairg, Scotland and Gothenberg, Sweden)	7·28	136 m (31%)	141 m (24%)	20 Jan.
52°N (Oxford)	8·32	104 m (21%)	113 m (18%)	10 Jan.

From Smithsonian Institute Almanac, U.S. Weather Bureau.

Nottingham there is a difference in onset of breeding of about a fortnight (Morris, 1958). The mole is not highly mobile and in different habitats there may be a differential response to the proximate stimulus. This could be tested experimentally by translocation.

The period when mammals will mate does not invariably coincide with oestrus in the female. In the fox a high proportion of apparently anoestrous vixens will mate during the month preceding the period when the first ovulations are commonly found, (13th January, Kent and Surrey, and 27th January in mid-Wales: H. G. Lloyd, unpublished) while in the rabbit Brambell (1944) found evidence of mating in anoestrous females during the 3 months prior to the onset of the breeding season. Whether or not this pre-season behaviour varies according to environmental circumstances is not known nor is there any information relating this behaviour to the time of onset of breeding in individuals.

VARIATIONS IN REPRODUCTION IN WILD RABBITS

I will now illustrate some of these points by examples of variations in different parts of the reproductive process in groups of individuals, and in populations, from wild rabbits in Britain. I shall not make much reference to the detailed and exciting studies on rabbits in Australia and New Zealand, mainly because there is too much information for inclusion in a short review. Myers (1966) has briefly reviewed the effects of population density on rabbits, Casperton (1968) has described some environmental influences upon the physiology of wild rabbits, and Sadleir (1969) has collated information on the effects of environment upon breeding of mammals.

Autopsies were made on rabbits taken from the following places, and some comparisons are made with data collected by Brambell (1944).

Wales 1957

Rabbits, which were exceedingly scarce, were collected from very widely scattered areas of Caernarvonshire and Anglesey, mainly from agricultural pastures.

Skokholm, Pembrokeshire

This island of 262 acres is mainly composed of rabbit resistant species of vegetation: the grasses *Agrostis tenuis*, *Festuca rubra* and *Holcus lanatus* predominate. Bracken is widespread, some heath occurs and large areas of thrift (*Armeria maritima*) extend along the spray-covered seaward sides of the island. The soil is light and rabbit warrens are extensive. Rabbits have been on the island for at least 700 years.

Skokholm 1959. The rabbit population was at "normal" climax size, of about 5000 adult animals in spring, but prolonged dry weather from April to September inclusive produced severe drought conditions (6·8 in. of rain fell compared with 21·8 for 1958 and the mean temperature was 61·8°F compared with 58·1° for 1958). Starvation conditions prevailed and the population fell to no more than 150 animals in February 1960. The 70 Soay sheep were destroyed on humanitarian grounds and not replaced.

Skokholm 1960. No animals were killed for autopsy. Pastures improved from August 1960 but thrift did not recover until early 1961.

Skokholm 1961. Pasture improvement was very marked, breeding population of about 900 rabbits.

Skokholm 1962. Pasture deterioration occurred and superficially its condition was not unlike 1959. Breeding population numbered about 3500.

Skokholm 1963. Prolonged and exceptionally cold weather killed or dehydrated the vegetation which was unprotected by a snow blanket. Conditions similar to summer drought prevailed and about 50% of over-wintering rabbits died in February and March 1963. A high proportion of juveniles died during the winter. Fifty-seven rabbits out of 224 tagged in October 1961 were recovered alive after April 1963, whereas only 6 juveniles out of 175 marked in August 1962 were recovered alive.

The Skerries

The Skerries are a group of 3 islands totalling 6 acres off Holyhead, Anglesey. They are connected at low tide. The vegetation is predominantly *Festuca rubra* and *Agrostis tenuis*. The sward is short at all times of the year and of very poor quality. The population was estimated to be about 100 adults in March 1964.

The Inner Farne Island

The Inner Farne Island is 9 acres in extent and in the winter and spring is composed mainly of *Holcus lanatus, Agrostis tenuis* and *Festuca rubra,* with much bare soil, probably as a result of scorching by bird faeces. The quality of vegetation, like that of the Skerries, was poor and not comparable in quality to that of Skokholm even in 1959. The total number of rabbits was about 100 in April 1969.

Skomer 1959

This population on the 900+ acre island suffered myxomatosis in

1954–1955 but rabbits were sufficiently numerous in 1959 to enable about 40 rabbits to be shot per day in the spring. This population was sampled only twice during the early part of the season. The composition of the vegetation was similar to Skokholm, but in 1959 it was more productive than Skokholm.

Except in Wales 1957, these populations were not subject to predation by mammals, but avian predators (mainly the Greater black-backed gull (*Larus marinus*)) were present on all the islands.

Onset, duration and productivity of breeding

In monotocous breeding species—such as deer—there is little variation in the period when pregnancies occur; in polytocous, post-partum oestrus breeders—such as wild rabbits—variation is often seen. It is necessary, when considering onset of the breeding season, to distinguish a base-line for time of onset of breeding in order to detect changes between and within populations of any species. In some seasonal breeders the onset of breeding in the female population can be spread over several weeks and out-of-season breeding may also occur. In rabbits, for example, 5% of the female population have been shown to be pregnant during the winter (Brambell, 1944). Each species has to be considered according to its pattern of breeding in this respect. In most populations of British rabbits, the proportion of animals pregnant increases quite sharply over a comparatively short period of time, and the basis that has been adopted for comparison between years (Lloyd, 1963) is the period when 50% of the female population are pregnant—at the onset and termination of the season. Using this criterion it will be seen from Fig. 1 that the onset of breeding in rabbits is subject to variation. In Fig. 1 the estimates of the densities of the population shown, from which the samples were drawn, are in some cases subjective (b, g and h) but in the remainder the indices of density are based upon capture/recapture data, and upon visual spotlight counts at night when the populations were very sparse (e, and to some extent g). Pre-myxomatosis populations of high density in N. Wales (h, based upon Brambell, 1944) showed onset of breeding (in 50% of does) in the second week of February but in low density populations in the same area in the third week of February in 1957. The season may however have been even earlier in 1957, because collection of samples did not begin until mid-February, and some of the lactating pregnant animals may have been in their third pregnancy of the season. The decrease in the proportion of pregnant does in late February and early March 1942 was due to suppression of post-partum oestrus in

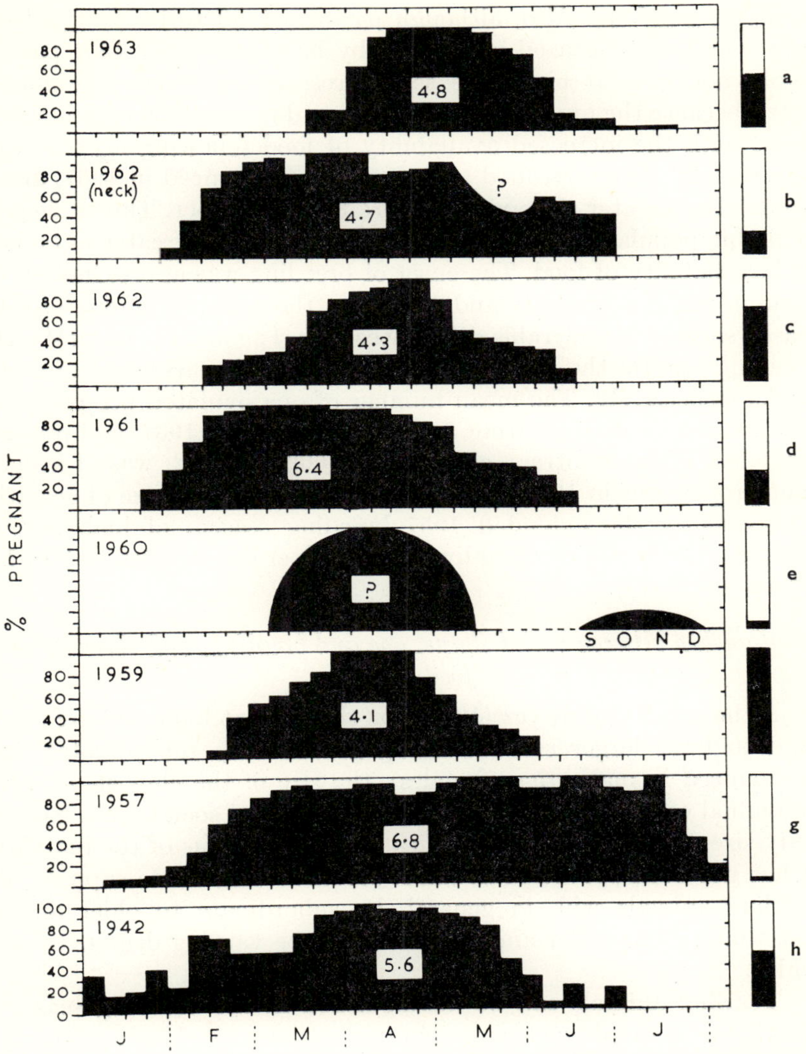

FIG. 1. Breeding season in British wild rabbits, showing the proportion of pregnant females, the mean ovulation rate (in centre of histogram), and the relative population density (on right). a–f: Skokholm; g: Mainland Wales (own data); h: Mainland Wales (after Brambell, 1944).

does which dropped their litters during a cold spell which occurred then. Had this not occurred breeding would probably have reached its peak sooner than early April. Agriculturally, the pastures where rabbits were collected in 1942 and 1957 were similar in type, but the productivity of the 1957 pastures would have been greater due to the decrease in rabbit numbers. Also, although pastures tend to be overgrazed in winter and under-grazed in summer by farm stock, the 1957 pastures would not have received the same pressure as in pre-myxomatosis years, because the number of farm stock had not by that time increased apace with the increased availability of food following myxomatosis. Superficially therefore, food abundance was enhanced in 1957, and its availability greater since rabbits were few in number. Thus a decrease in rabbit population density was accompanied by greater abundance and availability of food. The onset of breeding was approximately the same in the 2 years 1942 and 1957, but the duration of the breeding season showed considerable differences (Fig. 1; g and h); using the 50% breeding criteria the lengths of season in the 2 years were 15 and 22 weeks respectively. The mean number of ova ovulated per pregnancy also showed an increase from 5·6 in 1942, to 6·8 in 1957, and pre-natal mortality of entire litters, estimated at 60% in 1942, was of insignificant proportions in 1957 and occurred at a different stage of gestation. Consequently the extended 1957 breeding season, of high fertility, yielded a productivity of 30 live offspring per doe against 10·3 in 1942 (Brambell, 1944).

Relationship between social status, age and breeding, and observations on food resources

Although Mykytowycz (1965) has shown that high-ranking rabbits tend to have larger scent marking glands than subordinates (more pronounced in males than females), the size of the chin gland in any individual cannot reliably be used as evidence of social rank. Nor does body size necessarily give an indication of the status of the individual. It has not been proved although it seems reasonable to suppose, that yearling animals will, in general, be subordinate to older animals, and that among older animals there will be varying degrees of social rank.

Yearling animals can be distinguished in the breeding season on Skokholm by dried eye lens weights; sometimes by closure of the tibial epiphyses; and, to some extent, by body weight. If the reproductive performances are examined according to the two age-classes—yearling and older does—some differences in reproduction become apparent. Not only is the mean number of ova ovulated directly related to body weight in any season (Table II), and shows variation in this relationship in different populations and at different times of the breeding season, but also yearling animals in years of high population density do not contribute to the total breeding productivity at the same high rate as older does. Yearlings become pregnant later and cease breeding earlier

Table II

Mean ovulation rate related to body weight in wild rabbits

Body wt. in g	1941 Welsh mainland: Brambell, 1944	1942 Welsh mainland: Brambell, 1944	1957 Skokholm	1959 Skokholm	1959 Skomer	1961 Skokholm	1962 Skokholm (Main island)	1962 Skokholm (Neck)	1963 Skokholm
1600–1649									
1550–1599									
1500–1549		6·1	7·9						
1450–1499	6·1								
1400–1449									
1350–1399	5·2	5·9	6·7	4·4					6·1
1300–1349					6·8		5·0		
1250–1299	5·0	5·8	7·1			6·6		5·3	5·9
1200–1249				4·2	6·5		4·4		
1150–1199	4·8	5·6	7·0			6·7		5·2	5·0
1100–1149				4·0	6·6		4·7		
1050–1099	4·4	5·4	5·8			6·1		4·2	4·7
1000–1049				4·0	5·7		4·3		
950– 999			5·3					4·2	4·1
900– 949	3·7			4·0	4·9	5·5	3·9		
850– 899		4·3							
800– 849							3·8		3·2
750– 799				3·0					
700– 749									
Mean no. of ova	4·89	5·64	6·8	4·1	6·05	6·4	4·3	4·7	4·8
Mean body wt. in g	1214	1212	1223	1082	1078	1135	1043	1070	1092
Population density index	5	5	0·1	10	3	1·5	7	2	5

TABLE III

Skokholm Rabbits, 1959. *Proportion of total number of does becoming pregnant each week at the onset of breeding season, and proportion ceasing to breed at termination of season; the mean number of corpora lutea and mean cleaned body weights of each weekly group are shown*

	Becoming pregnant								Becoming anoestrous						
Proportion	0·05	0·35	0·13	0·07	0·09	0·08	0·21		0·29	0·13	0·22	0·10	0·02	0·08	0·16
Mean body weight (g)	1171	1164	1066	928	901	902	772		950	1063	1062	1076	1137	1132	1125
Mean number of corpora lutea	3·6	3·7	3·7	3·9	3·8	3·0	4·1		4·5	4·0	5·2	5·1	4·2	4·5	4·5

Date: February 14 — 21 — 28 — March 7 — 14 — 21 — 28 — April 4 —18 — 25 — 2 — May 9 — 16 — 23 — 30— June 6

TABLE IVA

Comparison of mean ovulation rates and body-weights in rabbits under 15 months and over 20 months, Skokholm

	<15 months		>20 months	
	Mean body wt. (g)	No. corpora lutea	Mean body wt. (g)	No. corpora lutea
1961				
March	1160	6·77	1147	5·27
April	1135	7·0	1101	6·12
May	1100	4·77	1129	6·0
Total	1138	6·54	1128	5·68
1962				
March	—	—	1201	4·33
April	1003	4·1	1157	4·22
May	1007	4·4	1166	5·23
Total	1009	4·22	1171	4·75
1962 (Neck)				
March	1060	4·0	—	—
April	1017	4·58	—	—
May	1057	5·85	—	—
Total	1046	4·5	—	—
1963				
April	957	3·5	1032	4·03
May	1081	5·21	1162	5·53
Total	1040	4·61	1064	4·9

TABLE IVB

Adjusted mean ovulation rates and body-weights in rabbits under 15 months and over 20 months. Skokholm

	<15 months		>20 months	
	Mean body wt. (g)	No. corpora lutea	Mean body wt. (g)	No. corpora lutea
1961	1135	6·18	1126	5·8
1962	1005	4·2	1175	4·66
1962 (Neck)	1044	4·81	—	—
1963	1019	4·35	1065	4·78

in the season than older does (Table III), and the mean number of ova shed per pregnancy is less (Table IVA and IVB). Almost the entire sample of pregnant animals examined *post mortem* can contribute to these data but only a proportion of the total sample will yield data for birth litter-sizes since only those animals in the last 5 days of pregnancy are used for this purpose. Unfortunately, the sample sizes then become too small to yield useful data on the actual birth litter-sizes of the two groups. Table V shows the increase in range of number of ova shed in populations of low density compared with those of high density in the same location (Skokholm and Wales).

Table IVB is based on Table IVA and shows the monthly mean number of ova ovulated per pregnancy and mean body weights. In some years, for example 1962 (Neck of Skokholm), the monthly samples were very different in size, many more does being examined in March than May. Since the mean number of ova per pregnancy tend to increase as the season progresses, and reach a peak at mid-season, the means of the total sample are biased in favour of particular months. Thus in Table IVB the mean of the monthly means has been taken as a basis for comparison. The sample taken from the small 9-acre promontory on the Neck of Skokholm is interesting because 89% of the animals examined were yearlings. These were immigrants into the area after a very high proportion of the Neck residents had been killed in the previous autumn. Not only were the mean body weights of these immigrant yearlings greater than the body weights of yearlings on the main part of the island, but the (adjusted) mean numbers of ova ovulated were at least as high as those of the older and heavier does on the main part of the island. Food availability on the Neck in 1962 was more favourable than elsewhere on the island—even though much of the grass on the Neck was composed of unpalatable *Holcus lanatus*—and it is consequently difficult to ascribe the increased fecundity of the Neck immigrants to improved food conditions, lower population density or decreased competition from older animals, or to any combination of these factors. Although improved, the fecundity of the does on the Neck was not as high as on the main island in 1961, in the year following the crash.

An attempt to determine the effect of increased food availability or quality on breeding, was made in 1962 and 1963. In March of the first year Nitro Chalk (15% N) was broadcast at a rate of 1 cwt to the acre over 10 acres of *Agrostis tenuis, Festuca rubra* and *Holcus lanatus* pasture. Although the colour improved, there was no noticeable growth of the sward, no detectable change in body weight of the rabbits nor in their fecundity, compared with rabbits taken in other areas. In 1963,

TABLE V

Number of corpora lutea per pregnancy

Number of corpora lutea	Number of examples										
14	—	—	1	—	—	—	—	—	—	—	—
13	—	—	0	—	—	1	—	—	—	—	—
11	—	—	1	—	—	0	—	—	—	—	—
10	—	—	6	—	1	1	—	—	—	—	—
9	1	8	18	—	1	6	—	—	—	—	1
8	6	41	26	—	7	7	1	2	2	—	2
7	28	64	51	1	12	10	3	1	9	7	6
6	61	123	40	4	26	25	8	6	19	9	4
5	66	130	25	44	13	15	33	6	34	9	2
4	59	69	9	80	7	4	69	17	29	3	3
3	54	17	2	38	1	1	19	16	9	0	—
2	1	2	1	1	0	0	0	0	1	0	—
1	0	1	0	2	0	0	0	0	1	0	—
	1941 Wales (Brambell 1944)	1942	1957 Wales	1959 Skok-holm	1959 Skomer (April and June only)	1961 Skok-holm	1962 Skok-holm	1962 Skok-holm Neck	1963 Skok-holm	1964 Skerries May June July	1968 Inner Farne Island (April only)
Population density index	5	5	0·1	10	3	1·5	7	2	5	10	6

during February, March and April, 10 cwts of a compound fertilizer (N 10%, P_2O_5 4%, K 10%), 8 cwt of Super Phosphate (P_2O_5 18%), and 10 cwt of Urea (N 45%) were spread in 3 equal monthly doses over the same area. The pasture developed luxuriantly. Rabbits, except those living in warrens in the fertilized area, abandoned it as a feeding ground and those feeding on the short grazed grasses round the warrens did not show any detectable difference in body condition or fecundity. One possible reason why rabbits shunned the area may have been due to the strong growth of *H. lanatus*, but rabbits will feed on this grass when it is at a young stage so other reasons must be sought. Only one major difference in reproductive behaviour was noted; compared with previous years the number of nesting stops prepared in the fertilized area was considerably greater and fewer stops were found in the surrounding unfertilized areas. No other attempts to manipulate the environment were made and it is regrettable that the nutritive values of stomach contents of autopsied rabbits of different age classes, in different years, were not estimated.

DISCUSSION

Food availability, population density, social rank/age, are mutually inter-related factors which are difficult to separate in field investigations of rabbits. It is interesting to note that on the 2 islands where food quality and quantity were poorest (Skerries and Farne Island), the mean body weights of breeding animals were as high as in the Welsh mainland samples of 1942 and 1957 and in the Skokholm sample for the low density year, 1961; yet the lowest reproductive productivity per pregnancy and per season, in all the series examined was observed in the Skerries sample (Table VI). Excluding considerations of food availability, it seems from data shown in Table VI that productivity is fundamentally related to population density, the relationship being inverse, but each group examined showing a different pattern. It is surprising, for example, that the extent of pre-natal mortality of entire litters in rabbits on the islands with high density populations is less than in the samples collected by Brambell in 1942. Furthermore the most frequent occurrences of pre-natal mortality on Skokholm occurred not at the highest density, but at a slightly lower density in 1962. Another significant feature of increased productivity associated with low population density was the occurrence of autumn and winter breeding in about 25% of does of all age classes on Skokholm in 1960. Breeding ceased in late May in 1960, but it recommenced in September

TABLE VI

Reproductive productivity and other variations

	1942 Wales (Brambell)	1957 Wales	Skokholm					1959 Skomer (April May and June only)	1964 Skerries	1968 Inner Farne Island (April only)
			1959	1961	1962	1962 (Neck)	1963			
Population density index	5	1 to 0·1	10	1·5	7	2	5	3	10	6
Mean cleaned body wt.	1128	1190	1082	1135	1043	1070	1092	1078	1129	1112
Duration of breeding in weeks (Over 50% pregnant)	15	22	11	15	10½	17	10	12	12	?
Mean no. corpora lutea	5·6	6·8	4·1	6·4	4·3	4·7	4·8	6·0	5·7	6·2
Implantation litter size	5·0	6·0	3·7	5·2	3·8	4·2	4·3	5·1	5·0	5·5
Birth litter size	5·1	5·9	3·4	4·8	3·9	4·7	4·3	4·6	2·8	4·6
Extent of pre-natal loss of entire litters	60%	Very low	5%	Very low	10–20%	Sample too small for estimation	Not detected	?	Not detected	Sample too small
Mean no. of young born per doe	10·3	30	8	16·3	9	12 to 16	9·5	12	7	?
Time interval in days between observation of 1st pregnancy and period when 90% does pregnant	—	—	41	27	60	33	27	—	20	—
Ratio of yearlings: older rabbits in breeding population ♀♀	—	—	—	2·5 : 1	3·3 : 1	9·5 : 1	0·55 : 1	—	0·81 : 1	—
♂♂	—	—	—	5·7 : 1	4·4 : 1	—	0·53 : 1	—	—	—

after pasture improvement. This did not occur in any other year between 1955 and 1968.

Although it is not possible to show a causal relationship between the variations in the different reproductive phenomena and changes in environmental components, it is clear that where comparisons of populations in different years are possible (Wales 1942, 1957; Skokholm 1959 to 1963) that higher productivity is associated with low population densities in expanding populations. The adaptation is clearly one for quick replenishment following the removal of some reproductive inhibition occurring at high densities. The mortality rates of offspring estimated crudely were, however, enormously high even in expanding Skokholm populations; from 1961 to spring 1962 at least 51% of the juveniles did not reach maturity (Lloyd, 1970); and between 2 high density years mortality of young animals was not less than 80%.

The genetical composition of the 6 populations examined must in each case be different. Some, such as those on Skomer and Skokholm, have bred in isolation possibly since their introduction in the 12th or 13th century. The Skerries or Farne groups of rabbits living on small islands are more likely to suffer extermination from climatic catastrophes, and reintroductions by trappers and lighthousemen are highly probable; indeed it is known that domestic varieties were used to reinforce the populations from time to time. On the Inner Farne islands the population in April 1969 was composed of 54% black rabbits as against less than 2% on Skokholm from 1959 to 1963. Albino rabbits are unknown on Skokholm but are common on Skomer. Some of the reproductive variations between populations may owe much to genetical differences but it is clear that for rabbits as for many other species *e.g. Mus musculus* (Southwick, 1958) and *Microtus californicus* (Houlihan, 1963) population density is a fundamental factor governing reproductive productivity.

In his brief review of the ecology, behaviour, and physiology of wild rabbits in Australia, Myers (1966) noted that in confined rabbit populations the size of the area occupied by the communities may have an adverse effect upon reproduction, in addition to the inhibitory effect of increasing density. The population densities contrived (equivalent to 200 rabbits to the acre) are about ten-fold those encountered in wild rabbit populations in Britain, but some of the variations in the reproduction of rabbits at the same densities on large and small islands may be due to this effect of available space. In high density years the Skokholm rabbit population numbers about 6000 breeding animals composed of yearlings and older animals in the ratio of about 3 : 1. To maintain a steady level of breeding animals from year to year at this ratio requires an annual mortality rate of about 80% of all young born. If the data available for the Skerries are typical of years without intervention of climatic catastrophes, the annual mortality rate of young rabbits for that island is calculated to be about 88%.

The rabbit is a highly successful and adaptable species and the apparent over-production and consequent enormously high losses cannot be detrimental to the communities. The advantage of the ability to increase productivity when populations are at a temporary low level is clear and has adaptive significance; possibly the significance of the inevitably high mortality of young may lie in the requirement of genetical diversification to endow the population with the ability to overcome small, frequent but irregular changes in the environment.

References

Brambell, F. W. R. (1944). The reproduction of the wild rabbit *Oryctolagus cuniculus* (L.). *Proc. zool. Soc. Lond.* **114**, 1–45.

Casperton, K. (1968). Influence of environment upon some physiological parameters of the rabbit *Oryctolagus cuniculus* (L.) in natural populations. *Proc. ecol. Soc. Aust.* **3**, 113–119.

Chitty, D. and Chitty, H. (1960). Population trends among the voles at Lake Vyrnwy 1932–60. Mammal investigation methods. Symp. Theriologicum. Brno 1960.

Einarson, A. E. (1956). Life of the mule deer. In *The Deer of N. America*. Taylor, W. P. (ed.). Pennsylvania, Stockpole Co.

Elton, C. S. (1924). Periodic fluctuations in the numbers of animals: their cause and effects. *Br. J. expt. Biol.* **2**, 132.

Hewitt, C. G. (1921). *Conservation of the wildlife of Canada*. New York: Scribners.

Houlihan, R. T. (1963). The relationship of population density to endocrine and metabolic changes in the Californian vole *Microtus californicus*. *Univ. Calif. Publs. Zool.* **65**, 327–362.

Jewell, P. A. (1966). Breeding season and recruitment in some British mammals confined on small islands. *Symp. zool. Soc. Lond.* No. 15, 89–116.

Lloyd, H. G. (1963). Intra-uterine mortality in the wild rabbit *Oryctolagus cuniculus* (L.), in populations of low density. *J. Anim. Ecol.* **32**, 549–563.

Lloyd, H. G. (1970). Post-myxomatosis rabbit population in England and Wales. O.E.P.P./E.P.P.O. Conference on rodents, Helsinki.

Morris, B. (1958). The yolk-sac of the mole (*Talpa europaea*). *Proc. zool. Soc. Lond.* **131**, 367–87.

Moule, G. R., Braden, A. W. H. and Lamond, D. R. (1963). The significance of oestrogens in pasture plants in relation to animal production. *Anim. Breed. Abstr.* **31**, 139–157.

Myers, K. (1966). The effects of density on sociality and health in mammals. *Proc. ecol. Soc. Aust.* **1**, 40–64.

Mykytowycz, R. (1965). Further observations on the function and histology of the submandibular (chin) gland in the rabbit, *Oryctolagus cuniculus* (L.). *Anim. Behav.* **13**, 400–412.

Newson, R. M. (1966). Reproduction in the feral coypu (*Myocastor coypus*). *Symp. zool. Soc. Lond.* No. 15, 323–334.

Poole, W. E. (1960). Breeding in the wild rabbit *Oryctolagus cuniculus* (L.) in relation to the environment. *C.S.I.R.O. Wildl. Res.* **5**, 21–43.

Sadleir, R. M. F. S. (1969). *The ecology of reproduction in wild and domestic mammals.* London: Methuen.

Shorten, M. (1954). *Squirrels.* London: Collins.

Southwick, C. H. (1958). Population characteristics of house mice living in corn ricks: density relationships. *Proc. zool. Soc. Lond.* **131**, 163–175.

Symp. zool. Soc. Lond. (1970) No. 26, 189–207.

SEASONAL AND AGE CHANGE IN SHREWS AS AN ADAPTIVE PROCESS

ZDZISŁAW PUCEK

Mammals Research Institute, Polish Academy of Sciences
Białowieża, Poland

SYNOPSIS

Seasonal morphological changes in shrews (Dehnel's phenomenon) are reviewed and some hypotheses concerning their nature are discussed.

Various studies of changes in the depth of the brain-case during post-natal life in different populations of the Common shrew (*Sorex araneus*) as well as in other species of the genus lead to the following conclusions. The winter decrease in the depth of the brain-case is much greater in northern and eastern populations than in southern and western ones. The duration of the spring jump in growth in the brain-case depth is shorter in eastern populations compared with western ones. Seasonal variations in brain-case are a characteristic phenomenon in all species of the genus *Sorex* L. at least in the Palaearctic. These changes result from two opposite processes (1) resorption of the occipital and parietal bones in autumn, and (2) growth of new bone tissue at the edges of the same bones in spring.

Measurements and weighings of pelvis and scapula show that no growth processes occur until the spring when a rapid jump in growth in adult shrews is observed.

Seasonal variations in body measurements, e.g. length, weight, were occasioned by changes in size and shape of the intervertebral discs, the nucleus pulposus of which flattens longitudinally during the autumn and early winter. Body weight changes observed in nature are inhibited in shrews kept under laboratory conditions. They take place, to a slighter degree, when animals are kept in cages out-of-doors.

A similar rhythm of seasonal changes is observed in the weight of the internal organs, although these change by different amounts. Seasonal changes in body weight as a whole and, in particular, of internal organs may be explained by tissue dehydration which takes place during the autumn and winter.

Morpho-physiological changes in glands of internal secretion and variations in metabolism are the basic physiological background of these drastic seasonal changes in shrews.

Seven phases of post-natal development can be distinguished in shrews: periods of intensive growth (nest development, spring jump in growth) are interspersed with phases of relative stability (summer), of autumn regression and of winter depression. Such a course of post-natal growth rate seems to be a property hitherto observed only in shrews. Unevenness of growth and development rates and the characteristic regressive changes in the shrew's organism may be the expression of hereditary, primarily structural adaptations to living in polar and circumpolar regions.

INTRODUCTION

The discovery by Dehnel (1949) of the phenomenon of seasonal changes in the brain-case of shrews has stimulated greater interest in these mammals. The process of shrinkage in the size of the skull depth has been termed "Dehnel's phenomenon" (Schubarth, 1958) or "Dehnel's concertina effect" (Crowcroft and Ingles, 1959). Some more recent authors take these terms to cover all processes of seasonal adaptation in shrews to the winter season (Mezhzherin, 1964). It seems, however, that only those processes of seasonal changes which are peculiar to shrews should be called "Dehnel's phenomenon".

Investigations of all aspects of seasonal and age changes in shrews, carried out during the past 20 years not only by our group, but also by other centres, have yielded abundant material. The purpose of this article is to review all these findings as well as to put forward some hypotheses on the causes and nature of such radical variations in shrews.

The age classification of the material must be clarified first. Most authors quoted here divide all the material into two main age-groups: (1) young adults (or "young" or "subadults")—i.e. animals caught from June of one calendar year to March of the following year, and (2) old adults (or "adults")—caught from April to late autumn, even as late as December, when the last shrews die off (*cf.* Dehnel, 1949; Borowski and Dehnel, 1953; Pruit, 1954; Rudd, 1955). The main criterion dividing the two groups is the intensification of growth at the attainment of sexual maturity. Morphological characters such as tooth wear, quality and colour of the fur and changes in skull sutures were found to be helpful in ageing shrews (Pearson, 1945; Dehnel, 1949; Conaway, 1952; Dunajeva, 1955; Pucek, 1955). By arranging the material month by month, beginning with June of the year of birth and continuing through the winter and spring to the late autumn of the following calendar year, one may analyse changes during the life cycle of shrews, i.e. from young adults to old adults. Many authors simplify the analysis by considering the yearly catch as covering one life-cycle although it is known that such samples consist of two generations, viz. young adults and overwintered old adults. But combining the samples from many years together gives good monthly averages for some generations or life-cycles. Since no young shrews enter the population after September, such a treatment of the material indicates very closely the real variability throughout an individual's life-cycle.

SKULL AND POSTCRANIAL SKELETON

Studies were concentrated primarily on changes in the brain-case depth during the post natal life-cycle in different populations of *Sorex araneus* over almost the whole geographical range of the species as well as in other species of the genus. Dehnel's (1949) findings were confirmed for other populations of *Sorex araneus* from Poland (Kubik,

1951; Serafiński, 1955; Pucek, 1955; Kowalska–Dyrcz, 1961, 1962), German D.R. (Schubarth, 1958), England (Crowcroft and Ingles, 1959), Soviet Union (Pucek, 1963; Viktorov, 1967), Finland (Skarén, 1964), Bulgaria and Czechoslovakia (Pucek, 1963; Pucek and Markov, 1964) and Austria (Spitzenberger, 1964). There are some consistent differences in the extent of the seasonal changes in the depth of the brain-

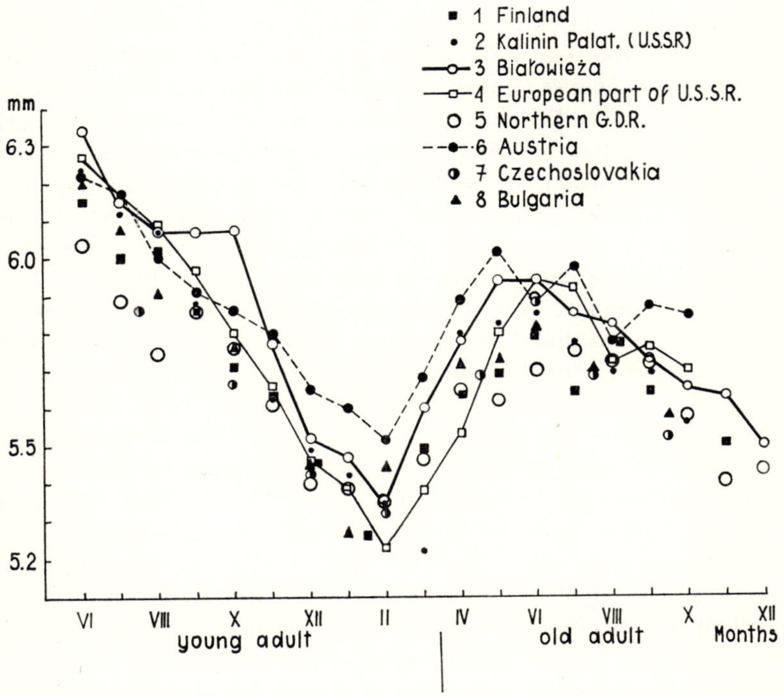

Fig. 1. Changes in height of the brain-case in *Sorex araneus* from some of the populations studied. 1. Finland (data from Skarén, 1964), 2. Kalinin Palatinate, USSR (unpublished data of L. V. Viktorov), 3. Białowieża Primeval Forest, Poland (data from Pucek, 1955), 4. European part of USSR (data from Pucek, 1963), 5. Northern part of German Democratic Republic (data from Schubarth, 1958), 6. Austria (data from Spitzenberger, 1964), 7. Czechoslovakia, 8. Bulgaria (data from Pucek and Markov, 1964).

case in different parts of the geographical range of this shrew in the Palaearctic, viz. (1) the decrease in this measurement from June to February is much greater in northern and eastern populations than in southern and western ones (Fig. 1). The maximum winter depression may vary from January to February (or even to the first half of March during harsh winters) according to latitude and severity of the climate. (2) The duration of the spring jump in growth in brain-case depth is shorter in eastern populations compared with western ones. According to studies by Viktorov (1967) in Kalinin Palatinate (U.S.S.R.) the rate of the spring jump in growth is 6·5 times faster than in the British Isles, 2·6 times faster than in Middle Europe, and 1·5–2 times faster than in Białowieża Forest.

Analogous changes were observed in some populations of *Sorex minutus* (Dehnel, 1949; Kubik, 1951; Serafiński, 1955; Caboń, 1956), *Sorex caecutiens* (Dehnel, 1949; Pucek, 1963), *Sorex unguiculatus* (Pucek, 1963; Skarén, 1964), *Sorex arcticus*, *Sorex vir* and perhaps in *S. raddei* and *S. asper* (Pucek, 1963). In the last two cases only a few specimens were studied but changes in the structure of the parietal bones indicate that seasonal changes must take place. These results lead to the conclusion that seasonal changes in brain-case depth are a characteristic phenomenon at any rate in all the Palaearctic species of the genus *Sorex*. Almost nothing is known about species of *Sorex* from North America and such studies would be of great interest.

Some preliminary observations show, however, that such seasonal changes are not confined to the genus *Sorex*. They have been observed also in *Neomys* (Dehnel, 1950), *Blarina* (Dapson, 1968) and, to some degree, also in rodents (Wasilewski, 1956 a, b, 1961; Pucek, 1963; Haitlinger, 1965). The only other cranial dimension which changes throughout the life-cycle of shrews, is the breadth of the brain-case, which decreases slightly during the winter (Pucek, 1955, 1963) and is also a little broader in old adult shrews than in young adults.

Further studies were concentrated on the mechanism of skull changes. Macroscopic and microscopic observations show that there are periodical changes in compactness of skull sutures, e.g. sutura sagittalis and lambdoides. Change in the depth of the brain-case arises from two opposing processes: (1) resorption of the occipital and parietal bones in autumn, occurring at the edges adjoining the sutura sagittalis and sutura lambdoides, and (2) growth of new bone tissue in the same areas in spring. These processes reduce the size of the ossa parietalia and of the os occipito-interparietale in the autumn and increase it in the spring (Pucek, 1955, 1957; Hyvärinen, 1968a, 1969). Resorption of bones is caused by osteoclasts, which were most numerous in microscopic sections during the early autumn and decreased through to February and even March (Pucek, 1955; Hyvärinen, 1969). It was also shown that osteogenesis is of a secondary character, i.e. new bone tissue replaces a cartilaginous model (Pucek, 1957).

Some changes in brain-case depth are observed in captive shrews. On the whole, the skull flattens less than it does in free-living animals. The processes of decrease and growth last longer in the laboratory.

This is a result of the inhibition of the regressive processes (resorption of bones of the neurocranium) in some individuals captured during the summer on the one hand, and on the other inhibition of the spring growth of bones in captive animals taken during the winter months (Pucek, 1964). Thus there were some individuals, caught during the summer or winter which even between May and August have low skulls and remain in a phase of resorption of neurocranium bones. Most individuals of this group show, however, "normal" skull growth during this period of time. These differences probably should be treated as pathological deviations of the normal and hereditary fixed processes.

Recently Dapson (1968) found that seasonal changes of cranial height take place also in *Blarina*, if the material is plotted by month of capture. But if the same data are graphed according to the age of the individuals (i.e. separately for cohorts born in different seasons) fluctuations apparent at the population level are not observed. Dapson (1968) concluded that the growth pattern in *Blarina* did not correspond to that in *Sorex*, and that the growth pattern described for Palaearctic shrews was an artifact. The second part of this conclusion is not valid. No shrews are born after September, and the most intensive changes in brain-case height are observed between October and February. Resorption of parietal and occipito-interparietal bones takes place during this period in all individuals, both at the population and individual levels. There is some individual variation in the timing of these changes, perhaps depending on the time of leaving the nest (June to September inclusive) but such variation cannot influence the general character of the changes in populations so evenly aged as those of shrews. The maximum age variability in trappable populations of young adult *Sorex* in autumn is only 4 months.

Measurements and weighings of the main bones of the post-cranial skeleton (e.g. pelvis, scapula) show great differences only between subadults and old adults, which have lived through the winter. No growth takes place until the spring when a rapid jump in growth in overwintered shrews is observed, both in the skull and the post-cranial skeleton. Then also, sexual differences are developed (Dolgov, 1961). It is interesting to note that measurements of the pelvis and scapula of those individuals which attained maturity in the first calendar year of their life fall in the range of young adults (Dolgov, 1961). Sexual maturation did not stimulate growth in this case.

BODY MEASUREMENTS

It is a long time since Adams (1910) first recorded changes in the body length of the the Common shrew. But Dehnel (1949) was the first to demonstrate this conclusively and to show a winter decrease in this measurement in some species of European shrews. This observation was confirmed for other populations of the commoner species of shrews (Borowski and Dehnel, 1953; Siivonen, 1954; Skarén, 1964). According to these findings, monthly averages show a small decrease in body length before the onset of winter and an increase during the next spring, when the general growth of shrews is observed. These changes vary in intensity in different years, depending perhaps, on changes in habitat conditions (Borowski and Dehnel, 1953). The mechanism of seasonal changes in body length has been clearly described by Saure and Hyvärinen (1965) and Hyvärinen (1969), who state that the size and shape of the intervertebral discs change during the postnatal life of shrews. The nucleus pulposus flattens longitudinally during the autumn and early winter, being thinnest during mid-winter, and swelling again during the spring. The annulus fibrosus is clearly folded at the margins in mid-winter, corresponding to the flattening of the nucleus pulposus, and calcified cartilage is absorbed at the margins of the discs. In overwintered individuals the annulus fibrosus is not folded, and a thick layer of new cartilage growing towards the vertebrae is observed. These changes in the intervertebral discs of the lumbar vertebrae determine the variations described in the whole spinal column and also in the head and body length.

BODY WEIGHT

Differences in body weight between young adults (subadults) and old adults have been known since the studies by Adams (1910, 1912), and Middleton (1931). Stein (1938), Dehnel (1949, 1950), Borowski and Dehnel (1953), Siivonen (1954), Pucek (1955, 1964, 1965), Niethammer (1956) and others (Rudd, 1955; Jameson, 1955) show that such changes have, firstly, a seasonal character and, secondly, are connected with the spring jump in growth. The decrease in body weight in monthly samples of a population is as much as 10–40%, and the spring gain can be 100% or more. Differences according to environmental conditions were established for one population (Borowski and Dehnel, 1953) as well as according to geographical range (Niethammer, 1956; Mezhzherin, 1964) (northern populations of *Sorex araneus* showing higher winter weight losses than southern ones).

Similar changes have been found in the body-weight of free-living, marked and released shrews (Shillito, 1963; Croin Michielsen, 1966). The latter established that weight loss from summer to winter in two generations of *Sorex araneus* was 9·6 and 14·5% and in *S. minutus*—

3·3 and 6·1%. The spring gain from winter to sexual maturity was 80·1 and 83·1% in *S. araneus* and 67·2 and 51·7% in *S. minutus*. Thus, winter depression in body weight is greater in *S. araneus* than in *S. minutus*, so that *S. minutus* would be less well adapted to winter conditions (Serafiński, 1955; Croin Michielsen, 1966).

In shrews kept under laboratory conditions, which have rather stable environmental and food conditions, weight changes are inhibited. Shrews caught in summer and kept in captivity all winter, did not exhibit either the winter loss or spring jump in body weight. Their average body-weights remained rather stable, although a rapid increase during the first few weeks in the laboratory followed by extensive individual short-term fluctuations was observed (Pucek, 1964).

Young individuals caught during the winter were found to gain weight rapidly, especially during the first weeks of their captivity, and this gain continued into the spring as in wild shrews. However, growth began much earlier than in nature and did not reach the rate recorded for free living populations (Pucek, 1964). This is understandable, because these animals did not breed in the laboratory.

Nevertheless, in both summer and winter-caught animals, growth of the cranial bones was observed in some individuals and not in others. These differences between the two groups kept in the laboratory may be a pathological result of long confinement in unsuitable laboratory conditions. However, failure to reproduce and consequent disturbances in the hormonal balance may have inhibited the spring growth of individuals caught during the summer of the previous year (Pucek, 1964).

Recently published studies by Wołk (1969) indicate that a significant, though much smaller, loss of body weight takes place in winter, when animals are kept in cages outdoors and fed *ad libitum*.

WEIGHT OF INTERNAL ORGANS

Seasonal changes found in the volume of the brain-case (Pucek, 1955) suggested that analogous variations should take place in the volume of the brain. This was demonstrated for weight and volume of the brain in *Sorex minutus* (Caboń, 1956) and in *Sorex araneus* (Bielak and Pucek, 1960). Thus seasonal changes in skull volume and height may be primarily due to changes in the size of the brain.

Buchalczyk (1961) investigated the seasonal variations in the salivary glands of *S. araneus*. Specific changes were found (differing however from those in hibernating mammals) in brown adipose tissue (BAT). According to Buchalczyk and Korybska (1964), variations in this organ are connected with the rhythm of physiological changes in the shrew's whole organism. The highest weights of BAT were found immediately after leaving the nest and, during the period of sexual maturation in the spring.

In the above instances organ weights of animals fixed in alcohol were studied. Pucek (1965) used fresh material and found that a regular rhythm of seasonal and age changes is observed in the weight of all the internal organs. These fluctuations are generally synchronized, particularly in autumn (regression) and in spring (jump in growth), although the various organs change by different amounts. If June is taken as a starting point, the losses up to midwinter in brain, kidney and BAT weights are as great as 20–23%, in liver 34% and in spleen 77%. The exception is found in the heart, the weight of which does not change during this time. If we compare the lowest winter values of organ weights with the highest weights in old adults during the summer, the following gains were recorded: brain, 8%; BAT, 30%; heart, 59%; kidneys, 70%; liver, 123% and spleen 495% (Pucek, 1965). These differences between organs are better seen when expressed as percent increase or decrease month by month, compared with the average values for young adult shrews in June (Fig. 2).

The total weight of all these organs mentioned above changes from 1·08 g in June to 1·16 g in August, 0·83 g in February and 1·50 g in June–September of the following year. Figure 3 shows changes in the proportions of internal organs of the Common shrew, the total average weight of all organs in each monthly sample being taken as 100%. Differential changes of particular internal organs lead to alterations of internal proportions and organization of the shrew's body during postnatal life (Pucek, 1965; Mezhzherin and Melnikova, 1966).

The relative weights of some internal organs are usually treated as morphological indices of physiological functions ("morpho-physiological indicators" of Schwarz, 1958). The heart index should then indicate the level of locomotory activity, the liver being an index of the general condition of the animal, the kidneys of the level of metabolic processes, etc. It is not easy to evaluate changes found in these indices in shrews (Pucek, 1965) from this standpoint, e.g. the heart index is 27% higher during the winter than in young or adult shrews caught during the summer months. This is probably a result of decreasing body weight and of the stable absolute weight of the heart and not a change in locomotory activity. Locomotor, as well as metabolic, activity was shown to decrease during the autumn and winter by Gębczyński (1965). A similar situation is observed in the case of relative brain or kidney weights (*cf.* Pucek, 1965; Hyvärinen, 1968b).

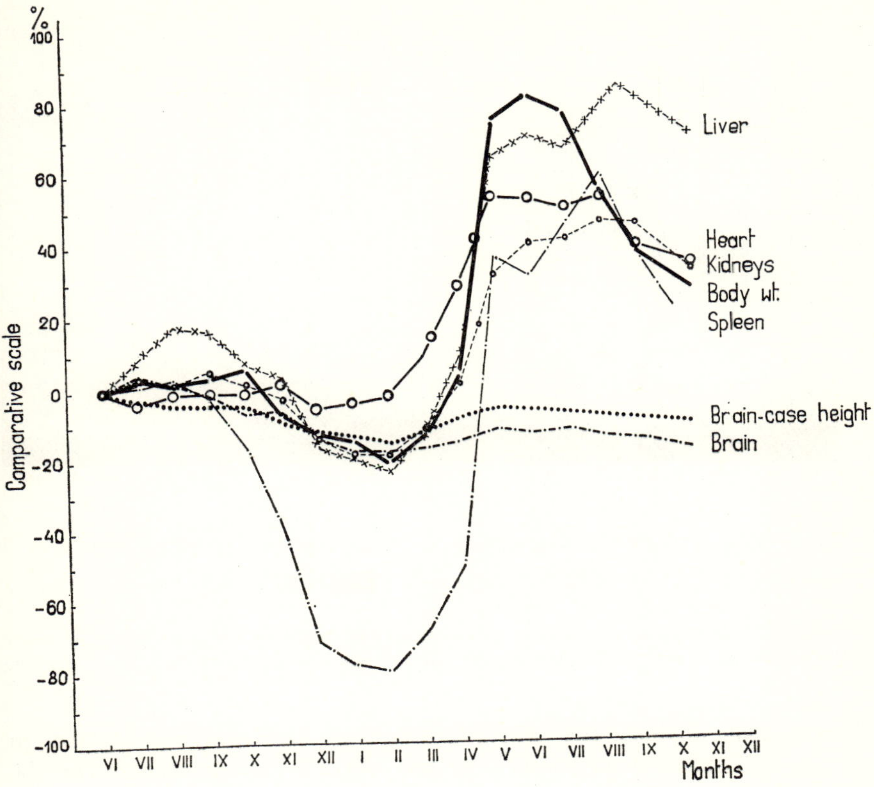

Fig. 2. Seasonal changes in the weight of internal organs in *Sorex araneus*, expressed as percentage increase or decrease of monthly averages. Average for June (young adult) taken as 100%, (based on data of Pucek, 1965).

More recent results throw some light on the possible mechanism of the seasonal decrease in body weight as a whole and, in particular, of the internal organs. Dehnel (1949) was the first to put forward the hypothesis, that seasonal changes in body weight may be due to dehydration of the shrew's body in winter. This hypothesis appears highly probable from the researches of Slonim (1961) and has recently been confirmed by the studies by M. Pucek (1965). She found a parallel decrease in water content of the brain tissue and in the weight of the whole organ during the winter. On the other hand brain tissue contained more lipids and dry fat-free material in winter than in summer. These findings were confirmed for brain tissue by Mezhzherin and Melnikova (1966) and by Mezhzherin and Finagin (1968) and for the whole body of a shrew by Górecki (1965) and Myrcha (1969). According

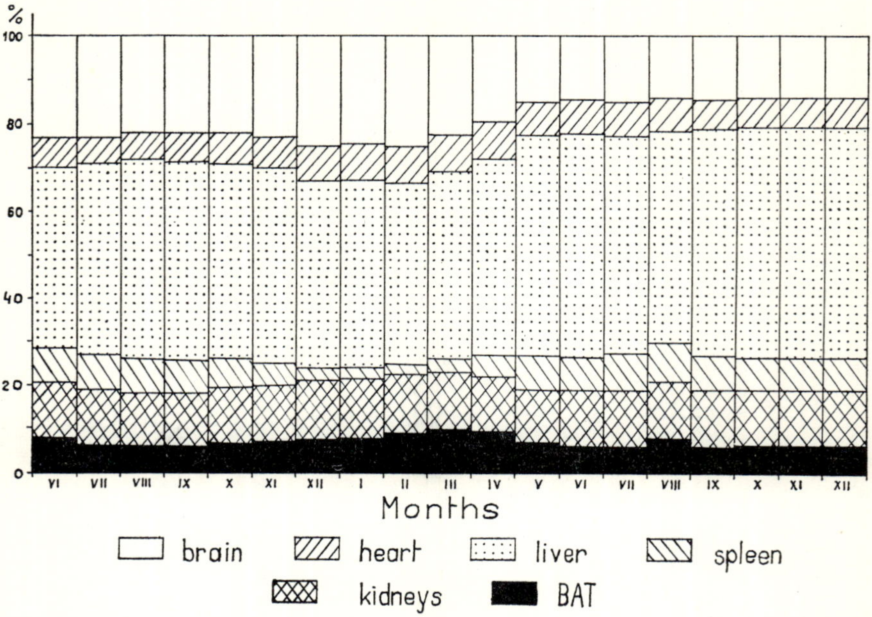

FIG. 3. Variation in internal proportions of the body in *Sorex araneus*. Average total weight of all internal organs in given month samples taken as 100%.

to the last author, the water content of the body is highest in spring and summer and lowest in the winter. Fat content shows the reverse changes, being highest in December–January. Losses in body water (in g) from summer to winter constitute 84% and 82% of the total loss of body weight of *S. araneus* and *S. minutus* respectively (Myrcha, 1969). So both the winter decrease and spring increase of body weight are mainly the result of changes in tissue hydration.

CHANGES IN ENDOCRINE SYSTEM AND PHYSIOLOGICAL INDICES

The radical morphological changes observed in shrews must be controlled by the respective physiological activities of the system of internal secretion. However, the seasonal adaptations to the winter season (the regulation of such processes as moult, reproduction, growth, etc.) influence the histological picture of the endocrine organs and create physiological backgrounds for morphological changes during the shrews' postnatal life.

The anterior pituitary has been studied by Hyvärinen (1967, 1969) who found it to be generally inactive in winter with low numbers of acidophil and PAS-positive cells. The increase in number of the acidophils in spring coincides with the general jump in growth, just as the increase in PAS-positive cells coincides with reproductive activity taking place at this time.

The weight of the adrenal glands is regarded by many authors as an index of their functioning and shrews show marked changes in the adrenal cortex (*cf.* Siuda, 1964; Pucek, 1965), which increases slightly from June to August or September, then decreases rapidly, attaining a minimum level in January to March. Adrenal weight and thickness of the cortex increase in spring rapidly in females but significantly less in males. The adrenal cortex undergoes these changes at the expense of the zona fasciculata and zona reticularis. The zona glomerulosa does not exhibit seasonal changes in its thickness (Siuda, 1964).

Parathyroid gland activity, measured by Hyvärinen (1969) by the size of its nuclei, is rather high till December, coinciding with the resorption of the skull bones. Parathyroid cells are small in mid-winter and average cells show regular growth until May, but there is considerable individual variation. Spring increase in parathyroid activity cannot be explained merely by the resorption processes in the skeleton, although these are always present and concerned with the rebuilding of bones while they are growing.

The activity of the thyroid gland is intensified in autumn during the moulting and in spring during sexual maturation. Only a part of the gland is functioning during the winter season (Dzierżykraj-Rogalska, 1953).

The activity of the thymus is also partially connected with the autumn moult. It undergoes complete involution in young adults in September and October, when the autumn moult takes place, and at least 6 months before sexual maturity is attained. The process of involution of this gland in shrews is irreversible (Bazan, 1953, 1956; Schwarz, 1959; Pucek, 1960, 1965).

Hyvärinen (1969) studied the activity of alkaline phosphatase during the whole life cycle in *S. araneus*. He found a good correlation between the activity of this enzyme and growth and bone formation in the brain case.

The general picture of physiological processes in the shrew's body may be obtained from an analysis of seasonal changes in metabolic rate (Gębczyński, 1965, 1969). It was shown by this author that Average Daily Metabolic Rate (ADMR) is significantly lower in the Common shrew in autumn and winter than in summer and (especially) in early spring (old adults). However the winter decrease in metabolism is less in shrews than in rodents, so that physiological adaptations of shrews to winter conditions play a lesser role than morphological adaptations.

PHASES OF POSTNATAL DEVELOPMENT OF SHREWS OF THE GENUS *Sorex*

The observations mentioned above give a reasonably comprehensive picture of morpho-physiological changes taking place during the shrew's life cycle. This enables us to distinguish 7 phases of postnatal development in these mammals (Fig. 4).

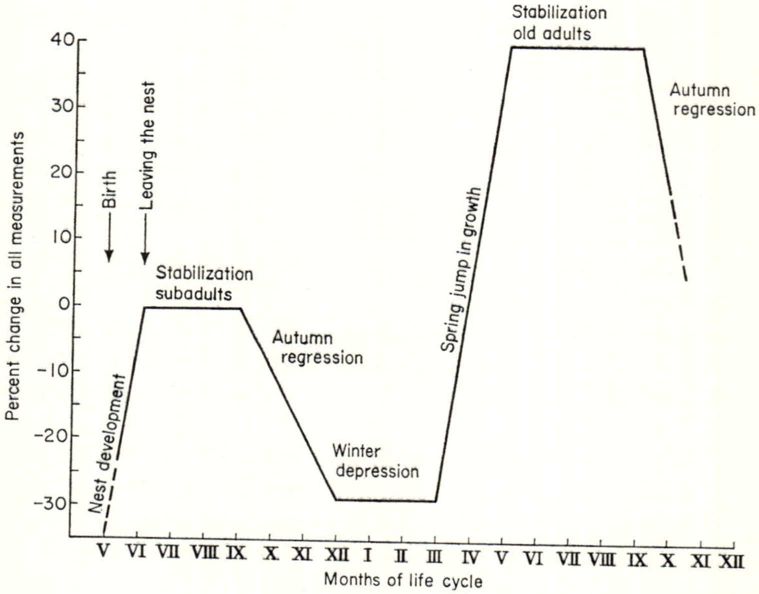

FIG. 4. Phases of postnatal development of shrews, shown diagrammatically. Scale of y-axis = average percent decrease or increase of all the characters studied in *Sorex araneus* (skull, body measurements, and weights of internal organs).

(I) Development in the nest lasts 21–23 days in *S. araneus* (Dehnel, 1952) and adult dimensions are attained by the time that the young leave the nest in June. From June to September it is possible to speak of a phase of relative summer stabilization (II). The weight of the body and such organs as the liver, spleen and kidneys increases slightly, but the skeleton does not grow. Other organs either do not change during this period (heart) or undergo regression (brain, skull, gonads, thymus) beginning almost immediately after leaving the nest.

From October to December there is a rapid autumn regression (III) in the morphological and physiological indices.

During the winter depression (IV) regressive changes continue until January or February.

The spring jump in growth (V) is clearly defined during the second half of March, and by May all the morpho-physiological indices have attained their highest values except that the relative weights of heart, kidneys and brain do not increase during this period.

(VI) In summer the physiological functions again become more or less stable. Senile changes can be observed (e.g. in the parathyroids) intensifying significantly in autumn during the phase of senile regression (VII). The reproductive functions cease and only the adrenal weight increases before winter. The shrews do not moult but gradually die off.

It will be seen from this short "history" of the shrew's life that the periods of intensive growth (development in the nest, spring jump in growth) are interspersed with phases of relative stabilization (summer) and phases of autumn regression and winter depression. Such a course of postnatal growth rate seems to be a property observed up to now only in shrews (Pucek, 1965).

The duration of and periods between different developmental stages in shrews differ from those in other mammals, and even from those in other representatives of *Soricidae* such as, for instance, the genus *Crocidura*. Intensive development in the nest enables young shrews and water shrews to attain the dimensions and proportions of fully grown individuals. The juvenile stage characteristic of mammals is therefore absent, as Dehnel showed earlier (1950). Shrews of the genera *Sorex* and *Neomys* may be considered as subadult after they leave the nest, i.e. full-grown but sexually immature. This stage generally lasts a relatively long time, in *Sorex* 6–9 months, that is, at least half of the animal's maximum life span. Sexual maturity is usually attained at the age of 7–10 months, in the spring of the following calendar year, i.e. during the second half, or even last third of its life. The adult stage, and senile stage which are difficult to distinguish, together last from 6–10 months.

UNEVEN RATE OF POSTNATAL DEVELOPMENT OF SHREWS AS AN ADAPTIVE PROCESS

In Dehnel's opinion (1950) the interval before sexual maturity, lasting in young shrews up to the spring of their second year, is an expression of their adaptation to life in polar and circumpolar regions. This hypothesis is borne out by more recent data which emphasize the unevenness of growth and of development rates, and the characteristic regressive changes in the shrew's organism (Pucek, 1965). The most important life processes are confined to the two summer periods when habitat conditions (e.g. weather, food supply) are favourable. Attainment of full physical development (body and skeleton) is principally confined to the short period of development in the nest. The second phase of growth with attainment of sexual maturity is "postponed", as it were, until the following calendar year. In the rare instances when shrews mature sexually in the first year of life, this occurs in individuals from the first litter immediately after they leave the nest (Bazan, 1956; Pucek, 1960).

Regressive processes in *Sorex* and *Neomys* may therefore be considered as being the expression of hereditary, primarily structural, adaptation to living in northern areas. In contrast, representatives of the genus *Crocidura* living in more southernly areas do not appear to exhibit similar seasonal changes. No variation in skull height and only a slight decrease in body weight were observed during the winter in *C. leucodon* by Buchalczyk (1960).

A consequence of the autumn regressive changes is a general reduction over the winter in body measurements. At the same time the further northwards one looks, the smaller are the dimensions of the forms inhabiting these areas (Mezhzherin, 1964, 1965). This creates a particularly unfavourable ratio of body surface to body volume, which involves the necessity of changes in the energy budget with season.

However, in comparison with rodents, shrews change their metabolic level (ADMR) in winter only slightly. Heat losses (expressed in kcal per $kg^{3/4}$ per day) in autumn and in winter are also only slightly (5·6%) higher than in young adults in summer (Gębczyński, 1965, 1969). On the other hand it was found that the density and length of the fur increases in winter (Borowski, 1958), producing more efficient insulation (Gębczyński and Olszewski, 1963). It seems that energy losses in shrews are thus reduced in a physical way to a greater extent than in other small mammals.

Reduction in body weight must naturally cause an increase in food requirements. The recent studies made by Myrcha (1967) and Wołk (1969) show that both in the different species of Insectivora and within the same species both stomach weight and food intake are inversely proportional to body weight. The increase in food intake attains its greatest values during the winter (Wołk, 1969) particularly at temperatures around freezing point (Mezhzherin, 1964; Mezhzherin and Melnikova, 1966). Shrews may be encountered at such unfavourable temperatures in nature in the layer below the snow cover (Judin, 1962;

Coulianos and Johnels, 1962). In the case of shrews, however, this increase in food requirements is probably the result of other processes. Experiments with *S. araneus*, kept outdoors, show that during the winter absolute food intake is only about 10% higher than in young adults in autumn or old adults in summer. However, the relative food intake (per 1 g of body wt.) is in mid-winter 25% higher than in young adults in autumn and 41% higher than in old adults in summer; thus this increase is attained owing to the reduction in body weight (*cf.* Wołk, 1969). Similarly, the absolute weight of the stomach does not undergo seasonal changes and therefore the increase in stomach index in winter is achieved also through the reduction in body weight (Myrcha, 1967). These findings may show that the absolute amount of food eaten by shrews daily in winter is similar to that in summer but much higher relative to body weight (*cf.* also Mezhzherin, 1964).

It may therefore be assumed that the shrews adapt themselves by means of specific morphological and physiological changes and alter the internal proportions of the organism to survive through the deteriorating environmental conditions in winter. They probably need less than the minimum winter level of their food supply and, though their food requirements increase relatively in winter, they will not increase absolutely owing to the decrease in the shrews' body weight.

These assumptions can be verified only by means of continued detailed ecological studies of shrews living under different conditions and of the abundance of their food supply.

References

Adams, L. E. (1910). A hypothesis as to the cause of the autumnal epidemic of the common and the lesser shrew, with some notes on their habits. *Mem. Manchr. Lit. Phil. Soc.* **54** (10), 1–13.

Adams, L. E. (1912). The duration of life of the common and the lesser shrew, with some notes on their habits. *Mem. Manchr. Lit. Phil. Soc.* **56** (7), 1–10.

Bazan, I. (1953). [Morphohistologische Veränderungen des Thymus im Lebenszyklus von *Sorex araneus* L.] *Annls Univ. Mariae Curie-Skłodowska* Sect. C. **7**, 253–304. (In Polish, with German and Russian summaries.)

Bazan, I. (1956). Untersuchungen über die Veränderlichkeit des Geschlechtsapparates und des Thymus der Wasserspitzmaus (*Neomys fodiens fodiens* Schreb). *Annls Univ. Mariae Curie-Skłodowska* Sect. C. **9**, 213–259.

Bielak, T. and Pucek, Z. (1960). Seasonal changes in the brain weight of the common shrew (*Sorex araneus araneus* Linnaeus, 1758). *Acta theriol.* **3**, 297–300.

Borowski, S. (1958). Variations in density of coat during the life cycle of *Sorex araneus araneus* L. *Acta theriol.* **2**, 286–289.

Borowski, S. and Dehnel, A. (1953). [Angaben zur Biologie der *Soricidae*]. *Annls*

Univ. Mariae Curie-Skłodowska Sect. C.**7**, 305–448. (In Polish; German and Russian summaries.)

Buchalczyk, A. (1961). Variation in weight of the internal organs of *Sorex araneus* Linnaeus, 1758. I. Salivary glands. *Acta theriol.* **5**, 229–252.

Buchalczyk, A. and Korybska, Z. (1964). Variation in the weight of the brown adipose tissue of *Sorex araneus* Linnaeus, 1758. *Acta theriol.* **9**, 193–215.

Buchalczyk, T. (1960). Variabilität der Feldspitzmaus, *Crocidura leucodon* (Hermann, 1780) in Ost-Polen. *Acta theriol.* **4**, 159–174.

Caboń, K. (1956). Untersuchungen über die saisonale Veränderlichkeit des Gehirnes bei der kleinen Spitzmaus (*Sorex minutus minutus* L.). *Annls Univ. Mariae Curie-Skłodowska* Sect. C.**10**, 93–115.

Conaway, C. H. (1952). Life history of the water shrew (*Sorex palustris navigator*). *Am. Midl. Nat.* **48**, 219–248.

Coulianos, C.-C. and Johnels, A. G. (1962). Note on the subnivean environment of small mammals. *Ark. Zoöl.* (2) **15**, 363–370.

Croin Michielsen, N. (1966). Intraspecific and interspecific competition in the shrews *Sorex araneus* L. and *S. minutus* L. *Archs néerl. Zool.* **17**, 73–174.

Crowcroft, P. and Ingles, J. M. (1959). Seasonal changes in the brain-case of the common shrew (*Sorex araneus* L.). *Nature, Lond.* **183**, 907–908.

Dapson, R. W. (1968). Growth patterns in a post-juvenile population of short-tailed shrews (*Blarina brevicauda*). *Am. Midl. Nat.* **79**, 118–129.

Dehnel, A. (1949). [Studies on the genus *Sorex* L.] *Annls Univ. Mariae Curie-Skłodowska* Sect. C.**4**, 17–102. (In Polish; English summary.)

Dehnel, A. (1950). [Studies on the genus *Neomys* Kaup.] *Annls Univ. Mariae Curie-Skłodowska* Sect. C.**5**, 1–63 (In Polish; English summary.)

Dehnel, A. (1952). [The biology of breeding of common shrew *S. araneus* L. in laboratory conditions.] *Annls Univ. Mariae Curie-Skłodowska* Sect. C.**6**, 359–376. (In Polish; English and Russian summaries.)

Dolgov, V. A. (1961). [Variation in some bones of postcranial skeleton of the shrews (*Mammalia, Soricidae*).] *Acta theriol.* **5**, 203–227. (In Russian; English and Polish summaries.)

Dunajeva, T. N. (1955). K izučeniju biologii razmnoženija obyknovennoj burozubki (*Sorex araneus* L.). *Byull. mosk. Obshch. Ispÿt Prir.* **60**, (6): 27–43. (In Russian.)

Dzierżykraj-Rogalska, I. (1953). [Histomorphologische Veränderungen der Schilddrüse in Lebenszyklus *S. a. araneus* L.] *Annls Univ. Mariae Curie-Skłodowska* Sect. C.**7**, 213–252. (In Polish; German and Russian summaries.)

Gębczyński, M. (1965). Seasonal and age changes in the metabolism and activity of *Sorex araneus* Linnaeus, 1758. *Acta theriol.* **10**, 303–331.

Gębczyński, M. (1969). Daily heat losses in the common shrew (*Soricidae*) in different seasons. *Proc. IVth Symp. Energy Metabolism*, 397–400. New Market upon Tyne: Oriel Press.

Gębczyński, K. and Olszewski, J. L. (1963). Katathermometric measurements of insulating properties of the fur in small mammals. *Acta theriol.* **7**, 369–371.

Górecki, A. (1965). Energy values of body in small mammals. *Acta theriol.* **10**, 333–352.

Haitlinger, R. (1965). Morphological analysis of the Wrocław population of *Clethrionomys glareolus* (Schreber, 1780). *Acta theriol.* **10**, 243–272.

Hyvärinen, H. (1967). Variation of the common Shrew (*Sorex araneus* L.) *Aquilo;* Ser. Zool. **5**, 35–40.

Hyvärinen, H. (1968a). On the mechanism and physiological background of the seasonal variation of the height of the skull in the common shrew (*Sorex araneus* L.), *Aquilo*, Ser. Zool. **6**, 1–6.

Hyvärinen, H. (1968b). On the seasonal variation of the activity of alkaline phosphatase in the kidney of the bank vole (*Clethrionomys glareolus* Schr.) and the common shrew (*Sorex araneus* L.), *Aquilo*, Ser. Zool. **6**, 7–11.

Hyvärinen, H. (1969). On the seasonal changes in the skeleton of the common shrew (*Sorex araneus* L.) and their physiological background. *Aquilo*, Ser. Zool. **7**, 1–32.

Jameson, E. W. (1955). Observations on the biology of *Sorex trowbridgei* in the Sierra Nevada, California. *J. Mammal.* **36**, 339–345.

Judin, B. C. (1962). Ekologija burozubok (rod *Sorex*) Zapadnoj Sibiri. *Trudy̆ biol. Inst. Sib. Otd. Akad. Nauk. S.S.S.R.* **8**, 33–134.

Kowalska-Dyrcz, A. (1961). Seasonal variations in *Sorex araneus* Linnaeus, 1758 in Poland. *Acta theriol.* **4**, 268–273.

Kowalska-Dyrcz, A. (1962). [Seasonal variability of the shrew (*Sorex araneus araneus* L.) in the region of Wrocław.] *Przegl. zool.* **1**, 5–22. (In Polish; English summary.)

Kubik, J. (1951). [Analysis of the Pulawy population of *Sorex araneus araneus* L. and *Sorex minutus minutus* L.] *Annls Univ. Mariae Curie-Skłodowska* Sect. C. **5**, 335–372.

Mezhzherin, V. A. (1964). [Dehnel's phenomenon and its possible explanation.] *Acta theriol.* **8**, 95–114. (In Russian; English and Polish summaries.)

Mezhzherin, V. A. (1965). [An essay on quaternary history and the origin of recent fauna of shrews (genus *Sorex, Insectivora, Mammalia*).] In: *Materialy po četvertičnomu periodu Ukrainy*, 164–174. Kiev. (In Russian; English summary.)

Mezhzherin, V. A. and Finagin, L. K. (1968). [Seasonal and age changes of cholesterol content in the brain of representatives of the genus *Sorex*.] *Vêst. Zool. (Kiev)* **2** (3), 29–32. (In Russian; English summary.)

Mezhzherin, V. A. and Melnikova, G. L. (1966). [Adaptive importance of seasonal changes in some morphophysiological indices in shrews.] *Acta theriol.* **11**, 503–521. (In Russian; English summary.)

Middleton, A. D. (1931). A contribution to the biology of the common shrew, *Sorex araneus* Linnaeus. *Proc. zool. Soc. Lond.* **1931**, 133–143.

Myrcha, A. (1967). Comparative studies on the morphology of the stomach in the *Insectivora*. *Acta theriol.* **12**, 223–244.

Myrcha, A. (1969). Seasonal changes in caloric value, body water and fat in some shrews. *Acta theriol.* **14**, 211–227.

Niethammer, J. (1956). Das Gewicht der Waldspitzmaus, *Sorex araneus* Linné, 1758, im Jahreslauf. *Säugetierk. Mitt.* **4**, 160–165.

Pearson, O. P. (1945). Longevity of the short-tailed shrew. *Am. Midl. Nat.* **34**, 531–546.

Pruit, W. O. (1954). Aging in the masked shrew, *Sorex cinereus cinereus* Kerr. *J. Mammal.* **35**, 35–39.

Pucek, M. (1965). Water contents and seasonal changes of the brain-weight in shrews. *Acta. theriol.* **10**, 353–367.

Pucek, Z. (1955). Untersuchungen über die Veränderlichkeit des Schädels im Lebenszyklus von *Sorex araneus* L. *Annls Univ. Mariae Curie-Skłodowska*, Sect. C.**9**, 163–211.

Pucek, Z. (1957). Histomorphologische Untersuchungen über die Winterdepression des Schädels bei *Sorex* L. und *Neomys* Kaup. *Annls Univ. Mariae Curie-Skłodowska*, Sect. C.**10**, 399–428.

Pucek, Z. (1960). Sexual maturation and variability of the reproductive system in young shrews (*Sorex* L.) in the first calendar year of life. *Acta theriol.* **3**, 269–296.

Pucek, Z. (1963). Seasonal changes in the braincase of some representatives of the genus *Sorex* from the Palearctic. *J. Mammal.* **44**, 523–536.

Pucek, Z. (1964). Morphological changes in shrews kept in captivity. *Acta theriol.* **8**, 137–166.

Pucek, Z. (1965). Seasonal and age changes in the weight of internal organs of shrews. *Acta theriol.* **10**, 369–438.

Pucek, Z. and Markov, G. (1964). Seasonal changes in the skull of the common shrew from Bulgaria. *Acta theriol.* **9**, 363–366.

Rudd, R. L. (1955). Age, sex and weight comparisons in three species of shrews. *J. Mammal.* **36**, 323–339.

Saure, L. and Hyvärinen, H. (1965). Seasonal changes in the histological structure of the spinal column of *Sorex araneus* (L.). *Nature, Lond.* **208**, 705–706.

Schubarth, H. (1958). Zur Variabilität von *Sorex araneus araneus* L. *Acta theriol.* **2**, 175–202.

Schwarz, S. S. (1958). Metod morfo-fiziologičeskih indikatorov v ekologii nazemnyh pozvonočnyh životnyh. *Zool. Zh.* **37**, 161–173.

Schwarz, S. S. (1959). Nekotorye biologičeskie osobennosti arktičeskoj burozubki (*Sorex arcticus* Kerr.). *Trudȳ Salehardskogo Sta.* **1**, 255–271.

Serafiński, W. (1955). [Morphological and ecological investigation on Polish species of the genus *Sorex* L. (*Insectivora, Soricidae*).] *Acta theriol.* **1**, 27–86. (In Polish; English summary.)

Shillito, J. F. (1963). Field observations on the growth, reproduction and activity of a woodland population of the common shrew *Sorex araneus* L. *Proc. zool. Soc. Lond.* **140**, 99–114.

Siivonen, L. (1954). Uber die Grössenvariationen der Säugetiere und die *Sorex macropygmaeus* Mill.-Frage in Fennoskandien. *Suomal-Tiedeakat. Toim.* (A) **4** (21), 1–24.

Siuda, S. (1964). Morphology of the adrenal cortex of *Sorex araneus* Linnaeus, 1758 during the life cycle. *Acta theriol.* **8**, 115–124.

Skarén, U. (1964). Variation in two shrews, *Sorex unguiculatus* Dobson and *S. a. araneus* L. *Annls. zool. Fenn.* **1**, 94–124.

Slonim, A. D. (1961). *Osnovȳ obshchei ȇkologicheskoi fiziologii mlekopitajushchikh.* Akad. Nauk SSSR.

Spitzenberger, F. (1964). *Zur Ökologie und Bionomie der Spitzmäuse* (*Soricidae, Mammalia*) *der Donauen oberhalb und unterhalb Wien.* Diss. Univ. Wien, (unpubl.).

Stein, G. H. W. (1938). Biologische Studien an deutschen Kleinsäugern. *Arch. Naturgesch.* N.F. **7**, 477–513.

Viktorov, L. V. (1967). Geografičeskije osobennosti "javlenija Dehnelja" u zemlerojek Evropy. *Byull. mosk. Obshch. Ispȳt. Prir.* **72**, 1, 159–160.

Wasilewski, W. (1956a). Untersuchungen über die morphologische Veränderlichkeit der Erdmaus (*Microtus agrestis* Linné). *Annls Univ. Mariae Curie-Skłodowska*, Sect. C.**9**, 261–305.

Wasilewski, W. (1956b). Untersuchungen über die Veränderlichkeit des *Microtus oeconomus* Pall. in Białowieża-Nationalpark. *Annls Univ. Mariae Curie-Skłodowska* Sect. C.**9**, 355–386.

Wasilewski, W. (1961). Angaben zur Biologie und Morphologie der Kurzohrmaus, *Pitymys subterraneus* (de Sélys-Longchamps, 1835). *Acta theriol.* **4**, 185–247.

Wołk, E. (1969). Body weight and daily food intake in captive shrews. *Acta theriol.* **14**, 35–47.

CHROMOSOMAL VARIATION

Symp. zool. Soc. Lond. (1970) No. 26, 211–222.

CHROMOSOMAL POLYMORPHISM IN MAMMALS

ANDRÉ MEYLAN

Vertebrate Section, Federal Agricultural Research Station
CH-1260 Nyon, Switzerland

SYNOPSIS

Following improvement in cytological techniques, studies on mammalian chromosome-have been expanded very considerably during the past 10 years. Variations in chromos some number and/or in chromosome morphology appear relatively frequently, but 2 things must be clearly distinguished: "chromosomal aberrations" producing abnormal or sterile individuals, and "chromosomal polymorphism" which segregates normally within a population.

Different types of chromosomal polymorphisms have been discovered in populations of eutherian mammals. The most frequent one is the Robertsonian or "centric fusion/fission" type. Pericentric inversions and supernumerary chromosomes occur in some species. Deletions have been found only in the X chromosome. Small variations in the size and shape of chromosomes have been observed in several species, but their interpretation is often difficult.

Populations of small terrestrial mammals seem to be very helpful for interpreting patterns of chromosomal polymorphism, but future work will have to use large samples and involve the analysis of diploid metaphases, meiosis and hybrids.

The study of chromosomal polymorphism in mammalian populations allows us to point a finger at one aspect of their evolution. In many cases, polymorphic forms can be regarded as "species *in statu nascendi*" (Matthey, 1969) and it is often difficult to distinguish intra- from inter-specific variation. New models of speciation will be needed to explain the observed situations (White, 1968).

INTRODUCTION

In a review of the problem of chromosomal polymorphism in mammals Matthey concluded in 1959 that the only case of true intra-specific variation was the one described by Sharman (1956) and by Ford, Hamerton and Sharman (1957) in the Common shrew *Sorex araneus*. During the past 10 years, cytological studies of mammals have been expanded so intensively that chromosomal polymorphism appears now, if not common, at least relatively frequent. This abundance of new data is the result of improvement in cytological techniques. The introduction of colchicine pretreatment in fixation as well as after tissue culture has made possible both a more accurate determination of the diploid number of each specimen, and also a study of the morphology of the chromosomes. Another important point is that these new techniques make practicable the study of large samples of animals. This is absolutely necessary if variation within populations is to be revealed.

The different problems about mammalian chromosomes have been recently reviewed in the excellent book "Comparative mammalian cytogenetics" edited by Benirschke (1969). The presentation of chromosomal polymorphism in mammals is a rather ticklish affair after the remarkable papers issued in this volume. Nevertheless, this small contribution will help those mammalogists who are not accustomed to cytological problems, to a better understanding of the scope of chromosomal polymorphism. It does not give an exhaustive bibliography about chromosomal variations in wild eutherian mammals, but merely references to the examples used.

Most of the chromosome complements of mammals are based exclusively on the examination of metaphases in somatic cells. This allows one to know the diploid number, and the size and shape of the chromosomes. The karyotype is then established by pairing autosomes of the same length and morphology and the sex chromosomes identified in males as a heteromorphic pair. It is only by comparing the chromosome complements of a number of specimens that differences can be detected. Furthermore, the standard cytological technique of examination under the optical microscope of fixed and stained material allows the recognition of only rather large chromosomal differences between individuals. Technical procedures may produce small variations in the number and morphology of chromosomes, and the "process of chromatid condensation may not always be uniform in all chromosomes at a given stage of the mitotic cycle" (Sasaki, 1961).

Better data on the chromosome complements of male mammals are produced when meiotic as well as diploid metaphases are analysed. The number and morphology of the bivalents at the first meiotic division can provide confirmation of the chromosome number, of the identification of the sex chromosomes, and of the proposed autosomal pairing. The presence of univalents, asymmetrical bivalents, trivalents, or multivalents can reveal more accurately the presence of chromosomal modifications. The study of the second meiotic division permits the confirmation of the observations made in metaphase I, and information as to whether balanced germ cells are produced. Unfortunately, the examination of male meiosis is seldom done by mammalian cytologists, even when chromosomal polymorphism has been observed in a population.

ABERRATION AND POLYMORPHISM

The general expression "chromosomal polymorphism" includes two different things which must be clearly separated. Certain chromosomal mutations are incompatible with the survival of the animal or the species by producing abnormal or sterile individuals. In natural populations of wild mammals, such chromosomal rearrangements are automatically eliminated by natural selection. But in laboratory or domestic mammals—as well as in human populations—abnormal individuals can survive. A considerable number of cases of such variation involving both autosomes and sex chromosomes have been described, mainly in mice. Some of them occur spontaneously, but many are produced by irradiation or by chemical action. Such examples can properly be called "chromosomal aberrations" and will not be reviewed here. Indeed, the only case of chromosomal aberration detected in a population of wild mammals is worth mentioning. Studying the chromosome complement of the Water vole, *Arvicola terrestris*, in Sweden, Fredga (1968) found a young female with 37 chromosomes instead of 36 in the normal karyotype. This case is interpreted as a trisomy of the smallest autosomes and the similarity to Down's syndrome in man is striking, even if the animal had no obvious phenotypic abnormalities.

The expression "chromosomal polymorphism" must be used only for variations compatible with the survival of the species or at least for the survival of individuals of a given population. Polymorphism, in some cases, can concern the species value itself (see below). Chromosomal mosaicism is exceptional (although it is not uncommon in man) and I shall omit any discussion of it in this short paper.

ROBERTSONIAN VARIATION

The most frequent type of variation observed in mammals is a special case of reciprocal translocation called Robertsonian. Figure 1 illustrates a typical reciprocal translocation. A break occurs in two non-homologous chromosomes A and B. One segment of A is transferred to B, and one of B to A. In our example, the reciprocal translocation has transformed two submetacentric chromosomes A and B—chromosomes with their centromeres in a submedian position—into one metacentric A′—with a median centromere—and one acrocentric B′—with a subterminal centromere. In the Robertsonian process (Fig. 2), breaks occur near the centromeres of two non-homologous acrocentrics, but in one (A) on the short arm, and in the second (B) on the long arm. The result of such a reciprocal translocation is the formation of a large metacentric chromosome (A′), and a very small one (B′) which disappears, having no genetical value.

The "double break and translocation" hypothesis to explain the transformation of two acrocentric chromosomes into one metacentric

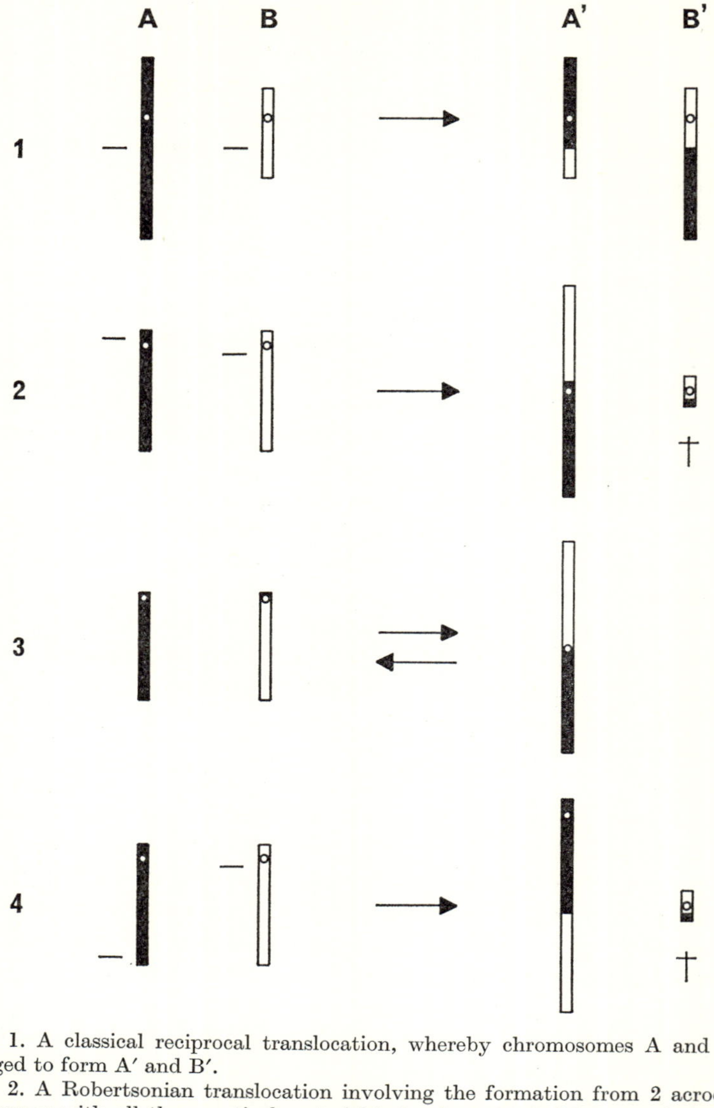

FIG. 1. A classical reciprocal translocation, whereby chromosomes A and B are rearranged to form A′ and B′.

FIG. 2. A Robertsonian translocation involving the formation from 2 acrocentric chromosomes with all the genetical material being incorporated into one large metacentric chromosome, the other translocation product being lost.

FIG. 3. An alternative mechanism for producing Robertsonian variation: reversible fusion (or fission) of the centromeres of 2 acrocentric chromosomes.

FIG. 4. Tandem fusion: the fusion of the long arms of 2 acrocentric chromosomes, and the loss of the reciprocal small chromosome. This has never been proved to occur in a mammalian population.

seems too complex to be a very common process of chromosomal rearrangement in mammals. Many authors prefer to regard the phenomenon as a direct "fusion" of two acrocentric or telocentric chromosomes (Fig. 3), hence the expression "centric fusion" for this process. The mechanism of this chromosomal mutation is not at all clear and it is dangerous to exclude the possibility of "centric fission" of one metacentric into two acrocentrics. Figure 4 illustrates the other extreme case of reciprocal translocation called "tandem fusion". This has never been proved to produce a chromosomal polymorphism in mammals.

The list of mammalian species in which Robertsonian polymorphism has been discovered is growing rapidly. For example, in the North American Short-tailed shrew, *Blarina brevicauda*, I observed two different chromosome numbers, 50 and 49 in a sample of 21 specimens (Meylan, 1967). In the karyotype with 49 chromosomes, a large submetacentric replaced two acrocentrics. Lee and Zimmerman (1969) have discovered a Short-tailed shrew with the other homozygous karyotype, in which there is a diploid number of 48 including two submetacentric autosomes. To date, a total of 100 Short-tailed shrews have been examined by Lee, Zimmerman and myself, 84 with 50, 15 with 49 and 1 with 48 chromosomes. In my own research on this species, it was only after having karyotyped 12 animals that I found an individual with a centric fusion. These data demonstrate the absolute necessity of analysing large samples of animals for a good understanding of polymorphism problems.

Finally, I must point out that the Robertsonian variation preserves the fundamental number or the number of chromosome arms.

An unusual case of centric fusion has been observed by Hsu (1969) in a population of the Least cotton rat, *Sigmodon minimus*. This involved a pair of homologous acrocentrics instead of non-homologous ones as in the common process. Seven specimens showed a diploid number of 30 and 6 a diploid number of only 29. The identity of the two elements which fused to form the metacentric was revealed in meiosis. The fusion of non-homologous acrocentrics should have given 13 bivalents and one trivalent in first meitoic divisions, but in this special case, the metaphases I showed 15 bivalents.

OTHER CHROMOSOMAL POLYMORPHISMS

Inversions are rearrangements occuring in the chromosomes themselves. In the pericentric inversion (Fig. 5), the breaking points are placed on each side of the centromere and the inversion can change the morphology of the chromosome. In our example, an acrocentric becomes metacentric after the pericentric inversion. If the breaking points are located on the same arm of the chromosome, the inversion, called a paracentric inversion (Fig. 6), does not change the shape of the element.

Several examples of polymorphism in solving pericentric inversions have been observed in mammalian populations, but none with paracentric inversions. This chromosomal variation has always been described from diploid metaphases and has never been confirmed by

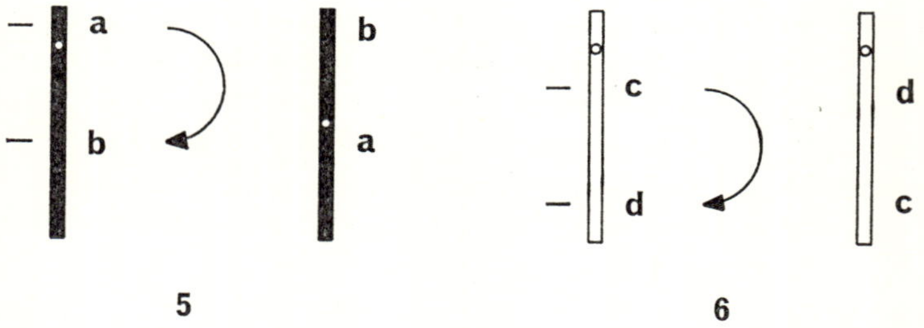

FIG. 5. A pericentric inversion, transforming an acrocentric chromosome into a metacentric one.

FIG. 6. A paracentric inversion, which involves no change in the position of the centromere in the rearranged chromosome.

observations of bivalent structure at the first meiotic division. The first case of polymorphism interpreted as the result of a pericentric inversion in an autosomal pair was described by Matthey (1966a) who found 3 different karyotypes in a sample of 8 African rodents, *Mastomys natalensis*.

Polymorphism caused by a variable number of supernumerary chromosomes has been described in the Red fox (*Vulpes vulpes*) by different authors (see Wurster and Benirschke, 1968). In this widely distributed species, the diploid number varies from 34–42 and the differences are due to a variable number of dot-like chromosomes. These very small elements are regarded as "inert" supernumerary chromosomes. A variation of the chromosome number in respect to supernumerary elements was also found in the rodents of the genus *Rheithrodontomys* by Blanks and Shellhammer (1968) and Shellhammer (1969).

Polymorphism involving a deletion, that is the loss of a part of a chromosome, has been found only in the X chromosome of some species. For example, in the field mouse, *Akodon azarea* (Bianchi and Contreras, 1967), the males have a diploid number of 38 with normal XY sex chromosomes, but the females show different sex chromosome constitutions, having normal XX sex chromosomes, with a deletion in one of the Xs or with the total loss of one X.

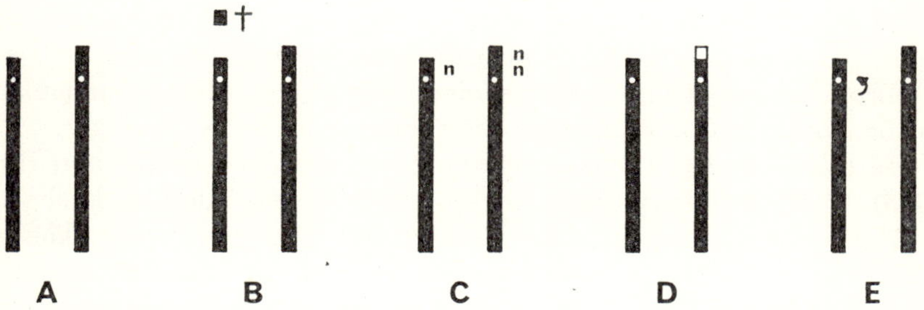

FIG. 7. Possible mechanisms for differences in the size or shape (i.e. position of the centromere) in a homologous pair of chromosomes: A. Heterogeneity of chromosome condensation. B. Small deletion. C. Duplication. D. Small reciprocal translocation. E. Pericentric inversion.

Small variations in the size or in the shape of well recognizable autosomal pairs have been observed in some species, but their interpretation is often a problem. Figure 7 illustrates the difficulties. If 2 homologous elements differ slightly in their length and/or in the position of their centromeres, this can be regarded (a) just as a result of the heterogeneity of chromosome condensation, (b) as a small deletion in one element (c) as a duplication, (d) as a tiny reciprocal translocation, or (e) as a pericentric inversion of a small chromosomal segment.

Polymorphisms involving a small variation of the short arm of acrocentric chromosomes have been found in *Rattus rattus*, *Rattus norvegicus*, and in the Guinea pig, *Cavia porcellus*. In this last species, Cohen and Pinsky (1966) explain variation of the first autosomal pair as a reciprocal translocation between 2 homologues leading to 3 different chromosome types and, consequently, to 6 different chromosomal constitutions.

The chromosomal polymorphisms occurring in mammalian populations are not always as simple as illustrated in the given examples. Robertsonian fusions and pericentric inversions, or more complex rearrangements can happen altogether. In several species of the genera *Erinaceus*, *Sorex*, *Thomomys*, *Perognathus*, *Sigmodon*, *Spalax*, *Mus*, *Acomys* and so on, polymorphisms involving different types of chromosomal mutations have been detected.

SIGNIFICANCE OF CHROMOSOMAL POLYMORPHISMS

In the last part of this rapid survey on the chromosomal polymorphism in mammals, it is necessary to try to understand the significance of the observed variations. As I am not a geneticist, I shall not discuss the consequences of different types of chromosomal mutations, but confine myself to some examples of the distribution patterns of chromosomal variation within mammalian populations, particularly to the evolutionary trends that they show.

In the European Common shrew, *Sorex araneus* (see Meylan, 1964, 1965), 2 different karyotypes have been observed. The first is characterized by 23 chromosomes in the male and 22 in the female. The sex difference in diploid numbers results in multiple sex chromosomes. The second form has 21 to 33 elements in the male karyotype and 20 to 32 in the female. This large variation in diploid number arises from Robertsonian fission of 6 metacentric autosomal pairs; the sex trivalent has exactly the same shape as in the first type. In 2 specimens of *Sorex arcticus* from Canada, I found a very closely related karyotype with 29 chromosomes in the male, 28 in the female, and the same sex trivalent. The common origin of these 3 *Sorex* types is indubitable and suggested by many other lines of evidence. It is believed that an ancestral Asiatic form of *Sorex* which had already acquired multiple sex chromosomes colonized both North America and Europe. In Europe, the population was split in two during the last glaciation and each population has been developing its own karyotype via rather complex chromosomal rearrangements. These 3 forms must be regarded now as distinct species, even if the 2 European ones cannot be separated morphologically. (The characters established by Ott, 1968, have no general value.) The chromosomal polymorphism of the *Sorex araneus—arcticus* group, apart from the Robertsonian variation in one European form, must be regarded as a consequence of allopatric speciation.

The model of speciation opposed to the allopatric model is the sympatric one. But Mayr (1963) considers sympatric speciation to be impossible and claims the necessity of geographical barriers. However, the recent discovery of 26 chromosomes in the Tobacco mouse, *Mus poschiavinus*, instead of 40 in *Mus musculus* is regarded by Gropp, Tettenborn and von Lehmann (1969) as a case of sympatric speciation. *Mus poschiavinus* differs from *Mus musculus* by centric fusions of 14 pairs of acrocentric autosomes giving 7 pairs of metacentrics. These Tobacco mice are living in the Poschiavo valley of the Swiss Alps besides normal House mice with 40 chromosomes and they do not interbreed under natural conditions. It has been possible to cross the *poschiavinus* mice with laboratory mice, but the fecundity of the hybrids with 33 chromosomes is considerably reduced owing to the difficulties of disjunction of the 7 trivalents at meiosis.

In the ground squirrel, *Spermophilus townsendi*, of the Western United States, Nadler (1968) has recognized 5 different karyotypes related mainly by Robertsonian centric fusions. Populations of ground squirrels with distinct karyotypes are well separated and they occupy different habitats. No hybrids were found in the contact zones studied and they are regarded as corresponding to different subspecies. The karyotype of *Spermophilus townsendi canus* has the highest diploid number and is considered by Nadler as the most primitive one. The other chromosomal forms of this species of ground squirrel are regarded as derived from an ancestral population similar in karyotype to *canus* via probable transient polymorphic systems related to the presumed directions of dispersal.

Another interesting and comparable case of polymorphism has been studied by Wahrman, Goitein and Nevo (1969) in the mole rat, *Spalax ehrenbergi*, from the Middle East. In this species, the diploid number varies from 52 to 60 and the number of chromosome arms from 76 to 84. Four karyotypes were observed with 60, 58, 54 and 52 chromosomes. The differences between these chromosome complements were interpreted as arising from centric fusions and pericentric inversions. The mole rats with distinct karyotypes occupy different areas. The chromosome number increases from northern to southern populations and, the distribution of the 4 forms coincides with regions characterized by different degrees of aridity. A few hybrids have been found between animals with 52 and 54 chromosomes, and between those with 58 and 60. In this case, the lowest diploid number is regarded as primitive and evolution has taken the form of an increase of chromosome number.

Chromosomal evolution in *Spermophilus townsendi* and *Spalax ehrenbergi* can be used to illustrate a model of speciation proposed by White (1968), the "stasipatric model". In this model, the chromosomal rearrangement is presumed to become established in a non-peripheral population, and thence to spread through the existing species population. The stasipatric model of speciation is considered by White to be distinct from the sympatric model. But it seems to me that this new model is just a dynamic explanation of sympatric speciation.

More complex are the polymorphic systems discovered in American mice of the genus *Peromyscus* by different cytologists and in African mice of the subgenus *Leggada* by Matthey (1966b). The chromosomal polymorphism in the *Peromyscus* group is regarded as a consequence of pericentric inversions, but the behaviour of the chromosomes during meiosis has never been investigated. The situation is so complicated at present that the last interpretation of phylogenetic relationships among this genus published by Hsu and Arrighi (1968) is highly conjectural. The situation in *Leggada* is most complex. Matthey has grouped the different forms according to the nature of the sex chromosomes (primitive, translocated or multiple), the fundamental numbers, and the presence of constant or polymorphic karyotypes. He concluded that it is necessary to classify the different forms in 3 main groups and that no simple evolutionary trend can be proposed.

CONCLUSIONS

In this period of explosive cytogenetic researches and of accumulation of new data on chromosomal polymorphisms, it is very difficult to recognize the significance of the different patterns of variation which have been observed. However, certain generalizations can be made.

The first point to note is that the most important cases of chromosomal variation have been found in small terrestrial mammals. This is probably the result of the ease with which large samples of animals can be collected and studied. But it seems to me that the patterns of the populations of small mammals, with limited exchanges in widely distributed forms as well as differences in population structure and dynamics, provide favourable conditions for the development of polymorphic systems.

The second point to note is that the samples have unfortunately never hitherto been big enough and the studies never continued for a sufficiently long period of time to be able to know if the polymorphisms are balanced or merely transient.

Nevertheless, the result of this rapid review of chromosomal polymorphism in eutherian mammals is that the study of chromosomal rearrangements allows us to pinpoint one aspect of their evolution. The species showing chromosomal variations are generally species which cause a lot of trouble to classical taxonomists. Matthey (1958 and later) has pointed out in a number of papers that the large chromosomal mutations by themselves do not involve phenotypic modifications. They lead progressively to intersterility but they have a lower evolutionary value than the gene mutations. Regarding their chromosome complements, polymorphic forms can be considered as "species *in statu nascendi*" (Matthey, 1969) and it is often difficult to distinguish intra- from inter-specific variation. In other words, as noticed by Hsu and Arrighi (1968), polymorphic forms do not reflect the results of evolution but allow us to observe the process itself. Finally, and according to White (1968), the study of chromosomal polymorphism in animals is showing that the classical models of speciation do not permit an explanation of the observed situations, and that "there are different kinds of mechanisms involved".

I hope that chromosomal polymorphism in mammalian populations will still interest many scientists for a long time. But researches must not be restricted to karyotyping diploid metaphases, but must be extended to the analysis of meiosis, and to the study of hybrids occurring in wild populations or produced under laboratory conditions.

References

Benirschke, K. (ed.) (1969). *Comparative mammalian cytogenetics*. New York: Springer-Verlag.

Bianchi, N. O. and Contreras, J. R. (1967). The chromosomes of the field mouse *Akodon azarae* (Cricetidae, Rodentia) with special reference to sex chromosome anomalies. *Cytogenetics* **6**, 306–313.

Blanks, G. A. and Shellhammer, H. S. (1968). Chromosome polymorphism in California populations of harvest mice. *J. Mammal.* **49**, 726–731.

Cohen, M. M. and Pinsky, L. (1966). Autosomal polymorphism via a translocation in the Guinea pig, *Cavia porcellus* L. *Cytogenetics* **5**, 120–132.

Ford, C. E., Hamerton, J. L. and Sharman, G. B. (1957). Chromosome polymorphism in the comon shrew. *Nature, Lond.* **180**, 392–393.

Fredga, K. (1968). Idiogram and trisomy of the water vole (*Avricola terrestris* L.), a favourable animal for cytogenetic research. *Chromosoma* **25**, 75–89.

Gropp, A., Tettenborn, U. and von Lehmann, E. (1969). Chromosomenuntersuchungen bei der Tabakmaus (*M. poschiavinus*) und bei Tabakmaus-Hybriden. *Experientia* **25**, 875–876.

Hsu, T. C. (1969). Robertsonian fusion between homologous chromosomes in a natural population of the least cotton rat, *Sigmodon minimus* (Rodentia, Cricetidae). *Experientia* **25**, 205.

Hsu, T. C. and Arrighi, F. E. (1968). Chromosomes of *Peromyscus* (Rodentia, Cricetidae). I. Evolutionary trends in 20 species. *Cytogenetics.* **7**, 417–446.

Lee, M. R. and Zimmerman, E. G. (1969). Robertsonian polymorphism in the cotton rat, *Sigmodon fulviventer*. *J. Mammal.* **50**, 333–339.

Matthey, R. (1958). Les chromosomes des mammifères euthériens. Liste critique et essai sur l'évolution chromosomique. *Arch. Julius Klaus-Stift. Vererb Forsch.* **33**, 253–297.

Matthey, R. (1959). Formules chromosomiques de *Muridae* et de *Spalacidae*. La question du polymorphisme chromosomique chez les mammifères. *Rev. suisse Zool.* **66**, 175–209.

Matthey, R. (1966a). Une inversion péricentrique à l'origine d'un polymorphisme chromosomique non-Robertsonien dans une population de *Mastomys* (Rodentia-Muridae). *Chromosoma* **18**, 188–200.

Matthey, R. (1966b). Le polymorphisme chromosomique des *Mus* africains du sous-genre *Leggada*. Revision générale portant sur l'analyse de 213 individus. *Rev. suisse Zool.* **73**, 585–607.

Matthey, R. (1969). Les chromosomes et l'évolution chromosomique des mammifères. In *Traîté de Zoologie* **16**, 855–909 and 999–1004. Grassé, P.-P. (Ed.). Paris: Masson.

Mayr, E. (1963). *Animal species and evolution*. Cambridge, Mass.: Harvard University Press.

Meylan, A. (1964). Le polymorphisme chromosomique de *Sorex araneus* L. (Mamm.-Insectivora). *Rev. suisse Zool.* **71**, 903–983.

Meylan, A. (1965). Répartition géographique des races chromosomiques de *Sorex araneus* L. en Europe (Mamm.-Insectivora). *Rev. suisse. Zool.* **72**, 636–646.

Meylan, A. (1967). Formules chromosomiques et polymorphisme robertsonien chez *Blarina brevicauda* (Say) (Mammalia: Insectivora). *Can. J. Genet. Cytol.* **45**, 1119–1127.

Nadler, F. (1968). The chromosomes of *Spermophilus townsendi* (Rodentia: Sciuridae) and report of a new subspecies. *Cytogenetics* **7**, 144–157.

Ott, J. (1968). Nachweis natürlicher reproduktiver Isolation zwischen *Sorex gemellus* sp. n. und *Sorex araneus* Linnaeus 1758 in der Schweiz (Mammalia, Insectivora). *Rev. suisse Zool.* **75**, 53–75.

Sasaki, M. (1961). Observations on the modification in size and shape of chromosomes due to technical procedure. *Chromosoma* **11**, 514–522.

Sharman, G. B. (1956). Chromosomes of the common shrew. *Nature, Lond.* **177**, 941–942.

Shellhammer, H. S. (1969). Supernumerary chromosomes of the harvest mouse: *Reithrodontomys megalotis*. *Chromosoma* **27**, 102–108.

Wahrman, J., Goitein, R. and Nevo, E. (1969). Mole rat *Spalax*: evolutionary significance of chromosome variation. *Science, N.Y.* **164**, 82–84.

White, M. J. D. (1968). Models of speciation. *Science, N.Y.*, **159**, 1065–1070.

Wurster, D. H. and Benirschke, K. (1968). Comparative cytogenetic studies in the Order *Carnivora*. *Chromosoma* **24**, 336–382.

Symp. zool. Soc. Lond. (1970) No. 26, 223–236.

CHROMOSOME POLYMORPHISM IN THE COMMON SHREW, *SOREX ARANEUS*

C. E. FORD

Medical Research Council, Radiobiology Unit Harwell, Didcot, Berkshire, England

and

J. L. HAMERTON*

Department of Genetics, The Children's Hospital of Winnipeg Manitoba, Canada

SYNOPSIS

The Common shrew, *Sorex araneus* L. in Britain has a basic diploid complement of 18 metacentric autosomes plus 3 sex chromosomes (XY_1Y_2) in the male and 2 in the female (XX). A Robertsonian system of chromosome polymorphism in which metacentric chromosomes 6, 7, 8 and, very rarely 4 (Fig. 1) are sometimes replaced by 2 acrocentric chromosomes was discovered at localities in Berkshire. One was an isolated thicket near the village of Chilton. Samples were trapped there in 3 successive summers and 160 animals were karyotyped. Six of 9 tests of goodness-of-fit to the Hardy-Weinberg proportions showed an excess of heterozygotes but none of the deviations were significant. Populations at other localities in Britain have been found to be monomorphic, or polymorphic in respect of, essentially, only one of the elements 6 to 8.

Shrews from Jersey and north France were found to have a distinct karyotype that differed from the basic karyotype of the British mainland animals by a minimum of 3 presumptive pericentric inversions and one tandem translocation. Both types were later identified in Switzerland by Meylan and it is now known that one type (Race B) has an essentially alpine, northern and eastern distribution whereas the other (Race A) occurs in lowland western Europe. The races overlap without hybridization at 2 localities in Switzerland and are considered as cryptic species. Robertsonian polymorphism is confined to Race B and in Europe includes also elements 3 and 5. The wide distribution suggests ancient origin and implies that the system is balanced. Maintainance by heterozygote advantage is assumed. Study of pregnant females and their embryos could provide unique information about mate selection and show whether assortative mating also contributes to the maintainance of the system.

INTRODUCTION

In 1949 Bovey reported that 2 male shrews, *Sorex araneus*, from lowland Switzerland, had a diploid complement of 23 chromosomes which included 3 presumptive sex chromosomes. This was the first instance of multiple sex chromosomes to be found in a eutherian species

* Formerly of: British Museum (Natural History), London, England and Paediatric Research Unit, Guy's Hospital Medical School, London, England.

but he had to leave open the question whether the system of sex determination was $XY_1Y_2 : XX$, with one more sex chromosome in the male, or $X_1X_2Y : X_1X_1X_2X_2$ with one more sex chromosome in the female. The issue was resolved a few years later by Sharman (1956) in favour of the XY_1Y_2 male. Sharman examined bone marrow preparations from animals trapped at Buckland in Berkshire, England (as well as testicular preparations from the males) and made the then suprising observation that the number of autosomes varied between one animal and another, though the total number of autosome arms appeared to be constant. This suggested that the species, or at least, that particular population, might display a Robertsonian system of chromosome polymorphism. Such systems, in which the alternative morphs are a metacentric chromosome, and 2 acrocentric chromosomes each corresponding to one arm of the metacentric, were well known among invertebrates (White 1954) but had previously been encountered only once in the class Mammalia (Wahrman and Zahavi, 1955).

THE CHILTON POPULATION

The year after Sharman made his observations a population of shrews was identified in a thicket of some 4 acres located near the village of Chilton, Berkshire and less than half a mile from our laboratory. At that time, the thicket was surrounded by arable land, the nearest cover suitable for shrews being about 200 yards away. Forty-four animals were trapped there during the summer of 1956 and Sharman's findings were confirmed and extended (Ford, Hamerton and Sharman, 1957). It was established that 3 of the smaller elements exhibited Robertsonian variation (Fig. 1) and 15 out of the $3^3 = 27$ possible different autosomal karyotypes were recorded. Despite the variation between animals, the karyotype was constant within a given animal. (Counts of less than the standard number were disregarded as artefact.) This rule was maintained in all shrews examined subsequently, apart from one animal in which a single spleen cell contained 2 abnormal chromosomes that were readily explicable as the complementary products of a reciprocal translocation, but not as the products of a new Robertsonian change (C. E. Ford, 1964). There is therefore every reason to believe that the variation between animals is a consequence of meiotic segregation and that recurrent chromosome mutation is at most a secondary factor.

Two very different questions are posed by any polymorphic system, genetic or chromosomal. One is concerned with the selective forces operating on the system and whether it is balanced or transient (E. B. Ford, 1964). The other is the historical problem, when and where did it originate and how did it spread? One of us (C. E. Ford) gave attention to the first of these questions. Trapping was continued at the Chilton thicket for a further 2 summers with the object of determining whether

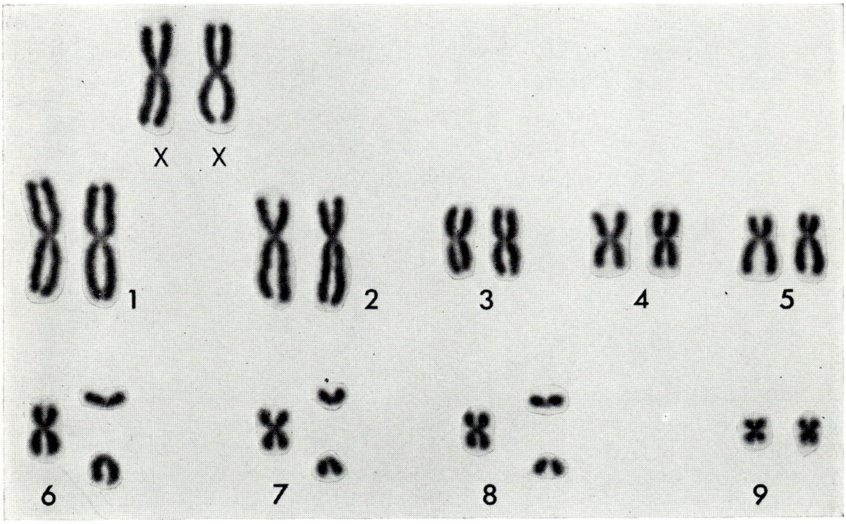

FIG. 1. Karyotype of *Sorex araneus* female trapped at Chilton, Berkshire. This animal was heterozygous for all 3 elements, 6, 7 and 8.

the frequency of the different karyotypic classes was in accordance with expectation from the Hardy-Weinberg distribution. In 1957, 57 animals were karyotyped and in 1958, 61 animals were karyotyped. These numbers represent about 90% of those trapped. Taking each element that occurs in alternative forms separately, the whole body of data provided 9 tests of agreement with the Hardy-Weinberg distribution. The relative isolation of the population, its small size and the high proportion of the young of each year that must have been taken call for caution in the interpretation of the tests. In fact, no single test revealed a significant deviation from expectation, but it is perhaps a pointer that in 6 of the 9 there was an excess of heterozygotes (Ford, 1970). All that can be said is that the data are not inconsistent with maintenance of the system by heterozygous advantage.

POPULATIONS IN OTHER BRITISH LOCALITIES

The obvious and perhaps only approach to the historical problem was to get information from different geographical localities. This aspect was taken up by J. L. Hamerton whose results are summarized in Table I.

Three points emerge. At many localities only a single karyotype was recorded, though had more animals been examined doubtless more populations would have been found polymorphic. Secondly, there are broad regional concordances. In south-east England, for example, from Kent to Hampshire, elements 7 and 8 are represented by the twin-acrocentric morph. In south Devon, animals from 4 out of 5 localities were homozygous metacentric for all variable elements, though twin-acrocentric 8 was identified again in the extreme west.

TABLE I

Shrews, Sorex araneus *from* 20 *British localities*

	Locality	Number of animals	Element			
			4	6	7	8
1	Bradfield Combust, Suffolk	21	0·02	M	M*	0·71
2	Epping Forest, Essex	1	M	0·50	0·50	M
3	Addington Park, Kent	2	M	M	A	A
4	Fetcham, Surrey	2	M	M	A	A
5	Beddington, Surrey	1	M	0·50	A	A?
6	Kirdford, Sussex	10	M	M	A	A
7	Alice Holt Forest, Hampshire	2	M	M	A	A
8	Bentley Station, Hampshire	8	M	0·08	A	A
9	Andover, Hampshire	20	M	M	A	A
10	Tewkesbury, Gloucestershire	6	0·08	0·25	0·17	A*
11	Mutter's Moor, Devon	3	M	M	M	0·33
12	Harpford Wood, Devon	1	M	M	M	M
13	Windy Cross, Devon	2	M	M	M	M
14	Exeter, Devon	3	M	M	M	M
15	Ladrum Bay, Devon	1	M	M	M	M
16	Molesworthy, Devon	2	0·50	M	M	A
17	Cape Cornwall, Cornwall	1	M	M	M	A
18	Anglesey, Wales	6	0·08	M	M*	A*
19	Peebles, Scotland	7	M	M	M	M
20	Kyle of Lochalsh, Scotland	3	M	M	M	M

M—Homozygous metacentric A—Homozygous acrocentric
Figures show the frequency of the twin-acrocentric morph when both morphs were present.

* Revised interpretation. Original records show both morphs present (see text).

And at 2 localities in Scotland all the animals were homozygous metacentric for all autosomes, like most of those from south Devon. Taken as a whole, Table I suggests an orderly increase in frequency of metacentric morphs from the south-east to the west and north. Thirdly, a twin-acrocentric morph of another element, number 4, was identified in single animals from 4 widely separated localities. It has also been found in the Berkshire populations and is evidently widespread but rare.

The alternative morphs of elements 4 and 6 can be identified at the microscope with confidence. Elements 7 and 8 are much less distinctive (Fig. 1) though there is little doubt about identity in favourable cells where they lie close enough together for direct comparison. Some inconsistencies in the primary data on elements 7 and 8 in the polymorphic populations came to light when the records were checked. In compiling Table I, therefore the convention was adopted that when 2 or 4 small acrocentric chromosomes corresponding to elements 7 and 8 were present they were considered to represent element 8, and that when 6 were present, element 7 was considered heterozygous and element 8 homozygous twin-acrocentric. This convention affects 4 entries in Table I, which are indicated by asterisks. It should be added that the observations were made on Feulgen squash preparations and that contemporary air-drying methods provide not only more metaphase spreads for examination, but they are of higher average quality. Identification of elements 7 and 8 by photography and measurement would be desirable in any future investigation.

SHREWS FROM JERSEY

As part of the geographical survey, on which he was advising, Dr. Peter Crowcroft trapped several shrews, presumptively *Sorex araneus*, on the Island of Jersey, British Channel Islands. These were examined by J. L. Hamerton who immediately discovered they had an entirely distinctive karyotype. The differences are conveniently displayed in a comparative ideogram (Fig. 2). This was constructed from measurements of the chromosomes of 10 cells of each type, animals with only metacentric chromosomes being selected to represent the British mainland form. The fewest assumptions that will account for the significant differences between the 2 karyotypes are 3 pericentric inversions and one tandem translocation with elimination of a centric fragment. (The combined mean lengths of chromosomes 1 and 10 of the Jersey form did not differ significantly from the mean length of chromosome 1 of the British mainland form.) In addition to these structural differences, which in fact are obvious on simple inspection of the karyotypes, there is a prominent secondary constriction located proximally in the long arm of chromosome 4 of the Jersey type which has never been seen in the corresponding chromosome of the British mainland form. These are the minimal differences; there may be more, but there cannot be less.

One previous difficulty was immediately resolved. We had been unable to reconcile Bovey's drawings of the chromosomes of his Swiss shrews (Bovey, 1949) with our knowledge of the basic karyotype of the shrews from mainland Britain. It was now obvious that Bovey's drawings agreed with the karyotype of the Jersey shrew and when shrews from Normandy also proved to have the same karyotype we began to think of distinct "Continental" and "British" karyotypes. It was also obvious that meiosis in hybrids between the 2 forms would almost certainly be disturbed to such a degree that few if any balanced haploid gametic nuclei would be produced. Two explanations seemed possible. Either the karyotype of the British shrews had diverged subsequent to the separation of Britain from the Continent, about 8000 years ago, which would have implied a remarkably rapid rate of karyotypic evolution; or the British form had existed on the Continent (and perhaps still did) but had been displaced at least from extensive

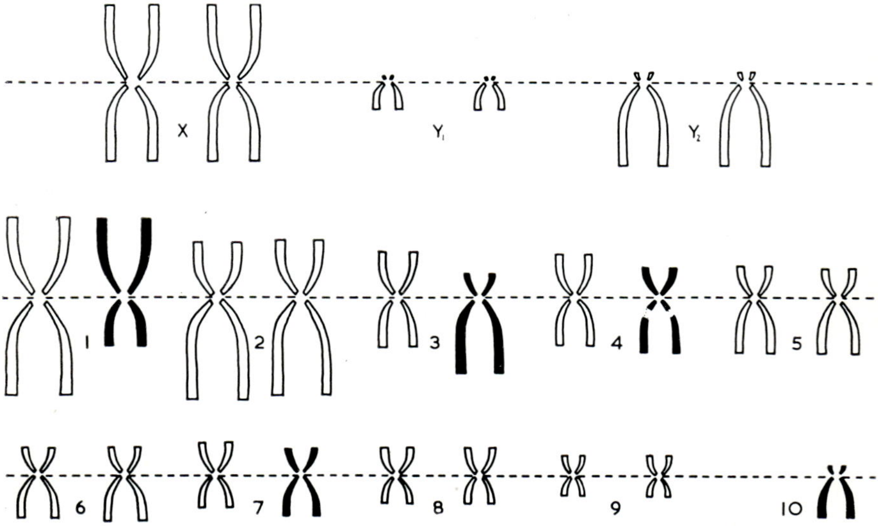

FIG. 2. *Sorex araneus*. Comparative ideogram of the British mainland form with the lowest chromosome number and the form from Jersey. The chromosome on the left of each pair represents the British form, that on the right, the Jersey form. Chromosomes in black indicate those of the Jersey form that differed significantly from those of the corresponding British form in mean length or arm ratio or both. (See text.)

areas by a karyotypically distinct cryptic species that had advanced as far as the Channel coast and Jersey.

At this point we communicated our results to Professor Matthey at Lausanne and shortly afterwards Meylan, working in his laboratory, found the "British" form at high altitude in Switzerland (Meylan, 1960). Meylan's subsequent extensive studies have demonstrated that the one form, which he calls Race A, does not exhibit Robertsonian polymorphism and occupies lowland western Europe; whereas the other, Race B, exhibits Robertsonian polymorphism, and has an alpine, northern and eastern distribution (Meylan 1964, 1965). Accordingly, the British mainland form is included in Race B and the animals from Jersey and Normandy in Race A.

Meylan also found a locality near Champéry (Valais, Switzerland) where animals of both races occur together without detected hybridization. This observation was confirmed by Ott (1968) who also found a second zone of overlap near Frenières on the opposite side of the Rhône valley, again without evident hybridization. Ott therefore formally proposed that Race A should be recognized as a new species, assuming that Linnaeus would have described a Swedish specimen of Race B as *Sorex araneus* since shrews of that race occur in Sweden (Meylan, 1965; K. Fredga, personal communication) whereas the nearest point at which animals of Race A have been recorded is just to the east of the Rhine delta (Meylan, 1965).

A further point that emerged from Meylan's work was that an additional element was involved in the polymorphism of some of the Swiss populations of Race B, making 5 in all. He also found that many populations of Race B were monomorphic and that in most polymorphic populations the alternative morphs of only one element were present. It is, of course, possible that if he had examined larger samples he would have found more populations to be polymorphic but his results are supported by our experience of the British populations.

It remains an open question, however, to what extent these monomorphic populations may reflect periodic extinction of small local populations and repopulation perhaps sometimes from single immigrant gravid females (Ford and Hamerton, 1958).

FURTHER OBSERVATIONS

The possibility, considered at one time, that the karyotype of the British shrews had undergone rapid evolutionary change suggested that it would be of considerable interest to examine samples from some

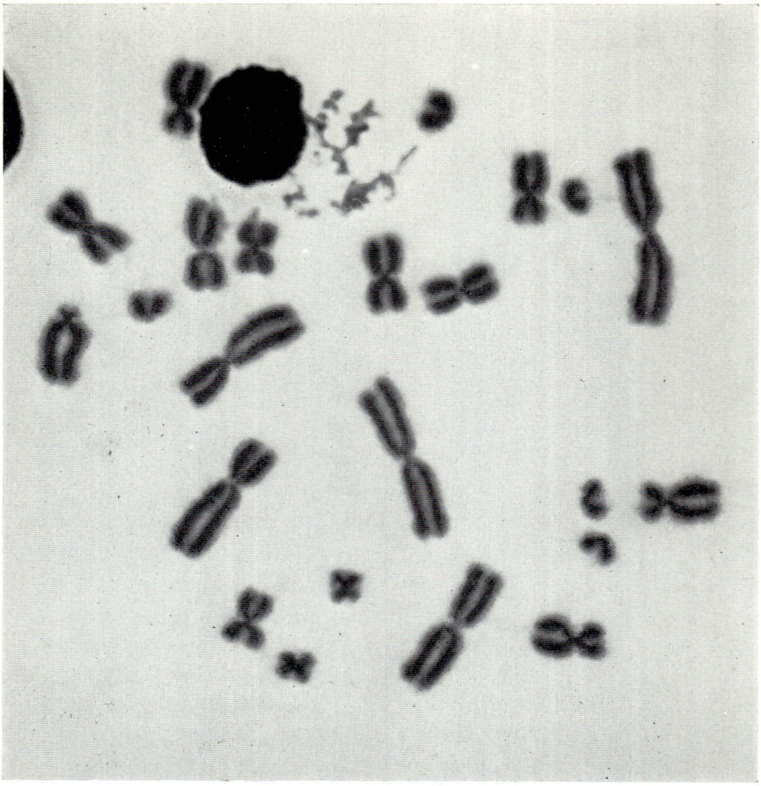

Fig. 3. Chromosomes at metaphase in a spleen cell from *Sorex araneus* male 232 trapped at Białowieża, Poland. ×2400.

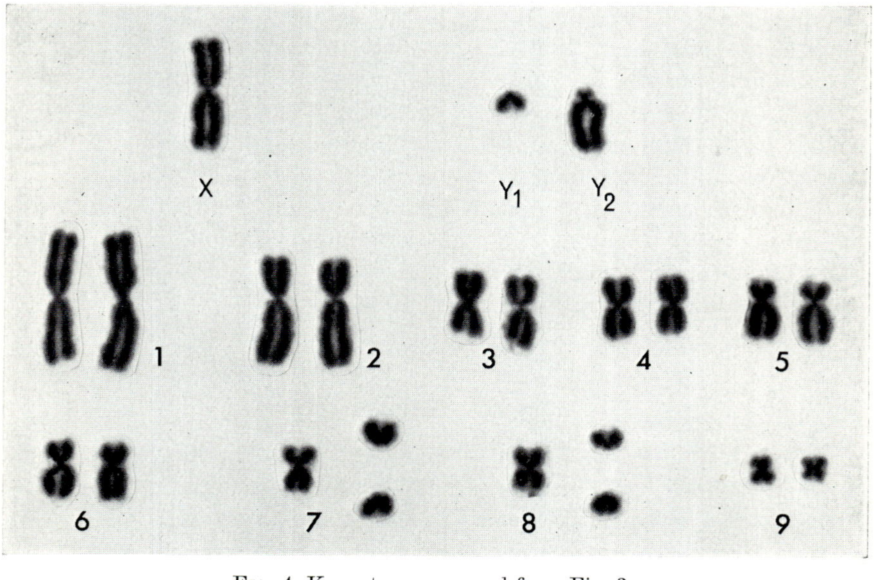

Fig. 4. Karyotype prepared from Fig. 3.

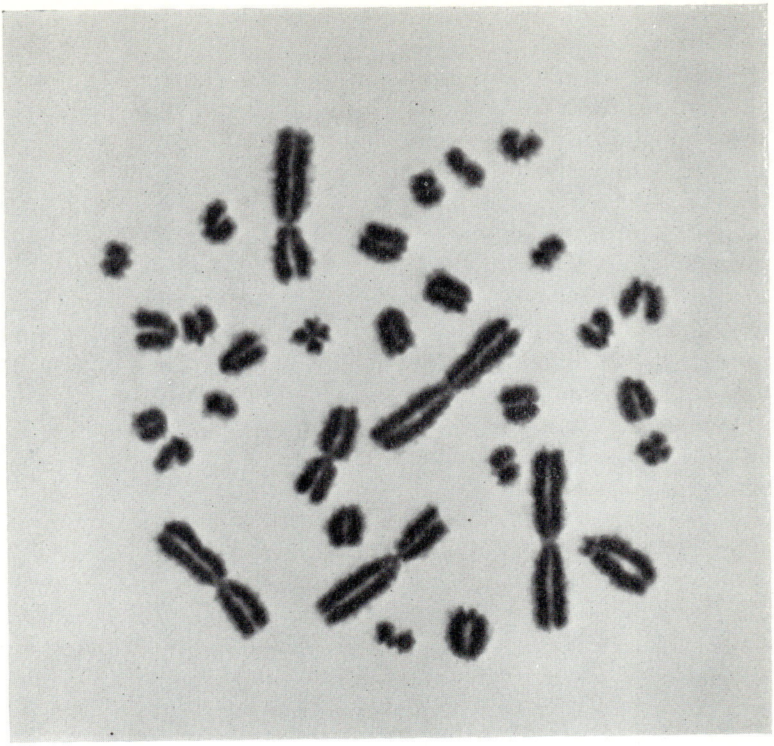

Fig. 5. Chromosomes at metaphase in a spleen cell from *Sorex araneus* male 233 trapped at Col de Voza, Haute Savoie, France. ×2400.

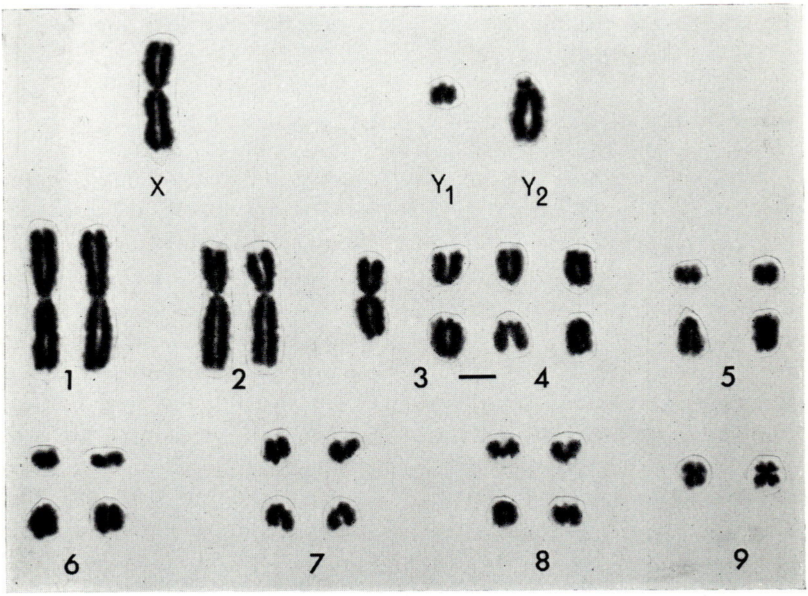

Fig. 6. Karyotype prepared from Fig. 5.

of the offshore islands. In the summer of 1963 samples of shrews were trapped on the Scottish islands of Islay, Jura and Gigha and also at a locality on the adjoining mainland by Ford and Graham (1964). Preparations were made and karyotypes established in the field from 70 animals. Twelve animals from Jura, 7 from Gigha and 11 from the mainland (Kintyre) had metacentric autosomes only. The sample of 40 animals from Islay, however, proved to be polymorphic, though element 8 only was involved, the twin-acrocentric morph being the more common with a frequency of 0·74. Direct inspection revealed no evident basic difference between the karyotypes of these animals and those of animals from elsewhere in Britain and this was confirmed by measurement of the chromosomes. There is therefore as yet no evidence for other than a single basic karyotype for *Sorex araneus* in Britain.

Two shrews trapped at Col de Voza, Haute Savoie, France at an altitude of 5500 feet and 3 taken at Białowieża in Eastern Poland proved to be of Race B and add new European localities for this race. Despite the small numbers they provided important additional information. The specimens from Białowieża demonstrated that the local population was polymorphic at least for elements 7 and 8 (Figs 3 and 4) thereby greatly extending the range over which the Robertsonian system is known to occur. One of the Col de Voza animals had a diploid number of 32 including 22 acrocentric autosomes (Figs 5 and 6), extending the Robertsonian system to a new element and providing the highest chromosome number yet encountered in the species.

DISCUSSION

It is clear that there is extensive karyotypic variation within the Linnaean species, *Sorex araneus* and that this variation takes 2 forms. Robertsonian changes are mainly responsible for the observed variation of the diploid number from 20 to 32. No less than 6 autosomal elements may be represented by alternative metacentric or twin-acrocentric morphs, though not more than 3 have yet been found to vary within a single population. Geographically the Robertsonian system is very widely distributed although the variation is extinguished in some localities and even whole regions to give monomorphic populations. Polymorphic populations have been identified from the Swiss and French Alps to Britain and Scandinavia and from western Germany to eastern Poland.

With such a widespread distribution and presumptively ancient origin it can hardly be doubted that the Robertsonian system is one of balanced polymorphism, though whether balanced or transient is strictly relevant only in relation to a particular population at a particular period of time. The question of how the system is maintained cannot be answered on the evidence we have, though there is a strong *a priori* presumption that it is by heterozygote advantage.

Despite the difficulties imposed by small samples from a small population, further investigations like those carried out on the Chilton population could provide valuable additional information. If dispersion of the young were limited so that there was an appreciable chance of siblings mating, the effect would be to increase the proportions of homozygotes in the population at the expense of the heterozygotes. Observed small deviations from the Hardy-Weinberg proportions in the direction of an excess of heterozygotes, as in 6 of the 9 tests provided by the Chilton data, would then be of increased importance. However, even if a significant excess of heterozygotes should eventually be demonstrated (after taking into account fluctuations of proportions introduced by the small numbers of parents of any one generation) it would not be proof of heterozygote advantage: it could be the result of assortative mating. Where the basis for preference is directly related to karyotype, assortative mating is perhaps not very likely, but it should be possible to find out something about it by examining the chromosomes of pregnant females and their young. Given the karyotypes of mother and embryos, it would often be possible to infer the karyotype of the sire absolutely. In the remaining cases the different possible sires could each be assigned a probability which would be dependent upon the number of embryos in the set and also on knowledge of the frequency of the different types of potential sire in the population. For example, if a female was homozygous metacentric for element 6, and she had 6 embryos with the same karyotype, the probability of getting such a set of embryos if the sire had been a heterozygote is only $0{\cdot}5^6$, but he would be identified absolutely as a heterozygote if there had been even a single heterozygote among the embryos.

The second type of chromosome variation is represented by the differences between the karyotypes of the 2 cryptic species that Meylan called Race A and Race B. As already said, the certain differences between the somatic chromosomes of these 2 forms could be accounted for by assuming 3 pericentric inversions and one tandem translocation. Study of meiosis in hybrids would be necessary to say whether or not these assumptions are valid. They might also reveal additional differences, not evident from comparison of the somatic karyotypes. Although in other mammalian species hybrids cells in the pachytene and later stages of meiosis may be absent, the closeness of the 2 shrew races allows the hope that if hybrids could be produced, cells in the required stages would be found.

Meylan (1964) has examined the possible evolutionary origins of the karyotypic differences between Race A and Race B. A question he did not consider, however, was the behaviour at meiosis of the chromosome rearrangements assumed to occur in the course of karyotypic evolution and their effect on fertility. Robertsonian changes present no problem of course. Their very presence in balanced polymorphic systems is sufficient evidence that if there are meiotic irregularities in heterozygotes the frequency is low enough not to have a significant effect on fertility. There is little information about their meiotic behaviour in mammals but it may be supposed that there is regular trivalent formation followed by convergent orientation on the spindle at metaphase.

It is possible that heterozygosity for a single pericentric inversion or even for a single tandem translocation (when the "recipient" chromosome is metacentric) could also be compatible with retention of (almost) full fertility. In the case of the pericentric inversion heterozygote, if chiasmata were largely excluded from the median (inverted) segment by distal localization, normal segregation would be the rule and the production of unbalanced cross-over products the exception. Frequent regular segregation in an animal heterozygous for a tandem translocation, like that represented by "British" chromosome 2, "Jersey" chromosome 2 and "Jersey" chromosome 10 in Fig. 2, is less likely, but could occur if, again, there was some measure of distal chiasma localization, combined with convergent orientation of the trivalent on the spindle. So new pericentric inversions and tandem translocations (of the kind considered) may not have to overcome high infertility barriers to become incorporated into the karyotype of the population. Again, the study of meiosis in Race A–Race B hybrids would be most informative. The general problem of chromosome rearrangements in relation to speciation in animals has recently been reviewed by White (1969).

In view of the very extensive Palaearctic distribution of *Sorex araneus* the possibility arises that it may be a complex that includes other karyotypically distinct cryptic species. Indeed, very recently 2 forms with karyotypes that cannot be reconciled with either of Meylan's races have been found, one in northern Sweden (K. Fredga, personal communication) and one in the Obi region of Siberia (Orlov and Kozlovsky, 1969). Preliminary information about the karyotypes of these forms suggests that an original set of acrocentric chromosomes have been combined into metacentric chromosomes by different series of Robertsonian changes.

Shrews have beautiful chromosomes, the preparative technique is easy and much remains to be done. Information about hybrid meiosis would be of outstanding value and studies of pregnant females and their embryos from polymorphic populations could give important information about the breeding system and relative fertility. At a more modest level there remain many parts of Europe from which simple identification of the karyotype in samples from the local population could at least help to fill in the still rather fragmentary distribution map of Races A and B and might reveal further unsuspected chromosome variation.

SUMMARY

Two karyotypically distinct races of the Common shrew, *Sorex araneus* L. have been identified in Europe. One, Meylan's Race A, occupies lowland western Europe and does not exhibit Robertsonian variation. The other, Meylan's Race B, exhibits Robertsonian polymorphism, and has an alpine, northern and eastern distribution. The minimal differences between the karyotypes are 3 presumptive pericentric inversions and 1 presumptive tandem translocation, and any hybrids between the 2 races would presumably be infertile. Zones of overlap without evidence of hybridization have been found in Switzerland, so the 2 races can be regarded as cryptic species.

The Robertsonian polymorphism of Race B involves 6 elements all of which can occur in the alternative metacentric and twin-acrocentric morphs. Some populations are nevertheless monomorphic. Others are polymorphic in respect of up to 3 elements.

A shrew population in an isolated thicket at Chilton, Berkshire, England was studied over a period of 3 years and 160 animals were karyotyped. This population was polymorphic in respect of elements 6, 7 and 8 (Fig. 1). The data provided 9 tests of goodness-of-fit to Hardy-Weinberg proportions. None of the deviations were significant at the 5% level but in 6 of the 9 there was an excess of heterozygotes.

The very wide distribution of polymorphic populations implies that the system is one of balanced polymorphism. The balance is presumed to be maintained by heterozygote advantage, but formal proof is lacking.

The study of meiosis in inter-racial hybrids and the identification of the karyotypes of pregnant females and their embryos could provide new information of great value.

Acknowledgements

Dr. Peter Crowcroft trapped many of the animals and gave much helpful advice. Mr. G. Breckon and Mrs. P. A. Henderson trapped the animals in the Chilton thicket. Dr. E. P. Evans and Dr. P. J. Ford obtained the animals from Białowieża and Col de Voza respectively. Mr. G. Breckon did the photography and prepared the karyotypes. We are grateful to all.

References

Bovey, R. (1949). Les chromosomes des Chiropterès et des Insectivores. *Revue suisse Zool.* **56**, 371–460.

Ford, C. E. (1964). Selection pressure in mammalian cell populations. In *Cytogenetics of Cells in Culture.* Harris, R. J. C. (ed.). *Symp. Int. Soc. Cell Biol.* **3**, 27–45. New York and London: Academic Press.

Ford, C. E. (1970). The population cytogenetics of other mammalian species. In *Human Population Cytogenetics.* Jacobs, Price and Law (eds). Edinburgh: University Press.

Ford, C. E. and Hamerton, J. L. (1958). A system of chromosomal polymorphism in the common shrew (*Sorex araneus* L.). *Int. Congr. Zool.* **15**, 177–179.

Ford, C. E., Hamerton, J. L. and Sharman, G. B. (1957). Chromosome polymorphism in the common shrew. *Nature, Lond.* **180**, 392–393.

Ford, E. B. (1964). *Ecological genetics.* London: Methuen.

Ford, P. J. and Graham, C. F. (1964). The chromosome number in the common shrew, *Sorex araneus* L. *Bull. Mammal. Soc. Br. Isl.* No. 22, 10–11.

Meylan, A. (1960). Contribution a l'étude du polymorphisme chromosomique chez *Sorex araneus* L. (Mamm. Insectivora). *Revue suisse Zool.* **67**, 258–261.

Meylan, A. (1964). Le polymorphisme chromosomique de *Sorex araneus* L. (Mamm. Insectivora). *Revue suisse Zool.* **71**, 903–983.

Meylan, A. (1965). Répartition géographique des races chromosomiques de *Sorex araneus* L. en Europe (Mamm. Insectivora). *Revue suisse Zool.* **72**, 636–646.

Orlov, V. N. and Kozlovsky, A. I. (1969). The chromosome complements of 2 geographically distant populations and their position in the general system of chromosomal polymorphism in the common shrew, *Sorex araneus* L. (Soricidae, Insectivora, Mammalia). *Citologia* **11**, 1129–1136.

Ott, J. (1968). Nachweis naturlicher reproduktiver isolation zwischen *Sorex gemellus* sp. n. und *Sorex araneus* Linnaeus 1758 in der Schweiz. *Revue suisse Zool.* **75**, 53–75.

Sharman, G. B. (1956). Chromosomes of the common shrew. *Nature, Lond.* **177**, 941–942.

Wahrman, J. and Zahavi, A. (1955). Cytological contributions to the phylogeny and classification of the rodent genus *Gerbillus. Nature, Lond.* **175**, 600–602.

White, M. J. D. (1954). *Animal cytology and evolution.* Cambridge: University Press.

White, M. J. D. (1969). Chromosomal rearrangements and speciation in animals. *A. Rev. Genet.* **3**, 75–98.

VISIBLE VARIATION

Symp. zool. Soc. Lond. (1970) No. 26, 239–250.

THE DETERMINATION AND DISTRIBUTION OF COAT COLOUR VARIATION IN THE HOUSE MOUSE

M. S. DEOL

Department of Animal Genetics, University College London, England

SYNOPSIS

Coat colour in the House mouse (*Mus musculus* L.) depends largely on the presence of the pigment melanin in the medulla of the hair. Melanin is produced by melanocytes in the bulb of the hair follicle, deposited on minute bodies called melanosomes and passed on to the growing hair at its base. There are two types of melanin: eumelanin, which is black, and phaeomelanin, which is yellow. The colour of the hair depends not only on the type of melanin and its quantity, but also on the structure and morphology of the melanosomes. All these factors are under genetical control. A large number of loci scattered over most of the chromosomal complement of the mouse are involved, with up to a dozen alleles at each. The rates of mutation at these loci are not low, but very few of the mutant genes have been found in wild populations. Natural selection appears to be the main reason for this. It is generally believed that selection acts through predators hunting by sight, favouring animals with the wild-type or black agouti coat because they are the least conspicuous. But it is probable that selection also operates on the pleiotropic effects of colour genes: most of the major coat colour loci seem to affect some other developmental pathway as well, and these effects might be deleterious.

PHYSICAL BASIS OF COAT COLOUR

For a clear understanding of coat colour variation in the mouse it is essential to be familiar with not only the nature and origin of the pigment concerned but also the structure and development of the coat.

The coat

The coat of the mouse consists of 2 main types of hair, overhair and underhair. The overhairs are long and usually straight, but may have a single constriction. The underhairs are short and kinked in several places. The underhairs make up more than four-fifths of the total, but because of their lack of prominence they do not play an important part in determining the overall colour of the animal. In fine structure the 2 types of hair are essentially similar: a wide central medulla is surrounded by a narrow cortex, which in turn is surrounded by a thin cuticle (Fig. 1a). The cells comprising all 3 layers are keratinized, and compactly cemented together. They are irregular in shape, loosely arranged and interspersed with air spaces in the medulla, solidly packed in the cortex and flattened into translucent scales in the cuticle (for details see Grüneberg, 1952).

The hair is formed in the hair follicle. This begins as an epidermal invagination into the dermis. The dermis forms a thickening directly

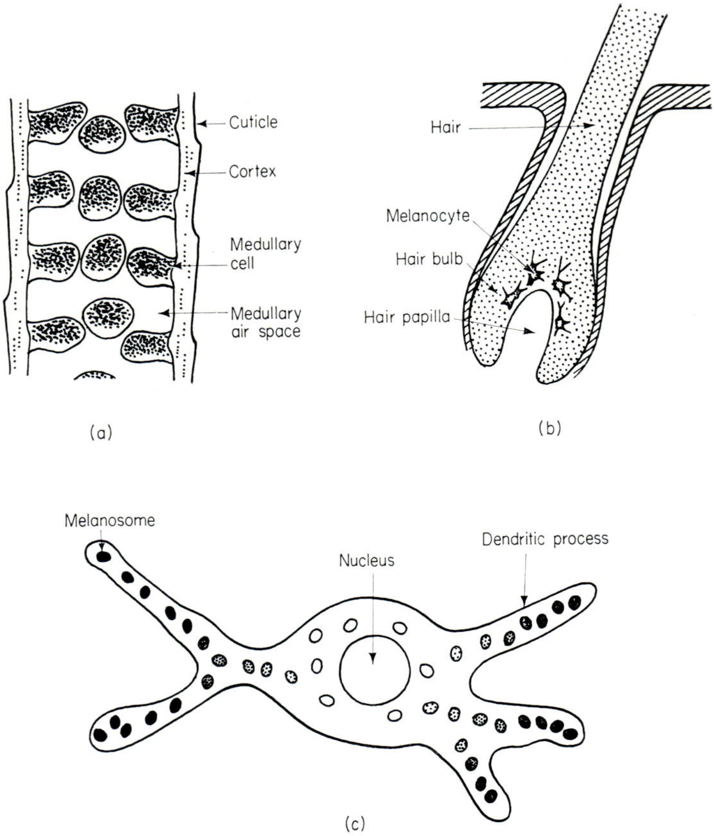

Fig. 1. (a) Longitudinal section of a hair. (b) Longitudinal section of a hair follicle. (c) Normal melanocyte with melanosomes at different stages of melanization. All drawings schematic.

underneath, and the blind end of the invagination comes to surround it partly. The dermal thickening develops into the hair papilla, and the surrounding part of the invagination into the hair bulb (Fig. 1b). The hair grows at the base of the follicle, and the whole coat is periodically renewed by moulting.

Melanin

The colouring matter in the hair is the pigment melanin, which is found in the medulla and the cortex. Melanin is of 2 kinds: eumelanin, which appears black or brown, and phaeomelanin, which appears yellow or reddish yellow. Eumelanin is an indole-quinone polymer of high molecular weight, and is always attached to some protein. It is formed from tyrosine. This colourless amino acid is converted into 3,4-dihydroxyphenylalanine, commonly called dopa, by the enzyme tyrosinase, a copper-containing oxidase. Dopa is next oxidized to dopa-quinone by the same enzyme. Further oxidations and intramolecular arrangements convert dopa-quinone into indole-5,6-quinone, which polymerizes to form eumelanin. No enzyme seems to be involved in these later steps. The mode of formation of phaeomelanin is less well understood, but it seems that tyrosinase again plays an essential role. It is possible that the precursor substance in this case might be tryptophan rather than tyrosine. There is no clear evidence for this, but a tryptophan-oxydizing enzyme is known to occur in the mouse skin.

The melanin polymers formed in this manner are not released into the cytoplasm of the cell in which they are synthesized, but are deposited on minute bodies called melanosomes, to the protein matrix of which they become attached by their quinone linkages (extensive references in Foster, 1965).

The melanosome

The melanosome is a specialized cellular organelle, generally oval in shape. It probably originates in the Golgi zone as a minute spherical vesicle, which develops an internal membranous structure as it enlarges and elongates. The internal membrane or membranes are heavily folded or wound so that in cross-section in an electron micrograph they appear like strands running lengthwise. Melanin is deposited on both sides of this membrane, not in a uniform layer, but at regularly spaced points. In all probability its synthesis also occurs at the same points. In normally pigmented animals these points are so numerous and the deposition of the pigment so heavy, that the internal membranous structure is completely obscured, and the melanosome appears as a uniformly pigmented body. Under the light microscope it is seen as the pigment granule.

The melanocyte

The synthesis of melanin and formation of the melanosomes take place within specialized cells, the melanocytes. They occur in a number of tissues, including the hair follicles. With the exception of those in the retina, which originate in the optic cup, the melanocytes are derived from the neural crest. The differentiation of a portion of the neural crest cells into melanoblasts (immature melanocytes) takes place at about the time of the closing of the neural tube, i.e. about the 9-day stage. It begins in the cranial region, and proceeds posteriorly. The melanoblasts then migrate to their destinations, those of the epidermis, with which we are primarily concerned, reaching theirs by the 12-day stage. When the follicles form a few days later, the melanoblasts migrate into them.

The melanocytes of the mature follicles occur in the bulb. They give the hair its colour not by becoming incorporated into the growing hair in any appreciable numbers, but by acting as unicellular glands. They secrete pigment granules or fully developed melanosomes into the cells of the growing hair at its base. When growth of hair in any cycle is complete, i.e. during the telogen stage, they become inactive, amelanotic and difficult to identify. They appear again soon after the beginning of the anagen stage of the next hair cycle. The fate of the follicular melanocytes after the completion of one cycle, and the origin of those taking part in the next cycle are not well understood.

The normal or wild type melanocytes in the mouse are stellate in shape, with numerous thick dendritic processes. The melanosomes are found scattered throughout the cell, including the processes (Fig. 1c). This form is described as nucleofugal (for details see Billingham and Silvers, 1960).

Factors affecting hair colour

The colour of the hair depends on the following factors:

1. The amount of pigment in the medullary cells, and to a lesser extent in the cortical cells, of the hair.
2. The type of pigment, whether eumelanin or phaeomelanin.
3. The distribution of pigment in the hair.
4. The size and shape of the melanosomes or pigment granules.
5. The distribution of pigment granules in the cells.

GENETIC CONTROL OF COAT COLOUR

We can see from the foregoing section that a gene affecting any step from the differentiation of the neural crest to the polymerization of melanin could influence coat colour. For instance, if a mutation results in a deficiency of the melanoblasts, or in an impairment of the melanocyte function, or the replacement of one type of pigment by another, or a change in the shape, size and distribution of the melanosomes, or an interference in melanin synthesis, the colour of the coat is almost certain to be changed. In fact, genes affecting coat colour in most of these ways are known. Over 35 independent loci are involved, covering 16 of the 20 chromosomes. The total number of mutant genes at these loci approaches 100. This is much greater than the number affecting any other trait, but it does not mean that coat colour is peculiarly subject to gene action. It is merely a reflection of the ease with which coat colour mutations are detected. A full account of all these loci is beyond the scope of this brief review (it may be found in Deol, 1963; Green, 1966; Searle, 1968). Only those loci will be considered that have a relatively high mutation rate and form the basis of coat colour differences in the most familiar strains of laboratory mice. Spotting genes will be treated as a single group, because although they occur at distinct loci they have certain important features in common.

Genes at the albino locus

This locus governs the synthesis of tyrosinase, and so determines the intensity of pigmentation. Both eumelanin and phaeomelanin are affected, but not always to the same extent. There are 6 known alleles at this locus, differing in the amount of tyrosinase produced. The most familiar ones are the wild type allele producing full colour (C), chinchilla (c^{ch}) and albino (c). The names of the genes are descriptive of the coat colour produced by them. Transplantation experiments have shown that the genes at this locus act within the melanocyte, i.e. the genotype of the melanocyte determines the amount of pigment it produces and not the genotype of the follicle in which it occurs. The spontaneous mutation rate at this locus is $10{\cdot}2 \times 10^{-6}$ (Schlager and Dickie, 1967).

Genes at the agouti locus

This locus controls the regional distribution of the 2 types of melanin in the hair by means of some switch mechanism which can change the type of pigment produced by the follicular melanocytes from eumelanin to phaeomelanin and *vice versa* in a very short time. Thirteen alleles are known at this locus, the most familiar ones being yellow (A^y), light-bellied agouti (A^w), grey-bellied agouti (A), which is the wild type one, and non-agouti (a). The agouti pattern consists in a sub-apical or apical yellow band in a black hair, and what the various alleles do is to change the width of this band, the extreme states being wholly yellow hair in $A^y/-$ mice and wholly black hair in a/a mice. These genes also determine the relative amounts of eumelanin and phaeomelanin on the back and the belly. The site of action of the genes at this locus is the hair follicle rather than the melanocyte. The switch mechanism operates through a change in the physiological milieu of the melanocytes. The same melanocyte can produce eumelanin or phaeomelanin depending on the physiological state of the follicle. The spontaneous mutation rate at this locus is $29{\cdot}7\times10^{-6}$. (It may be mentioned that this is a complex locus, there being some recombination between certain alleles.)

Genes at the brown locus

This locus is primarily concerned with the shape and structure of the melanosome. It probably is responsible for the synthesis of one of the protein sub-units of this organelle. Four alleles are known at this locus, the common ones being light (B^{lt}), black (B) and brown (b). Black, the wild type allele, leads to black oval melanosomes with an internal structure resembling a rolled membrane. The mutant allele b changes the shape of the melanosome to round and gives the internal structure the appearance of a distorted network. It is probable that secondary consequences of these alterations include changes in the activity of tyrosinase and in the number of the sites of melanin deposition or their effectiveness. This would account for not only the change in the colour of the eumelanin from black to brown but also in its texture, for while black eumelanin appears dense and finely grained the brown form appears loose and coarse. The B^{lt} allele, in addition, causes a slight clumping of the melanosomes, and leads to the disappearance of the melanocytes from the follicle before the hair has stopped growing, so that the base of the hair is unpigmented. Phaeomelanin does not seem to be affected to any appreciable extent. The genes at this locus act within the melanocyte. The spontaneous mutation rate is $3{\cdot}9\times10^{-6}$.

Genes at the dilute locus

This locus governs the distribution of the melanosomes in the melanocytes (and so also in the medullary cells of the hair). It has 5 known alleles. The wild-type allele is full colour (D), and the best known mutant one is dilute (d). In $D/-$ mice the melanosomes are scattered fairly evenly throughout the cell, but in d/d mice they occur in clumps, as a result of which the normal black agouti colour is diluted to bluish grey. The change in the distribution of the melanosomes appears to be a consequence of an alteration in their structure. This locus also affects the morphology of the melanocytes. In $D/-$ mice these cells have the typical nucleofugal appearance described in the preceding section. In *d*/*d* mice the melanocytes are nucleopetal: the dendritic processes are fewer and thinner, and the melanosomes tend to crowd round the nucleus, leaving the processes free. Both eumelanin and phaeomelanin are affected. The genes at this locus act within the melanocyte, although their action may be modified by the environment of the cell. The spontaneous mutation rate is $12{\cdot}5 \times 10^{-6}$.

Genes at the pink-eye locus

This locus is concerned with the structure of the melanosome. It has 7 known alleles. The wild-type allele is full colour (*P*) and the best known mutant one is pink-eye (*p*). The melanosomes in *p*/*p* mice have a shred-like appearance and a disorganized internal structure. It is believed that the abnormality lies in one of the protein sub-units of the melanosome, which either itself provides the binding sites for the melanin polymers or affects the capacity of the melanosome to bind melanin in some other way. The result is that the melanosomes carry much less pigment. There seem to be indications that the early stages of melanin synthesis are not affected. At any rate, there is no detectable deficiency of tyrosinase. There is also a slight tendency towards clumping of the melanosomes in *p*/*p* mice. Eumelanin is affected much more severely than phaeomelanin by the activity of the locus, so that the coat colour in *p*/*p* mice is yellow or orange. The genes at this locus act within the melanocyte. The spontaneous mutation rate is $0{\cdot}78 \times 10^{-6}$.

Spotting genes

At least 10 loci are concerned with white spotting, the number of known alleles varying from 2–8 per locus. The commonest spotting genes are piebald (*s*), dominant spotting (*W*), splotch (*Sp*) and patch (*Ph*). The effects of a particular gene may not be precisely the same in every animal, but in general its pattern of spots is distinguishable from that of others. They all have one important feature in common: no melanocytes, normal or abnormal, can be identified in the white regions. At its maximum, spotting leads to an all-white phenotype. This kind of white animal is to be regarded as having a single large spot and not as an albino, for in albino animals melanocytes are present in the follicles, although they are amelanotic on account of their failure to synthesize tyrosinase. Another difference between these 2 types of white mice is that while in albino animals no pigment is found in any part of the body, in all-white animals produced by spotting genes some pigment always occurs in the eye or the inner ear. As to the manner of origin of the white spots, there is evidence that some genes affect the neural crest, so that either the melanoblasts are not formed or they do not migrate in the normal manner or fail to differentiate into melanocytes on reaching their destination (Mayer, 1965). In other cases it appears that the neural crest is not affected, but some regional property of the skin is altered so that the melanoblasts cannot differentiate there (Mayer and Maltby, 1964). It is possible that some other factors also come into the picture. It may be significant that while there is only one locus for albinism there are so many for white spotting. Some types of spotting appear to be polygenic in origin. The spontaneous mutation rate at the splotch (*Sp*) locus is $0{\cdot}78 \times 10^{-6}$, and at the dominant spotting (*W*) locus $3{\cdot}52 \times 10^{-6}$.

Interaction between different loci

The colour of the coat is the product of all the loci that affect it, and the results of changes at a single locus cannot be discussed in isolation. If no reference is made to the other loci the assumption is that they carry wild-type alleles. If 2 or more loci are involved simultaneously the effects of each one may be quite different on account of their interaction. The results of all possible interactions can only be imagined, but we may here discuss a simple situation involving only 3 loci (agouti, brown and pink-eye) with no more than 2 alleles (A, a; B, b; P, p) from each.

Genotype	Coat colour
A/A; B/B; P/P	Agouti (wild type)
a/a; B/B; P/P	Black
A/A; b/b; P/P	Cinnamon
A/A; B/B; p/p	Orange or yellow
a/a; b/b; P/P	Chocolate
a/a; B/B; p/p	Blue-lilac
A/A; b/b; p/p	Orange or yellow
a/a; b/b; p/p	Cafe-au-lait

COAT COLOUR POLYMORPHISM IN WILD POPULATIONS

Colour genes observed in wild populations

The agouti locus is the only one with regard to which widespread polymorphism has been discovered in wild populations of mice (Philip 1938; Schwarz and Schwarz, 1943; Engles, 1948; Dunn, Beasley and Tinker, 1960; Petras, 1967). A wide variety of shades of the ventral fur, ranging from almost white to very dark, has been reported. However, it appears that only the light-bellied agouti (A^w) and dark-bellied agouti (A) alleles are concerned, and the different shades result from the action of different modifying genes. This type of polymorphism is so common that both A^w and A have equal claims to being regarded as the wild-type alleles for this locus. (Schwarz and Schwarz, 1943, have suggested that A^w is the wild-type allele for truly feral populations and A for populations commensal with man.) There have also been reports of differences in the colour of the back, some populations being appreciably darker and others lighter than normal (Jameson, 1898; Dunn *et al.*, 1960; Clegg, 1963). But again there is no evidence that any other allele is involved. In all probability, these variations are also the result of modifying genes affecting the expression of A^w and A alleles, perhaps by causing slight alterations in the width and position of the yellow band in the agouti hair. They could also be reflections of variations in hair structure or of a shift in the proportions of different types of hair.

Polymorphism due to genes at the pink-eye locus has also been observed in one population. Brown (1965) discovered pink-eyed mice with yellow fur in a granary on a farm in Missouri, U.S.A., which on crossing with appropriate laboratory stocks proved to be homozygous for the allele *p*. The proportion of the mutant mice in the population was 30·3%.

Spotting of various types, especially on the belly, has also been observed in wild populations. Littleford (1946), Dunn *et al.* (1960) and Petras (1967) have reported it from the U.S.A., Zimmermann (1941) from Germany, and Keeler (1933) from Turkey. It is not known which locus was involved in the populations investigated by Dunn *et al.* and Petras, or indeed whether the spotting was hereditary. In the other populations the piebald locus (allele *s*) seems to have been implicated. In the German population the frequency of spotted animals was 70%.

Colour genes and natural selection

Considering the large number of loci involved in the determination of coat colour and their relatively high rates of spontaneous mutation, the degree of coat colour polymorphism observed in wild populations is astonishingly low. This is generally attributed to natural selection. It is believed that selection operates through predators hunting by sight, favouring animals that are inconspicuous against their background. There is considerable evidence in support of this view. Dice (1947) exposed to owls buff and grey deer-mice, differing with regard to genes at a single locus, on soils of the same 2 colours alternately. Animals on matching soil always survived in larger numbers. In Brown's (1965) study of the population with *p*/*p* mice referred to above, cats were allowed access to the granary at a certain stage. They selectively eliminated the yellow coloured homozygotes, which appeared again after the cats had been removed. The prevalence of light-coloured mice in places with light-coloured soil, such as Bull Island in Dublin Bay (Jameson, 1898), points in the same direction.

However, it is possible that this aspect of selection may not be all. It is difficult to imagine that the agouti pattern provides the best protective coloration in all types of habitats and that all mutations at coat colour loci would result in making the animal appreciably more conspicuous: for instance, a cinnamon coat due to a mutation from B to b at the brown locus might be equally inconspicuous on some types of backgrounds. Moreover, as mentioned in the preceding sub-section, the coat colour is far from uniform in wild mice, but the variation always seems to be achieved through modifying genes rather than mutations at the known colour loci. These considerations suggest that natural selection may also operate on the pleiotropic effects of the colour genes. Of all the loci considered here, the albino locus is the only one which may possibly be concerned solely with pigmentation. The rest of them seem to affect some other developmental pathway as well, there being at least one allele at each locus with demonstrable side effects. At the agouti locus the yellow (A^y) allele, apart from being lethal in the homozygous condition, leads to gross obesity. At the brown locus the allele b causes increase in body size, at least on some genetic backgrounds (Grüneberg, 1952). The allele d at the dilute locus also increases body size. Another allele at the same locus, d^l (dilute-lethal), causes extensive demyelinization in the central nervous system. At the pink-eye locus the allele p^s (pink-sterile) causes abnormal behaviour and retardation, in addition to sterility. As to the spotting loci, mutations at many of them cause nervousness, megacolon and various degrees of deafness, which in cases of severe spotting may be total (Deol, 1970). As some of these mutations are known to affect the neural crest, from which a number of structures are derived, it is probable that they also cause other abnormalities not yet detected. The genes mentioned above have pleiotropic effects, often manifestly deleterious, that are obvious or easy to demonstrate. The possibility cannot be excluded that some other genes may also have similar effects which would become apparent only in the wild under competitive conditions.

In view of these considerations, it is conceivable that one of the reasons for the absence of extensive polymorphism in wild populations with regard to coat colour is that the vast majority of coat colour mutations have adverse effects on the physiological balance of the animal. The most favourable physiological state would then be associated, as far as coat colour loci are concerned, with the presence of the wild-type alleles at all of them, which would result in a black agouti coat. As black agouti is also the wild type coat colour in a number of other rodents and related species, the assumption that it is associated with the most favourable physiological state becomes stronger, for it could hardly be the most protective colour in all cases.

References

Billingham, R. E. and Silvers, W. K. (1960). The melanocytes of mammals *Q. Rev. Biol.* **35**, 1–40.

Brown, L. N. (1965). Selection in a population of house mice containing mutant individuals. *J. Mammal.* **46**, 461–465.

Clegg, T. M. (1963). Observations on an East Yorkshire population of the House Mouse (*Mus musculus* Linn.). *Naturalist, Hull*, No. 885, 39–40.

Deol, M. S. (1963). Inheritance of coat colour in laboratory rodents. In *Animals for Research*, W. Lane-Petter, (ed.) pp. 177–198. London: Academic Press.

Deol, M. S. (1970). The relationship between abnormalities of pigmentation and of the inner ear. *Proc. R. Soc.* (B.) **175,** 201–217.

Dice, L. R. (1947). Effectiveness of selection by owls of deer-mice (*Peromyscus maniculatus*) which contrast in color with their background. *Contr. Lab. vertebr. Biol. Univ. Mich.* **34**, 1–20.

Dunn, L. C., Beasley, A. B. and Tinker, H. (1960). Polymorphisms in populations of wild house mice. *J. Mammal.* **41**, 220–229.

Engles, W. L. (1948). White-bellied house mice. *J. Hered.* **39**, 94–96.

Foster, M. (1965). Mammalian pigment genetics. *Adv. Genet.* **13**, 311–339.

Green, M. C. (1966). Mutant genes and linkages. In *Biology of the Laboratory Mouse,* 2nd ed. E. L. Green, (ed.), pp. 87–150. New York: McGraw Hill.

Grüneberg, H. (1952). *The genetics of the mouse.* The Hague: Martinus Nijhoff.

Jameson, H. L. (1898). On a probable case of protective coloration in the house mouse (*Mus musculus*, Linn.). *J. Linn. Soc.* (*Zool*) **26**, 465–473.

Keeler, C. E. (1933). Akhissar spotting in the house mouse. *Proc. natn. Acad. Sci. U.S.A.*, **19**, 477–481.

Littleford, R. A. (1946). Occurrence of piebald spotting in a wild house mouse. *Am. Nat.* **80**, 283–288.

Mayer, T. C. (1965). The development of piebald spotting in mice. *Devl. Biol.* **11**, 319–34.

Mayer, T. C. and Maltby, E. L. (1964). An experimental investigation of pattern development in lethal spotting and belted mouse embryos. *Devl. Biol.* **9**, 260–286.

Petras, M. L. (1967). Studies of natural populations of *Mus*. III. Coat color polymorphisms. *Can. J. Genet. Cytol.* **9**, 287–296.

Philip, U. (1938). Mating systems in wild populations of *Dermestes vulpinus* and *Mus musculus*. *J. Genet.* **36**, 197–235.

Schlager, G. and Dickie, M. M. (1967). Spontaneous mutations and mutation rates in the house mouse. *Genetics, Princeton* **57**, 319–330.

Schwarz, E. and Schwarz, H. K. (1943). The wild and commensal stocks of the house mouse, *Mus musculus* Linnaeus. *J. Mammal.* **24**, 59–72.

Searle, A. G. (1968). *Comparative genetics of coat colour in mammals.* London: Logos Press.

Zimmermann, K. (1941). Some results of genetical analysis in populations of wild rodents. *Int. Conf. Genet.* **7**, 332.

Symp. zool. Soc. Lond. (1970) No. 26, 251–269.

HOMOLOGOUS MUTANTS IN MAMMALIAN COAT COLOUR VARIATION

ROY ROBINSON

St. Stephens Road Nursery, London, England

SYNOPSIS

Inter-species homology of gene loci gives a rationale for the interpretation of saltational variation of mammalian coat colour. The majority of colour mutants can be interpreted in terms of a few loci and the alleles thereof. Despite this, certain variants cannot be so easily explained. Some of these may be phenocopies and represent non-genetical variation; others will be unique mutants, i.e. mutants possessing no obvious counterpart in other species. However, the range of homologous mutants is steadily widening and it may be conjectured that few truly unique mutants will remain as the number of known mutants per species become greater.

The main categories of coat colour variation are outlined. Six loci are mainly involved, 3 to a greater extent than the others, as judged by the number of known mutant alleles which these possess. A number of other loci produce similar mutants in different species and this has afforded opportunity to broaden the discussion. A general synthesis involving mutants at 10 loci recurring among 73 species is attempted.

The existence of mimic genes creates difficulties but true mimics are rare. Critical comparison of mimic phenotypes usually discloses that differences, often subtle, do exist. Mimic genes will presumably exist in every species if only because each character is controlled by numerous inter-dependent loci. It may be useful to study the pattern of such mimicry between species, especially those which defy separation.

Morphism is displayed by many species but, in general, this seems to be of a simple kind. The majority are melanisms, followed by erythrisms. At the present time, little is known of the genetic basis to the various forms and this deserves to be rectified. This task would perform 3 useful functions: (1) provide genetic information on species not usually bred in the laboratory, (2) lead to a better understanding of the nature of genes involved in morphism, and (3), hopefully, provide greater understanding of the factors involved in the maintenance of morphism.

INTRODUCTION

Inter-species gene homology is not one of the central problems of genetics, but it certainly should rank as an important peripheral topic. It is important because it has relevance for several aspects of mammalian population genetics. Three of these may be specifically mentioned. It provides a guide to : (1) the nature of variant forms or uncommon segregants which are encountered from time to time, (2) the genetical basis of morphism, in which a variant form co-exists at a relatively high frequency with the wild-type, and (3) sub-specific or racial variation.

The concept of homologous genetical variation is not new. It occupied the attention of such diverse minds as Wright (1917), Vavilov

(1922), and Haldane (1927). However, it may come as a surprise to discover just how extensively the concept can be applied to coat colour variation in mammals. The application may lack the precision of work with *Drosophila* species (e.g. Spassky and Dobzhansky, 1950; Sturtevant 1929; Sturtevant and Tan, 1937; Sturtevant and Novitski, 1941) but this is compensated for by far greater breadth. Altogether, some 73 species can be sensibly compared (Table III), and this number can be greatly exceeded if every likely instance of homology is included.

The first step in gene homology is to categorize types of variation which recur in different species, without too much concern for the mode of inheritance. This course is practicable—rather than penetrating—and hence not altogether satisfactory. It would be desirable to pass beyond this stage and to formulate hypotheses of the probable genetical nature of the observed variation. The degree of confidence with which this can be accomplished depends upon the type of variation. However it should not be overlooked that the problem is fascinating *per se* and deserves to be pursued in its own right.

CRITERIA FOR HOMOLOGY

The presumptive basis of homology is the descent of loci capable of producing similar alleles in a variety of species and derived from a common ancestor. The more distant the relationship, as established by other criteria, the greater the presumed antiquity of the loci. There may be a correlation between the degree of importance of the locus for the mammalian genome and the frequency with which it occurs among species. This assumes, of course, that loci can be "lost" but not "gained" in the course of evolution. This assumption could very well be false, especially in view of the plausible concept of gain of genetic material by means of "gene duplication" (Ohno, 1967).

Several criteria of varying degrees of usefulness can be proposed for establishing homology. The most direct method would be that of interspecies crosses. If a recessive gene with a similar phenotype in two species, produces the same phenotype in the hybrid, this would be evidence for homology. The problem is more complicated, of course, for dominant genes. This method has proved successful for *Drosophila* (Sturtevant, 1929) but has limited application for mammals. However, it might be possible to transcend the species barrier by the aid of non-genetical techniques, but using genetically significant material (such as tissue transplants or *in vitro* physiological and chemical analysis).

Similarity of phenotype will always be the mainstay of homology even if the matching of some genes carries less conviction than others.

Two problems arise from this approach. The first is that exact matching cannot always be expected, and the second is the existence of mimic genes. As the number of known mutant genes increases, so, presumably, will the number and range of mimics. This aspect is likely to increase in importance and is discussed in detail below. At present, suffice to say that identity is often spurious, since differences between apparent mimics can frequently be detected upon close scrutiny.

At the opposite end of the scale is the situation where only one mutant of a particular phenotype is known for several species. The inference is that these species have only one locus capable of producing the mutant. Hence, the likelihood of homology is correspondingly enhanced. This criterion would not be worth mentioning but for the remarkable situation that only one brown mutant is known in a number of species. It will be interesting to see just how long this situation will hold. At this time, only 2 species have more than one gene producing a brown or brownish phenotype, for example, the Golden hamster and the American mink.

At first sight, the criterion of similar linkages between putative homologues would seem to be promising, especially when it is recalled that linkage between the homologues for albinism and pink-eyed dilute in 3 species (House mouse, Deer mouse, Norway rat) was an early genetical discovery. Alas, this is still the only reliable known instance, if the gratuitous cases of sex-linkages are excluded! The existence of mimics can confuse the issue of linkage, both positively and negatively. Since only the House mouse has been investigated for linkage in any depth, only the future will tell just how useful this criterion will be.

The final criterion seems to possess most potential. This is the existence of similar alleles at a locus. The greater the number of alleles, the greater will be the power of discrimination. At present, the agouti and albino loci (and to a lesser extent, the extension) are specially instructive: a large number of species have more than one mutant allele at these loci. In time, it may be possible to separate mimic genes by the nature of their alleles. Yet a warning must be sounded, because the House mouse has 2 loci with a series of mimicking alleles (variegated dominant spotting and steel). Mutant genes at these loci can be differentiated by appropriate analysis (see Searle, 1968*a*), but not in a manner which is dependent upon their allelic characteristics.

MIMIC GENES

One of the difficulties of matching genes is the existence of genes at different loci with similar phenotypes—so-called "mimic" genes.

At first sight, it may seem that these genes would make the matching of genes a hazardous business. However, on detailed examination, few genes turn out to be precise mimics of each other. Nevertheless, some element of doubt may remain if only because an exact inter-species match cannot always be expected. The same gene in different species is necessarily in a different genome and this implies that the phenotype may not be identical.

At this time, the existence of mimic genes can be regarded as a nuisance. Close mimicry may present difficulty but once resolved it will lead to greater insight into gene physiology, expression and possibly interaction. If the genetics of the House mouse is a valid yardstick, mimic genes will be common for all mammalian species. The extent and nature of mimic genes in certain groups and their apparent absence or variety in others will then become important. Perhaps this is looking ahead a great deal but the prospects will doubtless be as exciting (if not more so) as the discovery of a long conjectured but non-realized case of homology of today.

For example, mimicking genes may occur as the result of two processes which reduce to one. Most characters are the culmination of the effects of many loci and a mutation at any one would modify the expression. The majority would probably produce different modifications but this need not be so. The idea of a character being controlled by several loci could be advanced as a fact of observation or it could be made more precise. It could be associated with the concept of gene duplication, where a genome increases in scope and versatility by steady accretion of genetical material small enough to avoid serious disturbance to the chromosome mechanism or to genome function.

Duplication is one way by which the genome acquires new genetic material. That is, mutation *sensu stricto* is not the only mechanism by which genomes evolve. The gain of material by minute duplication could result in the presence of "spare" loci which in time could take over new functions. The loci could persist *in situ* to build up a cistron or it could be shifted to new sites by means of translocation. The duplication *per se* could increase the possibility of translocation by providing "weak points" in the chromosome. Translocation may in fact be part of the mechanism by which the original function of genes is modified. This possibility may not be too far-fetched because adjacent loci do influence one another and the juxtaposition of strange loci could assist in the acquisition of new properties. However, the "spare" loci presumably are still capable of mutation and mutant alleles arising thus could appear as mimics at the phenotypic level.

Therefore, in addition to the simple plotting of the pattern of mimicking genes between species, the distribution of these upon the chromosomes may turn out to be significant, particularly, should it be possible to trace certain groupings for different species with various degrees of relationship.

MORPHISM

The great majority of coat colour genes in wild populations occur as rare segregants but a number apparently flourish as constituents of polymorphism. This fact gives these mutants an added interest. Melanism is the most common morphism, followed by erythrism and leucism, the last probably requiring rather special conditions. Huxley (1955), and Searle (1968*a*) have reviewed most of the known morphisms in mammals and have shown how varied and opportunistic these may be. Local conditions determine the frequency of a particular morph and clines are a common feature.

Table I lists the more well known examples of morphism in mammals. The list is not exhaustive, and is selective in that only those cases thought to be basically monogenic are included. The various categories of melanism, erythrism, etc., are rather broadly interpreted to include, for example, umbrous agouti under melanism and wide-band agouti under erythrism. As the various morphisms become more intensively studied, so it is hoped that their genetical basis can be more accurately described. A number of the smaller North American rodents appear to have extremely local morphisms and the significance of these may repay investigation. It may be mentioned that in the towns and cities of man, populations of the domestic cat exhibit extensive morphism of coat colour and this is currently the subject of considerable attention (Todd, 1966; Dreux, 1967, 1968; Robinson and Silson, 1969).

From the present standpoint, morphism is of interest because it affords a ready means of obtaining information on the nature and genetics of colour variation for species other than those commonly bred in the laboratory. Certain technical problems may have to be surmounted but the effort is worth making. For example, it would be of first importance to discover if a case of melanism is due to a recessive or dominant gene. Given this information, coupled with minute examination of the phenotype, it should be possible to draw conclusions of the probable homology. Just to describe a dark morph as melanistic is insufficient. A completely black phenotype leaves little more to be said, but small details may show that it is really umbrous agouti. The darkening may be either predominantly in the dorsal region as a spinal

TABLE I

Summary of better known morphisms

B = brown form; E = erythrism; L = leucism; M = melanism

Species and vernacular name	Morphism	Local populations
Marsupialia		
Phalangeridae		
Trichosurus vulpecula (Brush opossum)	M	Tasmania
Lagomorpha		
Leporidae		
Lepus timidus (Alpine hare)	M	Ukraine
Oryctolagus cuniculus (Rabbit)	M	Tasmania, UK, etc.
Rodentia		
Sciuridae		
Sciurus carolinensis (Grey squirrel)	M	USA
Sciurus vulgaris (Red squirrel)	M	Finland and USSR
Marmota flaviventris (Golden-mantled marmot)	M	USA
Geomyidae		
Thomomys townsendii (Western pocket gopher)	M	USA
Cricetidae		
Cricetus cricetus (European hamster)	M	Germany and USSR
Ondatra zibethica (Muskrat)	M	Southern USA
Peromyscus maniculatus (Deer mouse)	E, M	USA
Peromyscus polionotus (Old-field mouse)	L	Florida, USA
Neotoma albigula (White-throated pack rat)	M	Southern USA
Onychomys leucogaster (Grasshopper mouse)	M	USA
Arvicola amphibius (Water vole)	M	Scotland
Muridae		
Rattus norvegicus (Norway rat)	M	USA and elsewhere
Rattus rattus (Black rat)	M	Northern Europe and America
Praomys morio (Harsh-furred rat)	M	Uganda
Carnivora		
Canidae		
Vulpes vulpes (Red fox)	M	N. America and Russia
Alopex lagopus (Arctic fox)	L	Greenland and Siberia
Mustelidae		
Putorius putorius (Polecat)	E	Wales and USSR
Ursidae		
Ursus americanus (American black bear)	B	USA
Felidae		
Herpailurus jaguaroundi (Jaguarondi)	E	S. America
Panthera pardus (Leopard)	M	S.E. Asia
Puma concolor (Puma)	E	USA
Viverridae		
Herpestes brachyurus (Short-tailed mongoose)	E	Malaysia

stripe or be an overall darkening. Further, the darkening may take the form of a narrowing of the agouti band or the replacement of banded by unbanded hairs. The erythristic and other morphisms also deserve critical examination. Most erythristic morphs will probably be due to *e* type morphs but the possibility of a wide-band morph should not be overlooked.

It may be remarked that widely separated morphisms, even if similar in appearances, may not be genetically identical. This is actually an appeal for ecologists, field naturalists and game wardens, who may uncover a case of morphism, to make a determined attempt to ascertain the heredity of the atypical form. Morphism is no longer a novel phenomenon and any study should be regarded as incomplete if either a detailed description or the mode of heredity cannot be given.

AGOUTI AND ITS VARIATIONS

One of the most ubiquitous coat colours in mammals is agouti. To many people this conjures up a picture of the drab coloured brown-grey of the House mouse or Brown rat. However, the agouti coat can exist in a wide variety of forms, often accompanied by black or white markings. The unifying feature is the presence of banded hair. Not all of the hairs need be banded (the stout guard hairs are often unbanded) so long as numerous undercoat fibres have bands. The agouti hair is black tipped but appears blue in the proximal portion (undercolour) where the pigment granules becomes less numerous. Interrupting the transition to blue is a band of yellow, usually variable in position, width and intensity for different parts of the body but typifying the agouti coat. The black pigment granules are eumelanin while the yellow granules are phaeomelanin, each with fundamentally different properties. The two types are produced by the same melanocytes, the switch from the production of one sort to the other occurring abruptly, smoothly and regularly during hair growth (Cleffmann, 1963).

Eumelanic pigment granules tend to appear an invariant black or deep brown-black. It is as if the pigment exceeds a threshold of maximum light absorption. On the other hand, phaeomelanin granules do not possess this property but tend to traverse the yellow-red spectrum of light reflection as their number increases. The outcome is considerable variation which ranges from pale cream to red. This variation usually follows the dorsal-ventral gradient but it may be subject to regional modification according to species.

The whole body surface has the potential to produce eumelanin—witness the almost universal occurrence of black forms. Usually, the dorsal-ventral gradient, and any other regional differences in pigment production, is overwhelmed. However, several observations indicate that it persists. Mutants which interfere with the production of pigment (such as pale chinchilla of the rabbit) produce a remarkable shaded phenotype showing the gradient even in conjunction with black. Those species which have a dark striped or spotted pattern overlaying the agouti show clearly the persistence of the regional difference. Bright sunlight or flash photography can often bring out ghost tabby or spotted patterns in the otherwise black domestic cat or leopard.

The regional distribution of phaeomelanin deposition is most evident in yellow or erythristic mutants. In these, the hairs are entirely yellow, being most intensely yellow sub-apically and paling to slightly bluish or white towards the skin. The dorsal-ventral gradient, normally partially obscured by eumelanin in the agouti, is now exposed. In those species with marked regional overlays of darker pigment, such as the tabby of the domestic cat, the overlap persists as an intense erythristic pattern (the orange or red tabby cat). Erythristic leopards and tigers have a comparable phenotype. Some agouti species with regional black markings fail to show regional variation of yellow pigmentation in erythristic forms. The usual reason for this is that the black markings are due to local absence (or narrowing) of the agouti band. This is a timely reminder that the darkness (or brightness) of an agouti phenotype is often dependent upon the width of the agouti band, without an increase in amount of deposited pigment.

Most rodents and lagomorphs are agouti as regards the type form. Many species in other orders are also agouti but there are notable exceptions. In some, the agouti banded hair has been much modified but is still discernible. A fair number of species have evolved a brown-grey phenotype without the use of an agouti band. It is possible that the effect is created by pigment granules with reduced pigment content. In others, it seems to have given way to a black or even to a dark fuscous brown phenotype. The displacement of agouti may take place by several mechanisms and any monogenic interpretation, bereft of breeding evidence, should be viewed with caution. For those people who do favour a monogenic interpretation, the genetic situations discussed for melanic morphism are likely possibilities (full melanism such as non-agouti or one of the umbrous agoutis).

Most reddish wild type phenotypes owe their appearance to increases in the width of the agouti band (so that the amount of black pigment is effectively reduced) or to increases in the amount of phaeomelanism. In almost every instance where it has been possible to study the heredity of intensity of yellow pigment, the answer has been polygenic. Mutations at the *E* locus are essentially "switch" genes changing eumelanin into phaeomelanin. It is doubtful if many erythristic species have incorporated a yellow (*e*) allele into the wild genome although it may be admitted that the process is not impossible. Erythristic morphism is a special case and may often, if not always, be due to *e* mutants. Even so, it may be advisable to ascertain if a "wide-band" mutant is not involved. These can closely simulate an *e* allele if the band is sufficiently wide.

In contrast to erythristic species are those devoid of phaeomelanin pigment. Complete elimination is ordinarily shown by the South American chinchillas, American mink and probably by a few other species. It is not always easy to be certain that the ability to produce phaeomelanin has been lost. An *e* mutant in such a species could produce either a dark eyed white animal or a light sandy or cream coloured phenotype indicative of rather feeble pigment production! The deciding factor would be the amount of pigment produced in the normal eumelanic areas (in certain sections of the individual hair fibres or regionally over the body). When this is converted to phaeomelanin by the action of an *e* type allele, the amount of pigment may become visible as a sandy or beige phenotype. Such phenotypes have been reported for *Chinchilla lanigera*.

Of more frequent occurrence are species which either fail to produce yellow pigment, or only produce it weakly, in some regions of the body. Again, there are problems of detection, should the coat have obscuring eumelanin pigmentation, although an *e* type phenotype would reveal the true situation. It is of interest that the presence of colourless or weakly coloured phaeomelanin granules or protogranules in the white areas may be demonstrable by suitable techniques. In some species, the so-called white areas can assume a pale yellowish tinge. That is, there is variability of expression which would be impossible if the ability to produce phaeomelanin was completely absent.

The preceding ideas may be profitably amplified by consideration of the arresting coat patterns possessed by certain North American ground squirrels and chipmunks. The colour is basically agouti but is interrupted by longitudinal stripes of dark and light pigmentation. Both the dark and light stripes are variable between species and probably within species. The dark stripes vary in intensity all the way to black while the light stripes vary from white to yellowish. As regards the eumelanin, the variation can be most simply explained by marked regional variation in the width of the agouti band. The band may be absent in the black stripes and of maximal width for the light, according to species. Some of the black stripes could be due to localized concentrations of all-black hairs, with no decrease in band width of intermingled agouti hairs. This possibility could be checked.

Widening of the agouti band is probably the cause of the yellow or light stripes although an alternative is that there are concentrations of completely non-eumelanic hairs. In some species, the stripes are yellowish, in others, these are very pale indeed, if not completely white. Concomitant with the evolution of light striping, the amount of phaeomelanin pigment in the same areas has declined.

The reticulated pattern of the giraffe has the appearance of being a light, rather than white, pattern. There is some variation, and this has been adopted as a ready means of racial distinction, although the variation may occur intra-racially as well as inter-racially (Guggisberg, 1969). Despite this, there is no disruption in its unique character. However, recently, a giraffe has been bred utterly devoid of reticulation (Ueno Zoo, 1968 and personal communication) and this is suggestive of monogenic control of the reticulation *in toto*.

The zebra is another example of a pattern of white markings peculiar to a closely related group of species. It is very distinctive and immediately recognizable. The Common zebra has variable markings, partly in the width of the white stripes but largely in their configuration. However, the British daily newspaper, *Daily Mirror* for 3rd January 1968, featured a specimen in which the stripes were reduced to short bars and dots. The report stated that it was the only animal of its kind in the herd.

The skunks, the badger and zorilla are examples of black animals with prominent and distinctive (aposematic) white markings. The markings are variable in the Striped skunk but are never entirely lost as a rule (Verts, 1967). However, non-spotted animals have been reported for the skunk and badger and, although these may represent extreme variants of a polygenic complex governing the markings, they may be indicative of a major locus for the development of the markings independent of their grade of expression. It has been suggested that the markings could be due to fixation of white spotting genes but this seems improbable because they occur in unusual positions and too regularly to be due to any of the well known white spotting mutants.

In summary, the above examples are suggestive that unique and distinctive patterns common to certain species groups are subject to genetic modification by genes with major effect. Where the markings can be removed at one stroke, this could be attributed to the action of a "switch-gene" since this behaviour is independent of variation in the markings themselves. The giraffe case affords an excellent example since racial variation of the reticulation is well known. Another system of control of a distinctive pattern is that of a series of alleles. Only one locus is concerned but the alleles involve distinctive patterns. At least three varieties of the tabby pattern in the domestic cat are basically produced in this manner.

It is a temptation to suppose that the white markings of certain species may be due to white spotting genes comparable to known laboratory mutants. This is feasible, of course, but the number of instances where this explanation can be offered with any degree of confidence is small. It is doubtful if the skunk pattern, for example, can be explained in this manner for reasons given previously. The Cape hunting dog and possibly some of the blotched type marsupials seem the likeliest possibilities but, even so, doubt remains.

DISCUSSION

It is not possible in a short article to cover all of the coat colour mutants which may arise in a species. To do so would produce a book

TABLE II

Meaning of symbols for gene loci cited in text and in Table III

Symbol	Designation	Remarks
A	Agouti	One of 2 loci (other *E*) mainly responsible for agouti. Main alternatives, black and tan or black.
B	Black	Black pigment. Main alternative brown.
C	Colour	Full manifestation of yellow and black pigment. Alternatives, absence of yellow (chinchilla), acromelanic albino or full albino.
D	Dilute	Dark pigment. Main alternative, blue.
E	Extension	Normal expression of agouti. Main alternatives, dominant black, bicolour or yellow.
P	Pink-eye	Normal pigmentation. Alternative, red eye and pale coat.
S	Piebald	Normal. Alternative, white spotting (trace to extensive white areas).
Si	Silver	Normal. Alternative, variable quantity of white hairs.
U	Umbrous	Normal. Alternative, darkening of agouti.
*W**	Wide-band	Normal. Alternative, increase in width of agouti band.

*In the House mouse, this symbol has been used to denote a "dominant white mutant".

TABLE III

Summary of known mutant homologues in mammals

The numerals indicate the number of probable homologous alleles for each locus for the species. Italics mean probable homology rather than reasonable certainty.

Species	*A*	*B*	*C*	*D*	*E*	*P*	*S*	*Si*	*U*	*W*
Insectivora										
Erinaceidae										
Erinaceus europaeus (Hedgehog)	—	—	1,*1*	—	—	—	—	—	—	—
Soricidae										
Blarina brevicauda (Short-tailed shrew)	—	—	1	—	1	—	1	—	—	—
Sorex cinereus (Masked shrew)	—	—	1,*2*	—	—	—	—	—	—	—
Sorex minutus (Pygmy shrew)	—	—	1	—	1	—	—	—	—	—
Sorex obscurus (Dusky shrew)	—	—	1	—	1	—	—	—	—	—
Talpidae										
Talpa europaea (Common mole)	—	1	1, *1*	—	1	*1*	—	—	—	—
Chiroptera										
Phyllostomatidae										
Carollia perspicillata (Short-tailed bat)	—	—	—	—	1	—	—	—	—	—
Vespertilionidae										
Myotis lucifugus (Little brown bat)	1	—	—	—	—	—	1	—	—	—
Myotis sodalis (Cluster bat)	—	—	—	—	—	—	1	—	—	—
Scotophilus castaneus (Lesser yellow bat)	—	*1*	—	—	*1*	—	—	—	—	—
Molossidae										
Tadarida brasiliensis (Mexican free-tailed bat)	—	—	1	—	—	—	1	—	—	—
Lagomorpha										
Leporidae										
Lepus europaeus (Brown hare)	—	*1*	2	—	1	—	1	—	—	—

Lepus timidus (Alpine hare)	1	—	1	—	1	—	—	—	—	—
Oryctolagus cuniculus (Domestic rabbit)	3	1	3	1	3	1	1	1	—	1
Sylvilagus floridanus (Eastern Cottontail)	1	—	—	—	1	—	—	1	—	—
Rodentia										
Sciuridae										
Citellus richardsonii (Richardson ground squirrel)	—	—	1	—	1	—	—	—	—	—
Citellus tridecemlineatus (Thirteen striped squirrel)	1	—	*1*	—	—	—	—	—	—	—
Marmota monax (Woodchuck)	1	—	1	—	—	*1*	—	—	—	—
Sciurus carolinensis (Grey squirrel)	1, *1*	—	1	—	1	—	—	—	—	—
Sciurus niger (Fox squirrel)	1	—	1	—	—	—	1	—	—	—
Sciurus vulgaris (Red squirrel)	1	*1*	1	—	1	—	1	—	—	—
Tamiasciurus hudsonicus (American red squirrel)	1	—	1	—	2	—	1	1	—	—
Tamias striatus (Eastern chipmunk)	1	—	1	—	—	—	—	—	—	—
Geomyidae										
Geomys bursarius (Eastern pocket gopher)	—	—	1	—	—	—	1	—	—	—
Thomomys bottae (Valley pocket gopher)	1	*1*	1	*1*	1	*1*	1	—	—	—
Thomomys talpoides (Northern pocket gopher)	1	—	—	—	1	—	—	—	—	—
Cricetidae										
Clethrionomys glareolus (Bank vole)	*1*	—	—	—	1	—	—	—	—	—
Cricetus cricetus (European hamster)	*1*	—	—	—	1, *1*	—	1	—	—	—
Mesocricetus auratus (Golden hamster)	—	1	1	—	1	1	1	—	—	—
Microtus agrestis (Short-tailed vole)	*1*	—	—	—	—	1	1	—	—	—
Microtus arvalis (Common vole)	1	—	2	—	—	1	1	1	—	—
Microtus californicus (Californian meadow mouse)	—	—	1	—	1	—	—	—	—	—
Microtus ochrogaster (Prairie vole)	1	—	1	—	—	—	1	—	—	—
Microtus pennsylvanicus (Meadow mouse)	1	1	1	—	1, *1*	1	1	1	—	1
Ondatra zibethica (Muskrat)	2	1	1	—	*1*	1	—	1	—	—
Peromyscus maniculatus (Deer mouse)	*1*	1	1	1	1	1	1	1	—	1
Pitymys pinetorum (Pine vole)	—	—	1	*1*	1, *1*	—	—	—	—	—
Sigmodon hispidus (Cotton rat)	—	*1*	1	—	—	—	*1*	—	—	—

TABLE 3 (*continued*)

Species	*A*	*B*	*C*	*D*	*E*	*P*	*S*	*Si*	*U*	*W*
Muridae										
Apodemus sylvaticus (Wood mouse)	2	—	1	—	*1*	—	—	—	—	—
Mus musculus (House mouse)	4	1	3	1	2	1	1	1	1	—
Rattus norvegicus (Norway rat)	1	1	2	1	1	1	1	1	—	—
Rattus rattus (Black rat)	2	*1*	1	*1*	2	*1*	—	—	—	—
Caviidae										
Cavia porcellus (Guinea pig)	2	1	2	—	2	1	1	1	—	—
Chinchillidae										
Chinchilla lanigera (Chinchilla)	1	*1*	—	*1*	—	—	—	—	—	—
Capromyidae										
Myocastor coypus (Copyu)	1	*1*	2	1	1, *1*	1	—	—	—	—
Carnivora										
Canidae										
Canis domesticus (Dog)	1	1	1, *1*	—	2, *1*	*1*	1	1	—	—
Canis latrans (Coyote)	1	—	1	—	—	—	1	—	—	—
Vulpes vulpes (Red fox)	*1*	—	1	—	*1*	—	—	—	—	—
Ursidae										
Euarctos americanus (Black bear)	1	1	1	*1*	—	—	1	*1*	—	—
Procyonidae										
Procyon lotor (Raccoon)	1	*1*	1	—	1	*1*	—	—	*1*	—
Mustelidae										
Gulo luscus (Wolverine)	1	—	1	—	*1*	—	—	—	—	—
Lutra canadensis (American otter)	—	—	1	—	—	—	—	*1*	—	—
Martes americana (American marten)	1	—	*1*	1	1	—	—	*1*	—	—
Martes martes (Pine marten)	—	*1*	—	1	—	—	—	—	—	—
Martes pennanti (Fisher)	—	*1*	1	—	*1*	—	1	—	—	—
Martes zibellina (Russian sable)	*1*	*1*	1, *1*	—	*1*	—	1	*1*	—	—

Meles meles (Badger)	1	—	1	—	1	—	—	—	—	—
Mephitis mephitis (Striped skunk)	—	1	1	1	1	*1*	—	—	—	—
Mustela vison (American mink)	1	—	1, *1*	*1*	—	*1*	—	—	—	—
Putorius furo (Ferret)	—	*1*	—	—	1	—	—	—	—	—
Putorius putorius (Polecat)	—	—	—	—	1	—	—	—	—	—
Felidae										
Felis catus (Domestic cat)	1	1	2	1	*1*	*1*	1	*1*	—	—
Lynx rufus (Bob cat)	1	—	*1*	*1*	—	—	—	—	—	—
Panthera pardus (Leopard)	1	—	*2*	—	*1*	—	—	—	—	—
Panthera tigris (Tiger)	1	—	1, *1*	—	1	—	—	—	—	—
Puma concolor (Puma)	1	—	1	—	*1*	—	—	—	—	—
Perissodactyla										
Equidae										
Equus asinus (Ass)	1	1	—	—	—	—	1	—	—	—
Equus caballus (Horse)	1, *2*	1	*1*	—	2	—	1	*1*	—	—
Artiodactyla										
Suidae										
Sus scrofa (Pig)	1	—	*1*	1	3	*1*	1	—	—	—
Bovidae										
Bos taurus (Cattle)	*1*	*1*	2	*1*	2	*1*	1	—	—	—
Bubalus bubalus (Waterbuffalo)	*1*	1	—	*1*	1	—	1	—	—	—
Capra hircus (Goat)	2	—	—	—	*2*	—	1	—	—	
Ovis aries (Sheep)	1	*1*	—	—	1	—	1	1	—	—

the size of Searle's (1968a), and this should be consulted by anyone seeking more examples. The House mouse has the greatest range of known mutant forms and it is obvious that most comparisons should commence with the species (Grüneberg, 1952; Green, 1966; Wolfe and Coleman, 1966). For carnivores, Poland (1892) is an interesting source. Unfortunately, detailed description alone is often not sufficient to identify comparable mutants but it would be noted that Wolfe and Coleman (1966) have contributed 16 coloured photographs of mutant mice which may be of value in this respect.

Table III presents a survey of reported mutants (or simulated phenotypes) in mammals. The survey is incomplete in that only selected loci (described in Table II) are considered but there are sufficient to indicate that homologous phenotypes (if not homologous mutants) are to be found in all of the main orders. Rodents and lagomorphs provide the most reliable genetical information because many are bred in the laboratory. Matching of mutants in these is easy because most of the recurring mutants are to be found, but also complex because of mimic genes and the presence of genes which seem peculiar to a species.

None of the species of bats in the table have many known variants but it is interesting that as a group they are producing forms comparable to known mutants of rodents. The ungulates present the most problems for homology. It is conceivable, of course, that they may contain genes not present in other orders. On the other hand, (1) it is difficult to perform critical experiments with many species, and (2) there is a lack of unanimity among specialists as to which genes may or may not be present (*cf*. Odriozola, 1951; Castle, 1954; Rendel, 1957; Lauvergne, 1966).

Gaps in the distribution of mutant alleles between species have promoted speculation whether these may have significance. Such speculation is understandable but it seems that many may be upset by subsequent events. For example, until recently, in the House mouse there was no recessive yellow nor dominant black known. Now both genes have been recognized (Searle 1968*b*). Similarly no acromelanic allele was reported in mice until 1961, although it had been known in other rodents for many decades. Correspondingly, full albinism has not yet been described in the Guinea pig and Golden hamster.

Sex-linked orange in the domestic cat was a unique gene until 1966 when a comparable mutant was found in the Golden hamster (Robinson, 1966). Lethal yellow at the agouti locus in the House mouse is unique or nearly so (the suggested occurrence of a similar gene in *Microtus arvalis* is doubtful), but the occurrence of viable yellow in the same species has a possible counterpart in the sable gene of the dog. In both species, a yellow phenotype is produced with variable amount of dark hairs. The dog contains an oddity at the agouti locus, a gene producing a black phenotype which is dominant to other members of the series instead of being recessive as in other species. It is possible to continue with lesser known examples, but pointless until a larger number of mutants have been recorded for a greater range of species.

However, it does seem possible that certain loci—as distinct from certain alleles—may turn out to be restricted in distribution. One probable instance has emerged already and this is the tabby locus. If the genetics of the domestic cat can be taken to be a guide to coat colour variation in the Felidae, one locus seems to be primarily responsible for the various striped, rosetted and spotted patterns of the family (R. Robinson, unpublished). The interesting aspect is that these patterns seem rare outside the family (although there is some expression of this type of pattern in jackel, genet and raccoon species in related families. Similarly, the giraffe and zebra patterns may be primarily determined by one locus having a restricted distribution. The important point is that the vast range of mammalian coat colours is slowly being brought into some sort of rational order which is not simply descriptive but based upon a genetic appraisal.

It may be wondered if one or more mutant genes may have become established as the wild type of a particular species. Only one instance would seem *prima facie* convincing. The chinchillas (*Chinchilla brevicauda* and *C. lanigera*) have a coat devoid of phaeomelanin and it is possible that these are homozygous for one of the chinchilla alleles of the laboratory. Against this is the fact that genes at other loci can remove phaeomelanin, so the case for a chinchilla allele is not conclusive The *lanigera* chinchilla is being bred in captivity. Several mutants have already been discovered and it is possible that this intriguing situation will be resolved. Another interesting possibility is that of the Cape hunting dog (*Lycaon pictus*) with a phenotype of irregular patches of black, yellow and white. Here, not one but 2 genes occurring as mutants in other species are *prima facie* involved. This case has peculiar features and it is in the balance whether this is a true explanation. Schwarz (1937) proposes that the Sumatran hare (*Nesolagus netscheri*) and the South American opossum (*Chironectes minimus*) may carry a black/yellow bicolour gene. Nachsteim (1958) suggests that a similar gene may be a species character of the European hamster (*Cricetus cricetus*). However, these cases can be more plausibly explained in terms of regional modifications of agouti to produce dark markings.

References

Castle, W. E. (1954). Coat colour inheritance in horses and other mammals. *Genetics, Princeton* **39**, 35–44.

Cleffmann, G. (1963). Agouti pigment cells *in situ* and *in vitro*. *Ann. N.Y. Acad. Sci.* **100**, 749–761.

Dreux, P. (1967). Gene frequencies in the cat population of Paris. *J. Hered.* **58**, 89–92.

Dreux, P. (1968). Gene frequencies in the cat population of a French rural district. *J. Hered.* **59**, 37–39.

Green, M. C. (1966). Mutant genes and linkages. In *Biology of the laboratory mouse*. Green, E. L. (ed.). New York: McGraw-Hill.

Grüneberg, H. (1952). *The genetics of the mouse*. The Hague: Nijhoff.

Guggisberg, C. A. W. (1969). *Giraffes*. London: Barker.

Haldane, J. B. S. (1927). Comparative genetics of colour in rodents and carnivora. *Biol. Rev.* **2**, 199–212.

Huxley, J. S. (1955). Morphism and evolution. *Heredity* **9**, 1–52.

Lauvergne, J. J. (1966). Genetique de la couleur du pelage des boivins domestiques. *Biblphia. genet.* **20**, 1–68.

Nachtsheim, H. (1958). Problems of comparative genetics in mammals. *Int. Congr. Genet.* **10**, **1**, 187–198.

Odriozola, M. (1951). *A los colores de caballo*. Madrid.

Ohno, S. (1967). *Sex chromosomes and sex-linked genes*. Berlin: Springer Verlag.

Poland, H. (1892). *Fur bearing animals in nature and in commerce*. London: Gurney and Jackson.

Rendel, J. (1957). Nedarvning av farg och teckning hos husdjur. *K. Skogs-O. LantbrAkad. Tidskr.* **96**, 207–263.

Robinson, R. (1966). Sex-linked yellow in the Syrian hamster. *Nature, Lond.* **212**, 824–825.

Robinson, R. and Silson, M. (1969). Mutant gene frequencies in cats of Southern England. *Thereot. Appl. Genet.* **39**, 326–329.

Schwarz, E. (1937). "Japanese" pattern in wild mammals. *Proc. zool. Soc. Lond.* **107**A, 127.

Searle, A. G. (1968*a*). *Comparative genetics of coat colour in mammals*. London: Logos Press.

Searle, A. G. (1968*b*). An extension series in the mouse. *J. Hered.* **59**, 341–342.

Spassky, B. and Dobzhansky, T. (1950). Comparative genetics of *Drosophila willistoni*. *Heredity* **4**, 201–215.

Sturtevant, A. H. (1929). Genetics of *Drosophila simulans*. *Carnegie Instn Wash. Publ.* No. 399, 1–62.

Sturtevant, A. H. and Novitski, E. (1941). Homologies of chromosome elements in the genus *Drosophila*. *Genetics, Princeton* **26**, 517–541.

Sturtevant, A. H. and Tan, C. C. (1937). Comparative genetics of *Drosophila pseudo-obscura* and *melanogaster*. *J. Genet.* **34**, 415–432.

Todd, N. B. (1966). Gene frequencies in the cat population of New York City. *J. Hered.* **57**, 185–187.

Ueno Zoo (1968). Brown calf giraffe. *Int. Zoo Yb.* **8**, 134.

Vavilov, N. I. (1922). The law of homologous series in variation. *J. Genet.* **12**, 47–89.

Verts, B. J. (1967). *The biology of the striped skunk*. Urbana: University of Illinois Press.

Wolfe, H. G. and Coleman, D. L. (1966). Pigmentation. In *Biology of the laboratory mouse*. E. L. Green, (ed.). New York: McGraw-Hill.

Wright, S. (1917). Color inheritance in mammals. *J. Hered.* **8**, 224–235.

Symp. zool. Soc. Lond. (1970) No. 26, 271–282.

TAIL TIP AND OTHER ALBINISMS IN VOLES OF THE GENUS *ARVICOLA* CLAÉPÈDE 1799

D. MICHAEL STODDART*

University of Oxford, Department of Zoology
Animal Ecology Research Group, Oxford, England

SYNOPSIS

Study skins and field reports of voles of the genus *Arvicola* were examined for the presence of small partial albinisms. These were found to occur in 3 main sites on the body: the tip of the tail; the crown of the head; and the junction of the neck and chest (called "throat"). They were found to occur quite commonly, although all 3 albinisms were never found together in one population.

The presence of these characters and frequencies of their occurrence do not appear to bear any simple relation to latitude, altitude or winter temperature, and in this respect the phenomenon does not appear to help solve the problem of winter whitening in northern species.

Albinisms occur more commonly in Britain than in Eurasia, although the range of observations about the mean occurrence of white tail tips is less in Britain than in Eurasia, and for white crowns is more in Britain than in Eurasia. Variation in frequency of the characters between populations within one particular region was found to be considerable.

The genetic control of coat colour is complex, and may be influenced by the pleiotropic effects of genes responsible for other quite unrelated characters. Albinisms in the 3 areas concerned could arise from the action of one gene responsible for the internal environment of the melanoblasts migrating to the epidermis from the neural crest of the embryo, or alternatively, from the pleiotropic effects of 3 alleles, only 2 able to be represented in any phenotype. The great variation in frequencies observed can be explained if the populations are small inbreeding demes. Ecological research supports this suggestion.

It is tentatively suggested that Britain was colonized originally by either one pregnant female vole, or a small band of *émigrés* producing a "founder" effect.

INTRODUCTION

Most studies on variation between mammal populations have been based on examinations of skeletal material, particularly of the skull (Delany and Healy, 1964). The genetical control of coat colour in rodents has been well investigated for some laboratory species (Searle, 1968), but little is known about other wild rodents. This study was conducted with the object of identifying any trends in the frequency of occurrence of white patches in the pelage of voles of the genus *Arvicola* Lacépède 1799, and to see to what extent populations varied in respect of "partial albinism".

In 1963, Corbet published a short review of the frequency of tail tip albinism in 18 species of British mammals, and it appeared from

*Present address: Department of Zoology, University of London King's College, London, England.

TABLE I

A list of the samples examined, their locations, and the frequencies of occurrence (%) of the three major albinisms

Sample No. (a)	Location (b)	No. of specimens in the sample (c)	Frequency of occurrence of white tail tips (d)	Frequency of occurence of white crowns (e)	Frequency of occurrence of white throats (f)
1*	South Finland	6	0	0	0
2	Nordmore, Möre og Romsdal, Norway	8	62·5	0	0
3	Bovra, Opland, Norway	2	0	0	0
4	Sandefjord, Vestfold, Norway	2	0	0	0
5 →	Hollum, Mandal, Vest Agder, Norway	11	72·8	0	18·2
6	Stockholm and Uppsala, Sweden	3	0	0	0
7*	North Denmark	2	0	0	0
8* →	Netherlands	217	16·4	0	0
9 →	Slains, Aberdeenshire, Scotland (a)	18	33·3 } 31·0	89·0 } 92·3 +	0
10 →	Slains, Aberdeenshire, Scotland (b)	28	28·6 }	— }	—
11	Slains, Aberdeenshire, Scotland (c)	7	—	71·2 }	—
12 →	Slains, Aberdeenshire, Scotland (d)	112	—	95·5 }	0
13 →	Eilean Gamhna, Argyll, Scotland	14	†66·6	7·2	—
14 →	Northern England	20	40·0	4·55	—
15 →	Southern England	75	20·0	1·3	—
16 →	Norfolk, England	34	55·9	20·6	—
17 →	Oxfordshire, England	19	52·6	0	0
18* →	Nr. Stuttgart, S.W. Germany	74	2·7	0	0
19	Zubureč, N.E. Hungary	4	25·0	0	0
20	Nr. Linz, Austria	2	0	0	0
21 →	Pyrenees	15	0	6·7	6·7
22 →	Aragon and Castellon, Spain	61	0	1·6	0
23* →	S. Moravian lowlands, Czechoslovakia	150	0	0·7	0
24 →	Slovenia, Jougoslavia	25	0	0	0
25 →	Koraband, Kopaonik mts., Jougoslavia	10	20·0	0	0

26*→	Roumanian lowlands	62	0	0	0
27*→	Transylvanian alps	87	0	0	3·4
28 →	Foggia and Rome, Italy	12	0	0	25·0
29 →	Vaud and Berne, Switzerland	219	22·8	0·91	—
30 →	Bulle, Fribourg, Switzerland	157	33·6	0·70	—
31 →	East Switzerland	14	79·0	0	—
32 →	Swiss/French Alps	26	0	0	0
33	Vol'sk, Govt., Saratov, RSFSR	6	0	0	0
34	Estonia, SSR	5	40·0	0	0
35	Yakutsk, Yakutskaya, ASSR	2	50·0	0	0
36	Arkhangel'sk RSFSR	1	0	0	0
37	Alzamay, Irkutskaya, RSFSR	6	0	100·0	100·0
38	Irkutsk, Irkutskaya, RSFSR	3	0	33·3	66·6
39*	Altai mts., Altayskiy Kray, RSFSR			F	
40*	Barabinsk, Novosibirskaya, RSFSR		F		F
41*	Taz River, Tyumenskaya			F	
42*	Tomsk, Tomskaya, RSFSR			NU	
43*	Volga basin, Astrakhan		R	R	R
44*→	Pakhra River, Moscow	63	7·9	—	—
45*‡	Ossetia, Georgia				
46*	Semipalatinsk, Kazakhstan, SSR		S		
47 →	Semiretschenskoi, Kirgiziya, SSR	11	100·0	9·1	0

Note:—The above table contains information on two species and many subspecies of *Arvicola*.

Symbols used in Table I

- → Sample used in further analyses.
- * Samples not personally examined.
- † Only 3 specimens in this sample were available for examination of this character.
- ‡ Reported by Ognev (1964) that some specimens have pure white bellies.
- — No records available for this sample.
- F Frequent.
- NU Not uncommon, R Rare, S Some: Best possible estimates obtainable from published literature.
- \+ Samples 9 and 12 only.

this study (based on material deposited in the British Museum (Natural History), London), that this form of partial albinism occurred more frequently in the genus *Arvicola* than in any other. His data for the Common shrew (*Sorex araneus*) suggest that the occurrence of white tail tips is (just) significantly greater in the north of Britain than in the south. He says (p. 327) "Since the tip of the tail is presumably the coldest part of the body, this phenomenon seemed to have some relevance to the problem of winter whitening in northern species".

MATERIAL AND METHODS USED IN THE INVESTIGATION

Study skins and field reports from 47 localities throughout Eurasia were considered for the analyses. They are listed in Table I, and their geographical positions are shown in Fig. 1. I examined material from 32 locations, and information was sent to me from the remaining 15 locations, these being indicated in the table with an asterisk. The partial albinisms looked for were (1) white tail tip, (2) white crown, and (3) white "throat". Other albinisms also occur, e.g. white feet, and sometimes large white flank patches producing a piebald effect (Service, 1903), but they did so only infrequently in the samples. They were not considered in the analyses on the grounds that they may have been

FIG. 1. Map showing the locations of samples mentioned in Table I. (Lambert Azimuthal Equal Area Projection.)

TABLE II

Frequency of occurrence of partial albinisms

Albinism	% occurrence in samples examined	No. of samples examined
Tail tips	52·4	38
Crowns	41·9	38
"Throats"	20·0	30

selectively preserved in the collections on account of their obvious peculiarity. On the whole the other partial albinisms are inconspicuous, and thus the samples can be considered to be representative of the populations from which they are taken. The material I examined is lodged in the British Museum (Natural History), except the material from Slains, Aberdeenshire (sample nos. 9, 10, 11 and 12), Eilean Gamhna, Argyll (sample no. 13), Norfolk (sample no. 16) and Oxfordshire (sample no. 17) which are in my possession. The samples from Switzerland (nos. 29, 30 and 31) were examined by myself, and are in the Station Fédérale de Recherches Agronomiques, Nyon, Switzerland. The material was all examined in good light by eye, but with the aid of a x10 magnifying glass. Overall commoness of albinisms is shown in Table II.

Partial albinisms range quite widely in extent. A white tail tip may be anything from a few white hairs in the terminal hair pencil, to a whole white end, up to 1 cm in length. White crowns are usually small, occupying less than 4 mm^2, but occasionally may cover up to twice this area. The position of this albinism is apparently constant, as is the position of the "throat" albinism. This occurs at the base of the neck at the top of the chest, and may occupy 1 cm^2, although half this is more usual.

THE GENETICAL BASIS OF WHITE SPOTTING AND PARTIAL ALBINISM

Albinism is caused by the lack of the pigment melanin in the follicles of the hairs. The pigment granules are produced in the hair shafts by melanocytes, which in turn are derived from melanoblast cells present in the neural crest of the young embryo. Wendt-Wagener (1961) showed for rats that if the migration of the melanoblasts from the neural crest to the epidermis was delayed by one or two days only, the phenomenon of "hooding" resulted, with only the head and foreflanks bearing pigmented hairs. Searle (1968) quoting the work of Schaible (1963) describes 8 pigment centres in the mouse, 6 bilateral and 2 median. These are shown diagrammatically in Fig. 2. If the melanoblasts are held up in their migration, then white patches occur in those regions farthest from the centres of pigmentation, i.e. the tail tip, the throat and mid-ventral region, the feet and the crown of the head.

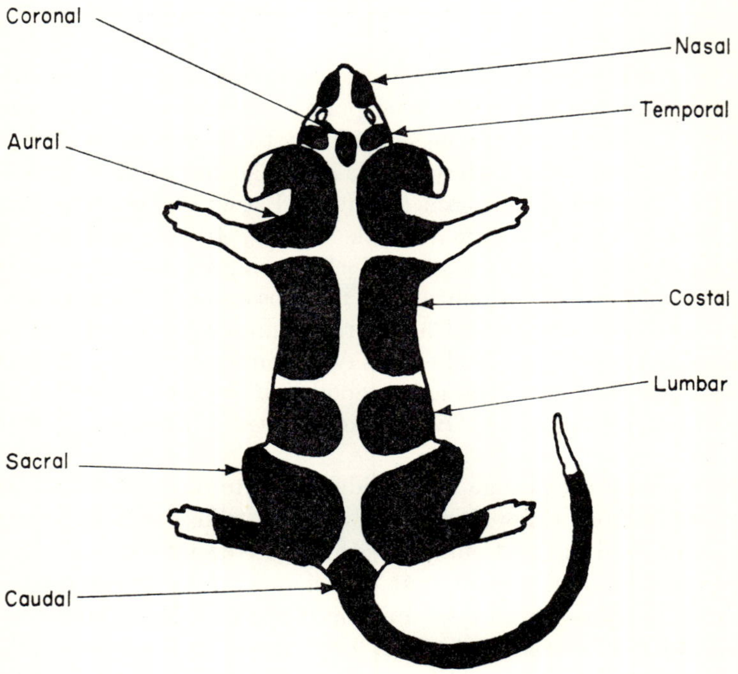

FIG. 2. Diagram of a laboratory mouse, showing 6 bilateral and 2 median centres of pigmentation (from Searle, 1968, after Schaible).

Several gene loci are responsible for irregular white spotting in mice, and many are pleiotropic in action. For example, *W* is lethal when homozygous, and produces white extremities and occasional haemopoietic disturbances when heterozygous. Other genes may be responsible for slowing down the rate of migration of the melanoblasts. A multiplicity of white spotting factors have been found to occur in guinea pigs, rats, rabbits and mice, and probably occurs in many other rodent species.

THE FREQUENCY OF OCCURRENCE OF PARTIAL ALBINISM

Results

In the following analyses only data based on samples of 10 specimens or more were used—these are indicated in Table I. Several of the data were most unsatisfactory in that the frequencies of partial albinism were merely recorded as "frequent", or "not uncommon" etc., (see sample nos. 39–43, 45 and 46). Nevertheless they are included in this paper for the sake of completeness, and may serve to indicate the regions from which sound data are required. In Table I are shown the frequency of occurrence of the 3 albinisms. They show that tail tip and crown albinisms are commoner in Britain than in Eurasia. These data are summarized in Table III.

TABLE III

Mean occurrence of tail tip and crown albinisms in Britain and Eurasia

Region	Mean occurrence of white tail tips %	Mean occurrence of white crowns %*
Britain	38·8 (n = 6) (range 20·0–55·9)	20·9 (n = 6) (range 0·0–92·3)
Eurasia	20·8 (n = 17) (range 0·0–100·0)	2·4 (n = 11) (range 0·0–9·1)

* Since this character is sometimes difficult to detect, I have used only data collected by myself.

An analysis of homogeneity for these characters using a χ^2 test shows that for tail tips Britain is more homogeneous than Eurasia, but less so for crowns (Table IV).

TABLE IV

*Homogeneity of regions for two albinistic characters**

Region	Tail tips		Crowns	
	χ^2 value	n–1 no. of observations	χ^2 value	n–1 no. of observations
Britain	14·7	5	288·9	4
Eurasia	122·6	11	8·7	3

* Instead of rejecting data giving an expected value of less than five, such data have been lumped together, and considered as one observation.

Variation between populations within each of the regions was found to be quite considerable. In Fig. 3 are shown the frequencies of occurrence of albinisms for several locations in Britain and 4 in Eurasia. These examples have been chosen since it is certain that all the specimens in each sample come from the same population and were collected at the same time.

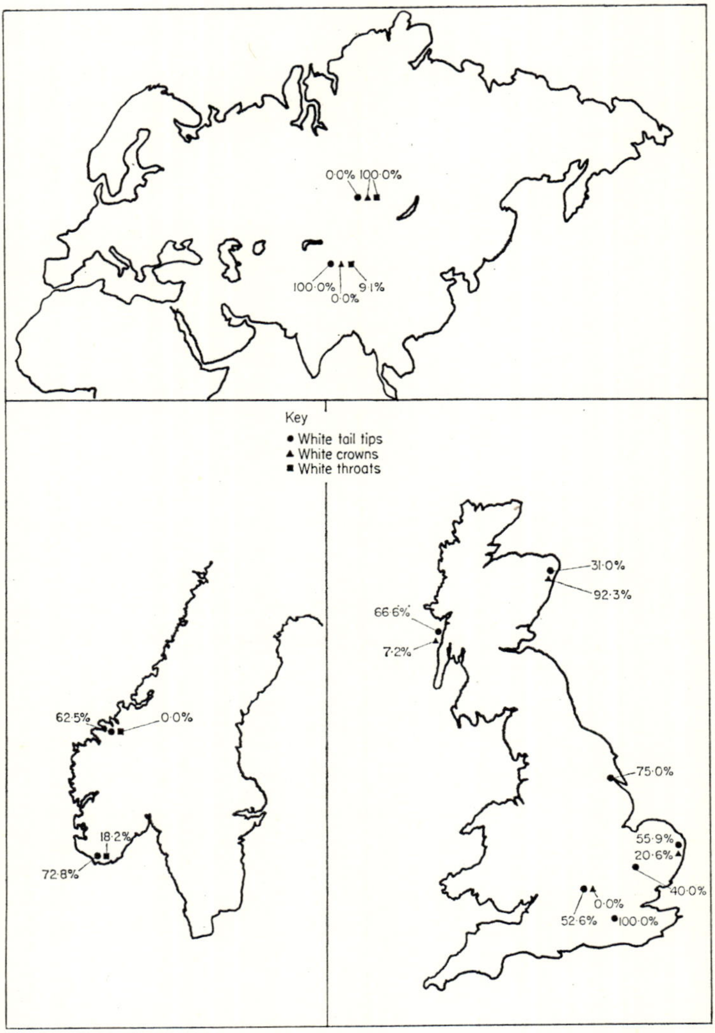

FIG. 3. Diagram showing the frequencies of occurrence of partial albinisms in populations.

DISCUSSION

There appears to be no correlation between the presence of partial albinisms and either high altitude or high latitude. Table I shows examples of frequent albinisms and low altitude (e.g. Norfolk, England), and rare albinisms and low altitude (e.g. Roumanian lowlands). Similarly examples occur of frequent albinisms and high latitude (e.g. Nordmore, Norway), and of rare albinisms and high latitude (e.g. South Finland). Likewise, winter temperature does not appear to be directly correlated with frequency of albinisms (e.g. frequent albinisms and low winter temperature, Switzerland, and frequent albinisms and high winter temperature, Scotland). It seems as if the occurrence of albinisms is genetically controlled, a suggestion that is backed up by laboratory breeding of animals taken from areas of high albinisms, which show the presence of these characters after several generations in captivity (A. Meylan, personal communication). It would be wrong to dismiss the effects of the environment on account of an absence of correlation of albinisms with the 3 environmental factors considered. However, Corbet's (1963) interpretation of white tail tips in *S. araneus* does not seem to hold for *Arvicola*.

The adaptive significance of these albinisms, *per se*, is not plain. There is no evidence from ecological or behavioural studies on the genus that visual signalling is in any way used as a means of communication. Indeed it would seem unlikely to be so in a genus with such poor eyesight and such a highly developed olfactory communication system. There are as yet no data available on the different rates of predation on populations with different frequencies of occurrence of white patches. If individuals with white patches were more at risk than those without, then the genes responsible would soon be eliminated from the population. Genes with absolutely no selective advantage will be gradually eliminated, but it is theoretically possible, as Sewell Wright has pointed out, for a gene conferring no selective advantage on the phenotype to increase in frequency under certain circumstances. For this to occur the breeding unit must be small. Segregation and recombination involve an element of chance, and the frequency of different alleles in a random mating population may vary considerably from generation to generation. Furthermore, small and isolated breeding units allow a rapid increase in homozygosity, due to consanguineous matings. *Arvicola* populations appear to be organized into small demes, and the large amount of genetic variation indicated above supports the idea that they are relatively isolated. The data from Norfolk (sample no. 16), in addition to the 3 albinisms mentioned, show a significant frequency

(11·8%) of white spots on the rump (see Fig. 2). This may be no more that a pleiotropic effect of a chance mutation, but only a considerable restriction of gene flow could have established the character at such a high frequency.

Studies of gene frequencies (or their effects) are useful tools in piecing together the past history of a genus (Delany and Healy, 1964). By considering the frequency of certain chromosomal translocations, Spurway (1953) has suggested that some populations of the newt *Triturus cristatus* may have arisen from just one founder, a gravid female, that managed to cross some ecological barrier and effect colonization. The data presented in this paper (and especially in Table III) are consistent with the suggestion that the genus *Arvicola* originally entered Britain in the form of one pregnant female, or at the most a small band of *émigrés* from one isolated population. The figures presented in Table I and Fig. 3 suggest some similarity in genetic composition between British and Scandinavian *Arvicola*, but clearly further speculation is foolish, and must await the collection of additional data.

The ecological mechanism effecting colonization in this manner has been shown to exist (Stoddart, 1968, 1970). I found several examples of dispersal by pregnant females, on one occasion to a distance of almost 3 km, as the crow flies. Each summer the number of breeding females in the population sharply dropped after the production of the first litter. (Members of this litter, incidentally, had a higher chance of survival than members of other litters born later in the breeding season.) I was able to locate only a few of the wanderers, since the area over which they dispersed was so great, but this was enough to establish the idea that movement is effected by pregnant females. Undoubtedly this is unusual for rodents, since this class of the population is more vulnerable to predation and other mishap than perhaps any other. This phenomenon need concern us no longer: suffice it to say that colonization and subsequent spread throughout Britain could occur in a fashion that is consistent with the genetic phenomena observed.

The mechanism of inheritance of partial albinisms in this genus awaits investigation. Inspection of Table I shows that no example is recorded of all 3 albinisms occurring in the one sample (see especially samples nos. 5, 9 and 37). This suggests that there are 3 alleles for these 3 features, and a maximum of 2 represented in the phenotype. No information is available on the change in frequency of albinistic characters over several years, for such an investigation would reveal whether once lost or suppressed, a particular albinism can be reestablished. A breeding programme designed for determining the mechanism of inheritance of these characters is in hand.

In conclusion, it appears that populations of *Arvicola* are small isolated demes showing a high degree of inbreeding. The three albinisms examined in this study appear to be evolutionarily insignificant in themselves, and are therefore likely to be the pleiotropic effects of genes that are being selected for some other advantage they give over the environment. The fact that albinisms are much more common in Britain than in Eurasia suggests that only a few individuals originally entered Britain, and it was quite by chance that they carried the genes capable of responding to new environmental needs, and producing their associated pleiotropisms. The founder principle, followed by coadaptation would strengthen this capability and the genome would spread throughout the land mass. Further colonists produced little genetic effect within the populations of residents. In Eurasia, the mutants do occur, and only locally are they able to become established. Clearly, a thorough genetical study of this genus will reveal the most interesting information, useful to both geneticists and ecologists.

Acknowledgements

I should like to thank many colleagues who have assisted in the collection of data for Table I, especially the following:

Dr. P. J. H. van Bree, Zoologisch Museum, Amsterdam, Holland.
Dr. G. B. Corbet, British Museum (Natural History), London, England
Dr. L. van Haaften, ITBON, Arnhem, Holland.
Dr. M. Hamar, Central Research Institute of Agriculture, Bucharest, Roumania.
Dr. A. M. Husson, Rijksmuseum van Natuurlijke Historie, Leiden, Holland.
Dr. A. Kaikusalo, Helsinki, Finland.
Dr. A. Meylan, Station Fédérale de Recherches Agronomiques, Nyon, Switzerland.
Dr. W. R. van Mourik, I.B.P. Biological Station, Oosterend, Holland.
Dr. P. A. Panteleyev, Academy of Sciences, Moscow, U.S.S.R.
Dr. J. A. Pedersen, Zoologisk Museum, Oslo, Norway.
Dr. G. Radek, Ludwigsburg, W. Germany.
Dr. J. Tast, Tampere, Finland.
Dr. H. Walhovd, Naturhistorisk Museum, Åarhus, Denmark.
Dr. J. Zejda, Institute of Vertebrate Zoology, Brno, Czechoslovakia.

I should also like to thank Professor J. W. S. Pringle for allowing me to work in the Animal Ecology Research Group, Department of Zoology,

Oxford University, and the Natural Environment Research Council for financial support. Logos Press Ltd. and Dr. A. G. Searle and Dr. R. H. Schaible kindly allowed me to reproduce Fig. 2. To them I also extend my thanks.

References

*Begg, C. M. M. (1959). *Introduction to genetics.* London: English University Press.

Corbet, G. B. (1963). The frequency of albinism of the tail tip in British mammals. *Proc. zool. Soc. Lond.* **140**, 327–330.

Delany, M. J. and Healy, M. J. R. (1964). Variation in the long-tailed field mouse (*Apodemus sylvaticus* (L.)) in north-west Scotland. II. Simultaneous examination of all characters. *Proc. R. Soc.* (B.) **161**, 200–207.

Ognev, S. I. (1950). *Mammals of U.S.S.R. and adjacent countries. 7 Rodents.* Acad. Sci. SSR, Moscow-Leningrad. Publ. Israel Program for Scientific Translations (1964).

Schaible, R. H. (1963). *Developmental genetics of spotting patterns in the mouse.* Unpubl. Ph.D. thesis, Iowa State University.

Searle, A. G. (1968). *Comparative genetics of coat colour in mammals.* London: Logos Press.

Service, R. (1903). Colour variation in Solway mammals. *Ann. Scot. nat. Hist.* **46**, 65–69.

*Sinnott, E. W., Dunn, L. C. and Dobzhansky, T. (1958). *Principles of genetics.* New York, Toronto, London: McGraw-Hill.

Spurway, H. (1953). Genetics of specific and subspecific differences in European newts. *Symp. Soc. exp. Biol.* No. **7**, 200–237.

Stoddart, D. M. (1968). *An ecological study of Arvicola terrestris (L.) with particular reference to population dispersion.* Unpubl. Ph.D. thesis, University of Aberdeen.

Stoddart, D. M. (1970). Individual range, dispersion and dispersal in a population of water voles (*Arvicola terrestris* (L.)). *J. Anim. Ecol.* **39**, 403–425.

Wendt-Wagener, G. (1961). Untersuchungen über die Ausbreitung der Melanoblasten bei einfarbig schwarzen Ratten und bei Hubenratten. *Z. Vererblehre* **92**, 63–68.

* Although not mentioned in the text, these standard genetics textbooks were used for reference.

Symp. zool. Soc. Lond. (1970) No. 26, 283–295.

VARIATION AND ECOLOGY OF ISLAND POPULATIONS OF THE LONG-TAILED FIELD MOUSE (*APODEMUS SYLVATICUS* (L.))

M. J. DELANY

Department of Zoology, University of Southampton Southampton, England

SYNOPSIS

Differences in size (as indicated by greatest length of the skull), colour of belly fur, length of pectoral stripe and colour of the dorso-lateral pelage are described in populations of *Apodemus sylvaticus* from the Hebridean, Scilly and Channel Islands as well as several localities on the British mainland. Island mice are generally larger than those from the mainland whilst even the differences between islands can be relatively great. The mean length of the pectoral stripe varies very much between populations: on some islands, e.g. Barra, mice have either no stripe or a very short one, whilst elsewhere, e.g. Alderney, all the mice may have a well-formed stripe. On the Channel Islands between 23·1% and 61·1% (according to the island) of the mice collected had a rich buff colouration to their belly fur. On 3 of the Hebridean Islands (Barra, North Uist and South Uist) many of the mice had paler dorso-lateral fur than was found in any other Scottish populations examined.

Field mice occur in a wide range of habitats in all the localities studied. However, they appear to be very much less abundant on the Outer Hebrides than on the Scottish mainland. There is also an indication that mice are slightly less numerous on the Isles of Scilly than in western Cornwall. Among the possibly more significant faunal differences, ground predators are absent from most islands, and on some islands the House mouse (*Mus musculus*) becomes a more frequent component of the fauna.

Some of the possible reasons for the morphological variations are discussed in relation to the environmental conditions the mice are known to experience on the islands.

INTRODUCTION

During the latter part of the last century and the early part of the present century several workers (e.g. de Winton, 1895; Barrett-Hamilton, 1899; Kinnear, 1906; Barrett-Hamilton and Hinton, 1913) drew attention to distinctive morphological features displayed by island populations of Long-tailed field mice (*Apodemus sylvaticus* (L.)) within the British Isles. These and other workers erected an elaborate taxonomic nomenclature which ultimately resulted in the description of 3 island species and 15 island subspecies. Many, although not all, of these descriptions were based on small samples, and the diagnoses of species and subspecies were frequently not sufficiently detailed to permit identification of individual animals. However, in spite of these shortcomings, there were clear indications that several islands supported populations of animals different from those of the adjacent mainland.

In 1957 an extensive re-examination of the status of island populations of *Apodemus* was initiated. Between then and 1963 collections of small mammals were made on the Inner and Outer Hebrides, Isles of Scilly, Channel Islands, Fair Isle and Foula in addition to various localities in mainland Britain. At the same time as the animals were collected detailed information was obtained on the ecological conditions under which they were living and trapping methods were standardized in most places so as to permit comparison between localities. These data were obtained since it was anticipated that they might shed some light on the selective processes to which mice on small islands were subjected; and also to note the extent to which conditions might differ from those on the mainland.

Several morphological characters, many of them skeletal, have already been examined in these mice and the data subjected to rigorous statistical analysis (Delany and Healy, 1964, 1967a, 1967b) with a view to determining the extent of morphological differentiation and affinity of these populations. However, the present paper selects only the more obvious (overt) characters of those examined and pools this information for the first time, for all the islands studied during the course of the research. It also briefly considers the results obtained from the ecological surveys conducted at the times of the collections. The attempt is thus to review this very broad field, since much of the more detailed information is already available elsewhere.

CHARACTERS EXAMINED

Size

One of the most striking and earliest recognized differences between island and mainland mice has been the tendency of the former to display larger overall size. This character can be expressed in various ways e.g. weight, head-and-body length, greatest length of skull (occipitonasal length), etc. For present purposes the greatest length of the skull has been selected as this is probably one of the more consistently accurate measures. The data on individual mice have been adjusted (Delany 1964) so as to bring each animal to a uniform age and the means for all the islands summarized in Table I. (It has not been possible to calculate the figures for Fair Isle, St. Kilda and Foula with the same statistical accuracy as those from other localities. These figures are based on those given by Miller, 1912; Barrett-Hamilton and Hinton, 1910–1921; Delany and Davis, 1961; and Delany, 1963.)

TABLE I

Means of greatest length of skull (mm) adjusted to uniform age

	25·0–25·4	25·5–25·9	26·0–26·4	26·5–26·9	27·0–27·4	27·5–27·9	28·0–28·4
Mainland							
Applecross	×						
Church Pl.	×						
Scorrier	×						
Cap Gris Nez	×						
Laga		×					
Islands							
Raasay	×						
South Uist		×					
Mull		×					
Lewis			×				
Barra			×				
Colonsay			×				
Jersey			×				
Guernsey			×				
Alderney			×				
Tresco			×				
North Uist				×			
Sark				×			
St. Mary's				×			
Herm					×		
Foula					×		
Rhum						×	
Fair Isle							×
St. Kilda							×

From Table I it is apparent that mainland mice, whether from Britain or France, are of small size, 4 of the 5 mainland populations examined having means falling within the smallest size range. The size of island mice varies considerably from one island to another with most island populations having means greater than any of the mainland populations. This size difference is not in general due to populations having similar ranges of mouse size within them with mice of a particular size in numerical preponderance. The larger mice on the majority of islands are bigger than the largest mouse on the mainland. It is also of note that of the main archipelagos examined the mice in all display trends to larger size although none of the 4 islands having the largest mice are in the Channel or Scilly Isles.

Belly markings

Two characters are readily distinguishable. They are the pectoral stripe or spot and the almost uniform buff coloration present on the underside of some individuals. The pectoral stripe may be absent or present in varying length from a mere speck to approaching 40 mm.

From Rood's (1965) (Fig. 1) comparison of pectoral stripe length in Isles of Scilly and Cornwall populations it is apparent that considerable differences can occur in the frequency distribution of individuals

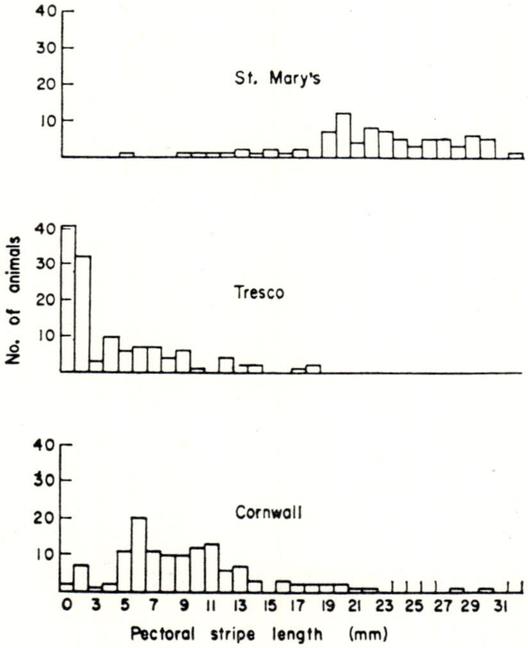

FIG. 1. Pectoral stripe length in samples of *Apodemus* from Isles of Scilly and Cornwall (from Rood, 1965).

with certain stripe lengths. The Cornwall population appears intermediate between the 2 islands. Comparison of mainland with a selection of island populations from elsewhere in Britain (Fig. 2) presents a similar picture. Mice from North Uist and Barra have short stripes whilst those from Rhum and Alderney have long ones. The Jersey mice are of interest as they have a considerable spread in stripe size as well as a relatively large number of mice with no stripe. The data in Fig. 1 have not been adjusted for age of mice (the necessary adjustment has been found to be small, partly due to mice in juvenile pelage being

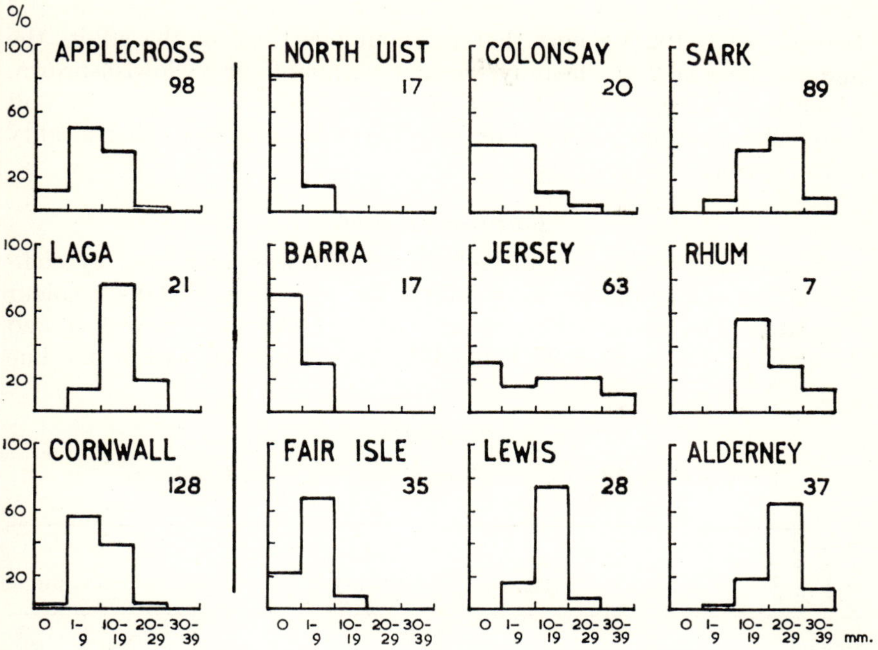

FIG. 2. Pectoral stripe length in samples of *Apodemus* from 3 mainland localities (column at extreme left) and 9 islands. The figures given for each locality are sample sizes.

omitted), nor for the size of the animal. As island mice are generally larger than mainland mice it may be appropriate to adjust stripe size slightly downwards for these populations.

Bishop (1962) reported the presence of an overall buff suffusion on some mice from the Channel Islands. The percentage of animals showing this character are summarized in Table II and can be seen to vary appreciably from island to island. Markings of this type have

TABLE II

Occurrence of buff suffusion

	Number examined	% with suffusion
Guernsey	99	41·4
Jersey	65	23·1
Sark	90	61·1
Alderney	37	45·9

been recorded also in mice from St. Kilda as Barrett-Hamilton and Hinton (1910–1921) report that in about one third of the adults the undersurface is "....heavily washed with buff or yellowish-brown, which tinge merges laterally in that of the flanks...." Extensive buff belly coloration has not been observed in mice from any other locality in the present study.

Dorso-lateral pelage

Examination of colour in Hebridean mice was facilitated by comparing the colour of the dorso-lateral fur with Munsell (1947) colour standards. The age-adjusted means (Delany, 1964) of the values and chroma are summarized in Table III. All colours fell within one hue

TABLE III

Colour variation in dorso-lateral pelage of Hebridean mice

chroma rotate through 90°			value		
	4·1–4·4	4·5–4·8	4·9–5·2	5·3–5·6	5·7–6·0
2·8–3·1					North Uist
3·2–3·5				South Uist	
3·6–3·9	Mull		Barra		
4·0–4·3	Lewis : Laga	Colonsay			
4·4–4·7		Applecross			
4·8–5·1			Rhum		
5·2–5·5		Raasay			

(10YR). The mice from North Uist have a pale, weakly coloured fur with those from South Uist and Barra showing a similar trend. In Raasay and Rhum, mice having a richer colour predominate, and in those from Mull, Lewis and Laga darker individuals are numerous. There is considerable overlap in the colour range of the populations with the Applecross animals almost completely covering the range of all others. Notable exceptions are certain mice from North and South Uist which have higher values than any from Applecross.

Of the mice from the Channel Islands Bishop (1962) reported, "Visual comparison of the dorsal aspect of the pelts showed no obvious trend to colour differentiation on any island, the range of variation being similar for each. In comparison with some skins of *Apodemus sylvaticus* and *A. flavicollis* from Southampton, the Channel Islands specimens showed some affinity to *A. flavicollis* in that the colours appeared to be generally brighter than in *A. sylvaticus*". Rood (1964) noted no obvious pelage colour differences in mice from Cornwall, Tresco and St. Mary's.

ECOLOGY

At each locality visited traps were placed in as many different types of habitat as possible. The results for the Hebrides, Channel Isles and Isles of Scilly are given by Delany (1961), Bishop and Delany (1963) and Rood (1964) respectively. On some islands (e.g. Barra), there is little variety of vegetation although it is evident from the combined results that *Apodemus* occurs in a very wide range of habitats.

During the summers of 1957–1960 trappings took place in north-west Scotland. Each year traps were set in various habitats at Applecross on the mainland and also on a selection of islands. Trapping was by means of trap lines with a 5 pace interval between traps, followed by the removal of traps after 3 nights in one place. This method was rigidly adhered to except in the Outer Hebrides in 1959 and 1960. As the trapping success was low there on the second and third night in 1958 (less than 1%), the traps were moved after only one night's trapping when visiting these islands in 1959 and 1960. In Tables IV

TABLE IV

Trapping success in deciduous woodland and moorland in North-west Scotland

Locality	% Success	Range	Summers trapped
	Woodland		
Applecross	20·1	9·3–31·3	4
Laga	13·5	11·8–14·7	2
Mull	9·4	5·5–14·0	2
Rhum	5·0		1
Colonsay	6·2		1
Lewis	2·6	0·3–3·4	2
Barra	2·4	0·1–2·8	2
	Moorland		
Applecross	6·9	4·2–11·1	4
Mull	2·5	0·0–5·0	2
Rhum	0·0		1
Colonsay	4·0		1
North Uist	0·7		1
South Uist	1·3	0·7–1·8	2

TABLE V

Trapping successes in agricultural land in North-west Scotland

Locality	% Success	Range	Summers trapped
Applecross	21·7	12·6–30·7	4
Laga	0·0		1
Mull	16·7	11·1–33·3	2
Raasay	5·1		1
Rhum	9·5		1
Colonsay	5·3		1
Lewis	2·4	1·1–2·9	2
North Uist	2·4		1
South Uist	1·8		1
Barra	2·5	0·6–3·1	2

and V comparisons have been made for trapping success over 3 nights and in order to include data from the Outer Isles a 1% catch has been assumed for the second and third nights. Agricultural land and deciduous woodland appear to be more favoured in all localities than moorland. Possibly more important is the very low numbers that have been obtained on the Outer Hebrides compared to the mainland, with the Inner Hebrides occupying an intermediate position.

Comparisons of trapping success in the Scillies and Cornwall at similar times of the year are given in Table VI. The trapping method

TABLE VI

Trapping success (%) *in Isles of Scilly and Cornwall* 1961–2

	St. Mary's	Tresco	Cornwall
	Bracken		
March/April 1961	2·0	2·4	3·0
May/June		2·0	3·1
November/December	1·7	3·7	5·0
August 1962		5·0	2·2
	Hedgerows		
May/June 1961	0·7		4·7
July	0·7		2·0
November/December	4·7		18·0

in this study was not the same as in the Hebridean survey (see Rood, 1964) thereby preventing comparison of the results between the 2 areas. However, it appears that with the exception of August 1962, field mice were caught more frequently in similar habitats on the mainland than on the islands. The magnitude of the differences is variable. No ecological comparisons were made between island and mainland in the Channel Island survey (Bishop and Delany, 1963), and it can be only broadly concluded that the mouse was present in moderate numbers in most habitats trapped.

Ground predators are generally poorly represented on the islands. In the Hebridean Islands examined the only species present were weasels (*Mustela nivalis*) on Raasay and Polecat ferrets (*Putorius furo*) and stoats (*M. erminea*) on Mull. Feral cats are probably widespread but not numerous. Stoats are present on Jersey and Guernsey whilst no ground predators have been recorded from the Isles of Scilly.

The frequency with which the House mouse (*Mus musculus*) has been caught away from human habitations on the islands is of note. Of the 14 islands surveyed in detail on which *Apodemus* occurs *Mus* has been caught on 9. In the Hebrides it was taken on the 4 Outer Isles and Colonsay. Elton (unpublished) in 1933 caught *Mus* more frequently than *Apodemus* in fields around Stornoway, Lewis whilst Delany (1961) obtained similar results in agricultural land on Colonsay, North Uist and South Uist. *Mus* has not been trapped on the mainland in any of the work reported here.

DISCUSSION

From the present analysis it appears that on small islands *Apodemus* shows a trend to larger size; a phenomenon apparently independent of the geographical situation of the island. In addition, fur colour and belly markings can vary appreciably from island to island although (unlike the case with size) generally within the overall limits displayed by the mainland population. None of the island mice are probably sufficiently distinct on morphological grounds to be regarded a distinct species. This has been confirmed by Jewell and Fullagar (1965), who found individuals of island races produced fertile offspring when crossed with each other and with mainland mice.

The information presented here considers only the overt morphological characters shown by the mice and does not attempt to analyse them together with data on less obvious features. Much work of this type has already been undertaken (Berry, Evans and Sennitt, 1967; Delany and Healy, 1964, 1967a, 1967b) and has proved most useful in suggesting affinities between certain islands and mainland populations. None of these analyses has been as embracing as the present survey, nor do they automatically shed any more light on the selective pressures responsible for the changes witnessed.

Corbet (1961) has suggested that, in the absence of ground predators, intra-specific competition increases which in turn favours selection for larger size. It is of note that on Mull, Raasay, Guernsey and Jersey, where ground predators are present the mice are among the smaller of the island races. It is anomalous that on Jersey and Raasay large races of Bank voles are present. Furthermore, a number of islands with mice of similar size (e.g. South Uist, Lewis and Tresco) to those of Mull, Raasay, Guernsey and Jersey do not have ground predators.

Many of the mice caught on the Uists and Barra were obtained from vegetation of the machair which has a soil with a large sand component. It is possible that the lightly coloured dorsal fur would blend better with this background. However, several of the mice from Lewis were also obtained from machair and they do not as a group display the paleness of the three more southern islands. Further more extensive studies on the Outer Hebrides could provide more conclusive information on the reasons for pelage colour variation in the Hebrides.

A noteworthy feature of these studies has been the apparently lower numbers of *Apodemus* on the islands compared, where possible, to the nearby mainland. This has been most striking for the Outer Hebrides although Boyd (1959) reports *Apodemus* to be relatively numerous on the remote and exposed island of St. Kilda. Delany (1961) has suggested that considerable cover may be afforded by the dry-stone storage chambers (cleitean) on St. Kilda. On the other Outer Hebrides these structures are not found, and stone walls, their possible substitute, are relatively few. There is thus probably little physical protection and a poor food supply for the mice in the winter months when much of the herbaceous vegetation is dead and close to the ground. Under these conditions survival could be particularly difficult and could account for the low numbers present.

The relationship with the House mouse is interesting. On islands where man is absent but has lived, and on which the Field mouse is present e.g. St. Kilda, the House mouse becomes extinct (Williamson and Boyd, 1960). On islands where man is either absent or present and the Field mouse is absent the House mouse can occupy a variety of habitats e.g. Treshnish (Darling and Boyd, 1964), May (Berry, 1965), St. Martins (Rood, 1964), Skokholm (Berry, 1968). Thus, the House mouse apparently cannot compete successfully with the Field mouse but is capable of occupying a variety of niches in its absence. It is now tentatively suggested that on the Outer Hebrides an intermediate state of affairs exists. The Field mouse is present in low numbers as a result of environmental pressures possibly unrelated to the House mouse. Under these conditions it is possible that with a reservoir in human habitations, the House mouse can temporarily if not permanently move out into fields as a result of the low numbers of the Field mouse. These conclusions would appear not inconsistent with Berry and Tricker's (1969) observations on the distribution of *Apodemus* and *Mus* on Foula. That House mice have been caught away from habitations, apparently as a less numerous member of the field fauna than in the Outer Hebrides, on Colonsay as well as on some of the Channel and Scilly Isles suggests that a similar, if incipient, situation is also prevailing on these islands. This supports the suggestion that on many small islands *Apodemus* may be experiencing greater, although far from fully understood, selection pressures than on the mainland.

The interpretation of variation in *Apodemus* is greatly hampered by lack of information on the history of the island faunas. The length of time mice have been on each island, their place of origin, the extent to which interchange has been experienced between one situation and another since the time of first colonization and the genetical composition of the various colonizers would all considerably help to elucidate the situation. Although there are apparently some features that island populations have in common, there can also be marked differences between them. In this connection it must be recognized that the contemporary characteristics of an island population are the result of its previous history in respect of both the selective pressures that have operated through time and the genetical composition of the original and possibly later immigrants The foregoing observations on *Apodemus* in the British Isles are paralleled by small rodents on islands elsewhere in the world For example, the studies of Foster (1963) on *Peromyscus* of the Queen Charlotte Islands and Alexander Archipelago of Canada as well as his review of the variation shown by this genus on the islands of California show a similar complex pattern of the variation that may be displayed by a small rodent.

References

Barrett-Hamilton, G. E. H. (1899). On the species of the Genus *Mus* inhabiting St. Kilda. *Proc. zool. Soc. Lond.* **1899**, 77–88.

Barrett-Hamilton, G. E. H. and Hinton, M. A. C. (1910–21). *A history of British mammals*. London: Gurney and Jackson.

Barrett-Hamilton, G. E. H. and Hinton, M. A. C. (1913). On a collection of mammals from the Inner Hebrides. *Proc. zool. Soc. Lond.* **1913**, 821–839.

Berry, R. J. (1965). Island house mice. *Animals* **6**, 438–442.

Berry, R. J. (1968). Ecology of an island population of the house mouse. *J. Anim. Ecol.* **37**, 445–470.

Berry, R. J., Evans, I. M. and Sennitt, B. F. C. (1967). The relationships and ecology of *Apodemus sylvaticus* from the Small Isles of the Inner Hebrides, Scotland. *J. Zool., Lond.* **152**, 333–346.

Berry, R. J. and Tricker, B. J. K. (1969). Competition and extinction: the mice of Foula, with notes on those of Fair Isle and St. Kilda. *J. Zool., Lond.* **158**, 247–265.

Bishop, I. R. (1962). *Studies on the life histories, ecology and systematics of small mammals inhabiting the Channel Islands.* Unpubl. M.Sc. thesis. Southampton University.

Bishop, I. R. and Delany, M. J. (1963). The ecological distribution of small mammals in the Channel Isles. *Mammalia* **27**, 99–110.

Boyd, J. M. (1959). Observations on the St. Kilda field-mouse *Apodemus sylvaticus hirtensis* Barrett-Hamilton. *Proc. zool. Soc. Lond.* **133**, 47–65.

Corbet, G. B. (1961). Origin of the insular races of small mammals and the "Lusitanian" fauna. *Nature, Lond.* **191**, 1037–1040.

Darling, F. F. and Boyd, J. M. (1964). *The Highlands and Islands.* London: Collins.

Delany, M. J. (1961). The ecological distribution of small mammals in north-west Scotland. *Proc. zool. Soc. Lond.* **137**, 107–126.

Delany, M. J. (1963). A collection of *Apodemus* from the Island of Foula, Shetland. *Proc. zool. Soc. Lond.* **140**, 319–320.

Delany, M. J. (1964). Variation in the long-tailed field-mouse (*Apodemus sylvaticus* (L.)) in north-west Scotland. I. Comparisons of individual characters. *Proc. R. Soc.* (B) **161**, 191–199.

Delany, M. J. and Davis, P. E. (1961). Observations on the ecology and life history of the Fair Isle field-mouse *Apodemus sylvaticus fridariensis* (Kinnear). *Proc. zool. Soc. Lond.* **136**, 439–452.

Delany, M. J. and Healy, M. J. R. (1964). Variation in the long-tailed field-mouse *Apodemus sylvaticus* (L.)) in north-west Scotland. II. Simultaneous examination of all characters. *Proc. R. Soc.* (B) **161**, 200–207.

Delany, M. J. and Healy, M. J. R. (1967*a*), Variation in the long-tailed field-mouse (*Apodemus sylvaticus* (L.)) in the Channel Isles. *Proc. R. Soc.* (B) **166,** 408–421.

Delany, M. J. and Healy, M. J. R. (1967*b*). Variation in the long-tailed field-mouse (*Apodemus sylvaticus*) in south-west England. *J. Zool. Lond.* **152**, 319–332.

Foster, J. B. (1963). *The evolution of the native land mammals of the Queen Charlotte Islands and the problem of insularity.* Unpubl. Ph.D. thesis. University of British Columbia.

Jewell, P. A. and Fullagar, P. J. (1965). Fertility among races of the field mouse (*Apodemus sylvaticus*) and their failure to form hybrids with the yellow-necked mouse (*Apodemus flavicollis*). *Evolution, Lancaster, Pa,* **19**, 175–181.

Kinnear, N. B. (1906). On the mammals of Fair Isle with a description of a new subspecies of *Mus sylvaticus. Ann. Scot. nat. Hist.* **1096**, 65–68.

Miller, G. S. (1912). *Catalogue of the mammals of Western Europe.* London: British Museum (Nat. Hist.).

Munsell Color Company Incorporated (1947). *Munsell book of color.* Baltimore: Munsell Color Co. Inc.

Rood, J. P. (1964). *Ecological studies on the small mammals of the Isles of Scilly.* Unpubl. Ph.D. thesis. Southampton University.

Rood, J. P. (1965). Observations on the life cycle and variation of the long-tailed field-mouse *Apodemus sylvaticus* on the Isles of Scilly and Cornwall. *J. Zool., Lond.* **147**, 99–107.

Williamson, K. and Boyd, J. M. (1960). *St. Kilda summer.* London: Hutchinson.

de Winton, W. E. (1895). The long tailed field mouse on the Outer Hebrides: a proposed new species. *Zoologist* **19**, 369–371.

VARIATION AND ECOLOGY

Symp. zool. Soc. Lond. (1970) No. 26, 299–325.

ECOLOGICAL STRUCTURE AND GENE FLOW IN SMALL MAMMALS

PAUL K. ANDERSON

Department of Biology, The University of Calgary
Calgary, Alberta, Canada

SYNOPSIS

Prevailing systematic and ecological concepts start from the premise that Mendelian populations are of relatively large size, and that species are collections of individuals which behave like gas molecules within containers formed by ecogeographical barriers of varying permeability. For rodents, at least, this view is contradicted by evidence that evolutionary and ecological processes are greatly modified by highly complex spatial and temporal structuring of local populations.

While it is likely that each species has a unique basic structural pattern and a repertoire of modifications, there is a common pattern involving highly isolated demes of very small size. These demes are primarily responsible for maintenance of the continuing species population in an area, and occupy secure microhabitats as "survival populations". Seasonal and other increases in carrying capacity of the larger environment are exploited by the production of colonizing phases, distinguished by a broad spectrum of morphological, physiological, behavioural, and genetical characteristics.

In heterogeneous natural environments, such structuring permits routine density regulation to work through group cohesion, exclusion of immigrants, and export of emigrants. Other suggested intrinsic density regulating mechanisms are secondary or pathological (induced by exceptional or experimental conditions). In particular, most experimental studies have failed to consider confinement to be a major variable. Consequently, the pathologies induced by lack of egress have been interpreted as normal regulatory mechanisms.

Small deme size, exclusion of immigrants, and competition in the establishment of new colonies through emigration, imply significant roles for genetic drift and interdeme selection in evolutionary processes. Gene flow may be so restricted and canalized by population structuring that it plays only a minor role in the maintenance of species integrity. Instead this last may depend largely on coadaptation of genomes, and convergent selection pressures. Further possible consequences of complex population structuring are the occurrence of well-defined ecotypes in mosaic patterns, and a potential for sympatric speciation.

Effective ecological, systematic, and genetical research is dependent upon the multidimensional structuring and the ecological and genetical properties of populations being considered as major components of the overall strategy by which species interact with the environment.

INTRODUCTION

The simple statistical concept of a population as any group of individuals specified by an investigator can be dangerously seductive when applied to the natural world, since its application may cause us to overlook the patterns in which population units exist. Such misapplication of valid statistical concepts may, perhaps, lie behind the general lack of appreciation of the complexity and significance of population structuring.

This paper deals with structure in populations of mouselike rodents. Its theme is that each species has a complex and functional pattern of population structuring. It is argued that such a basic pattern, with its adaptive modifications, lies at the heart of the ecological and evolutionary strategy of the species.

The broader aspects of population structure are limited by the ecology, in the widest sense, of rodents. For example, rodents are warm blooded animals of small size. This implies high metabolic rate and relatively large body surface. While small body size carries with it the ability to seek out favourable micro-climates, the large surface-to-mass ratio dictates a limited tolerance to extremes in ambient conditions, and in turn this means that weather becomes a critical factor in population dynamics (Frank, 1957; Fuller, 1969). Since physiological life span is generally under 10 years, and ecological life span under one year, seasonal periodicities and less regular fluctuations in the environment are significant in relation to the length of a rodent generation. In most rodent niches, the carrying capacity of the environment fluctuates rapidly. While some of these fluctuations are periodic and predictable, others are completely discontinuous in both their nature and occurrence. Both periodic and stochastic changes imply variation in the intensity and focus of selection over intervals which are significant in rodent terms.

With the limitations of small body size, restricted movement, and short life span, the exploitation of new ecological opportunities and the effective response to environmental stochasticism, require high reproductive rates and efficient dispersal mechanisms. Rodents are, essentially, colonizing forms (Lewontin, 1965). They are adapted to rapid exploitation of transient environmental surpluses, as well as to frequent catastrophe and local extinction.

The capacity for maintenance of small populations in isolated pockets of suitable habitat is important in this context. Here social organization has a considerable role to play. Formation of small and cohesive groups is facilitated by multiple births: rodents are born as social organisms, spending a crucial post-natal period of development in close association with at least the female parent and several siblings. This developmental experience establishes the social basis of population structure.

Many of these generalizations can be exemplified by a review of the structure of House mouse populations.

POPULATION STRUCTURE IN *Mus musculus*

The House mouse appears to have originated in steppe communities in Eurasia (Schwarz and Schwarz, 1943). It has demonstrated unusual abilities as a colonizing species, travelling with man to the far corners of the earth where it has been successful in establishing both commensal and feral populations in widely divergent environments.

What appears to be the fundamental pattern of House mouse population structure is most easily discernible where it happens to coincide with obvious environmental discontinuities. Such a situation exists in a number of habitats, among them the small grain storage buildings on the Canadian prairies (Anderson, 1965).

Each granary holds a single population unit. This is a behaviourally defined group of about 10 weaned mice, of which 4 to 7 are reproductively active. Among the latter, females outnumber males two to one. House mice are territorial (Crowcroft and Rowe, 1958; Anderson and Hill, 1965), and in this and similar situations group territories are formed (Eibl-Eibesfeldt, 1950; Hill, 1966). The groups are stable in size and composition, control their own density, and occupy habitats which are essentially predation-free and are supplied with unlimited food. The control of density is achieved through group territoriality (which excludes immigrants and isolates the group reproductively), combined with export of excess young. The latter serve a colonizing function, which, although it is probably costly in terms of individual survival, is of great ecological and genetical importance. Such a method of population regulation has significant consequences. Each granary population is a reproductively isolated interbreeding unit or deme. This can be demonstrated ethologically (Hill, 1966), and genetically as well as ecologically (granaries as little as one metre apart house demes with clearly different genetic identities (Anderson, 1964, 1965) and these may persist for several generations). The size of these independent demes is so small that there is a high degree of inbreeding. Populations differ from each other both because they are likely to be founded by a single pair, and because, through drift, genetic fixation occurs readily at many loci in succeeding generations. As a consequence, it can be postulated that interdeme selection, operating through interdeme competition for establishment of new colonies by exported emigrants, has an evolutionary role.

This general picture of the formation of small, genetically independent groups, showing evidence of genetic drift, is supported for analogous environmental situations by the work of Petras and Reimer (Petras, 1967*a*, 1967*b*; Reimer and Petras, 1967, 1968) and conforms to mathematical models developed to explain the anomalous gene frequencies observed in high-transmission-ratio *t*-alleles (Lewontin and Dunn, 1960; Lewontin, 1962; Dunn and Levene, 1961). Evidence for the smallness of effective population size, and consequent genetic drift, has been adduced for the superficially much larger populations of English corn ricks (Rowe, Taylor and Chudley, 1963; Berry, 1963; Berry and Searle, 1963).

In the granary situation, habitat discontinuities are well-defined and provide reference points for both the mice and the investigator. Can such ecological and genetic structuring persist in a continuous habitat? Justice (1962) was unable to detect evidence of family structure in feral *Mus musculus* inhabiting cultivated fields and pastures in Arizona. He felt that movement during the dispersal phase was restricted only by discontinuities in the habitat when densities were low to moderate, and that dispersing mice might cross exposed habitat when density was high. On the assumptions (not explicit) that such movement constitutes genetical dispersal and that there is no mate selection, he concluded that random mating was affected by both habitat discontinuity and density. Although he did not appreciate its full significance, Justice recognized that he was studying populations in highly unstable habitats, commenting only that individuals from "reservoir populations" might contribute more to the continuing gene pool than did individuals from those of the less stable environments.

The distinction between populations of unstable and stable habitats was long ago recognized and formalized by Russian investigators (Fenyuk, 1937) and more sophisticated implications of this concept will be reviewed later in this paper. Some aspects have been recently recognized and reiterated for the House mouse by Newsome (1969*a* and 1969*b*).

The most useful account of House mice living in relatively continuous environments, however, is that of Naumov (1940). He reported that in Eastern Europe and central Asia reproduction occurs in field and steppe environments only during the summer months. As successive litters reach the juvenile stage they disperse and seek their fortunes in new habitats. These individuals are the colonists in such unstable habitats as those studied by Justice (1962) and Newsome (1969*a*). In the autumn, however, the pairs of adults which have been breeding through the summer (in the more stable "survival" habitats) set to work to gather and deposit in a single heap, a twelve-month supply of weed seeds. The pile of accumulated seeds is covered with earth and a passage leads from it to a winter nest. The last litter of the season does not disperse, but stays in the winter burrow with the parents.

This means that they replace their own parents as owners of the seed store when old age and climatic vicissitudes take their toll, and thus continue the family line through sib matings on the same site. While the seed storing phenomenon does not occur throughout the ancestral range of *Mus musculus*, the populations of the "survival habitats" are analogous to those of the Alberta granaries. Further, the proven ability and tendency of House mice to discriminate in mate selection (Mainardi, 1963; Levine and Lascher, 1965) suggest that mating is not wholly random even among emigrants colonizing new habitats.

The implication that movement of dispersing juveniles cannot be equated directly with gene flow is further supported by evidence that strange mice are not readily integrated into established populations even under conditions of confinement (Andrzejewski, Petrusewicz and Walkowa, 1963), and succeed only very rarely in doing this in unconfined populations (Eibl-Eibesfeldt, 1950). The only direct study of gene dispersal (Anderson, Dunn and Beasley, 1964) was carried out in a continuous grassland habitat on an island. Here the introduced allele apparently failed to spread as fast as would have been expected if the island was occupied by a completely random breeding population. This is in full accord with the model described above, derived from Naumov's (1940) report. It thus appears that the granary pattern can be taken as characteristic of the structure of *Mus musculus* populations in both continuous and discontinuous habitats.

The extreme subdivision of *Mus musculus* populations approximates to the "island model" suggested by Wright (1943). In such a system, intrademe variation is greatly reduced, while interdeme variation is maximized as a consequence of genetic drift. How, in this situation, is the integrity of the species as a whole maintained? Much of the answer to this question may be contained in the recent study of Selander, Hunt and Yang (1969). These workers compared polymorphic loci in samples from populations of *Mus musculus domesticus* in the southern half of the peninsula of Jutland with samples from populations of *Mus musculus musculus* in the northern half of Jutland. These 2 forms are presumed to represent the terminal points of northern and southern dispersal routes from the centre of origin (Schwarz and Schwarz, 1943). They found marked differences in the frequency of alleles at 86% of loci. Their explanation for this differentiation in a homogeneous environment was that identical alleles have different fitnesses in the 2 populations due to the co-adapted nature of the genomes.

Now this coadaptation hypothesis could be examined through laboratory matings. If such studies indicate that the genomes do indeed resist reorganization, it will suggest that the co-adaption concept is substantially correct. The integrity of a species with breeding structure based on demes would receive alternative or additional support through stabilizing environmental selection, and Van Valen and Weiss (1966) have suggested that there is some evidence of such stabilizing selection in the reduction of variability with age in populations of *Rattus rattus*.

I have tried to sketch, in broad outline, a picture of the fundamental pattern of population structure in *Mus musculus*. It is clear that adaptive variations on this pattern can exist (Anderson, 1961). However, I give little credence to claims that adrenal stress and/or associated behavioural pathologies of confined populations, as interpreted by various authors (Strecker and Emlen, 1953; Southwick, 1955*a*, 1955*b*; Christian and Davis, 1964) have any relevance to unconfined populations of House mice whose population structure conforms to this basic pattern. The remainder of this paper will be devoted to exploration of the implications of the House mouse pattern and the degree to which various elements apply to other rodent species.

ENVIRONMENTAL HETEROGENEITY IN SPACE

All field biologists are aware that species occupy a range of habitats. Western ecologists have generally assumed that habitat variation forms more or less a continuum about some local optimum in respect to the adaptive mode of the local population. Russian investigators, on the other hand, have tended to place great emphasis on dichotomy in local habitats, classifying them as "survival stations" and "colonization stations" (Fenyuk, 1937; Naumov, 1940).

Survival stations are characterized by the continuous presence of the species in question. It is here that the habitat and the adaptive mode of the species population most closely match. Density need not be high, but both density and population composition are likely to be relatively stable. From the previous example of the granary *Mus* populations, and from ideas developed later in this paper under the heading of seasonal generations, the following list of additional characteristics can be assembled. Inter-specific competition is likely to be minimal, while intra-specific contacts are likely to be largely amicable and formalized. Most individuals inhabiting survival stations will be born there, mate with parents or sibs and have long life spans. This "survival component" is composed largely of animals likely to over-winter before reproducing. A high proportion live to sexual maturity, after which time the mortality rate tends to be relatively low (King, 1968). Selection is likely to operate through differential reproduction, rather than through differential mortality.

In contrast to the survival sites, colonization habitats are characterized by seasonal or otherwise intermittent occupancy (e.g. *Mus* in Australian wheatfields (Newsome, 1969*a*)), and by extreme fluctuations in density. In species with periodic density fluctuations, colonization habitats are unoccupied during "low" years (e.g. *Microtus oeconomus* studied by Tast, 1968), while in favourable seasons they may be the sites of maximum population density. These habitats are generally colonized by dispersing young born in spring and summer. There may be an excess of males (e.g. observations by Russell (1968) on pocket gophers of the genus *Pappogeomys*). While selective mating is not impossible, outbreeding and recombination are maximal in colonization habitats. The life span of individuals born in, or settling in, colonization habitats is relatively short and selection may operate largely through differential mortality.

The varying carrying capacity of colonization habitats, provides a basis for population eruptions. Major regional increases in rodent numbers are likely to involve coincidence of environmental conditions favouring early occupancy, heavy and prolonged reproduction, and unusual persistence of populations in such habitats. It is following occupation of secondary breeding areas that *Lemmus* highs occur in Finnish Lapland (O. Kalela, personal communication) and the relative condition and abundance of rodents on colonization and survival "stations" are major factors in the Russian system for prediction of rodent abundance in agricultural regions (Polyakov, 1958).

In many cases colonization habitats may be populated, but reproduction may fail to occur (e.g. Newsome, 1969*a*) and they may function as "behavioural sinks" in the sense outlined by Calhoun (1962). Calhoun noted that his experimental procedures led to the concentration of non-reproductive and socially subordinate individuals in certain areas of both indoor and outdoor enclosures. I encountered evidence of the same phenomenon among the feral House mice on Great Gull Island (unpublished data). The central part of the island, while superficially only slightly different from other grassland areas, was characterized by absence of reproduction, very high recruitment, and very low survival. I believe this part of the island represented, as did certain portions of Calhoun's enclosures, the marginal habitats of non-island environments. Such areas, with little or no reproduction and high mortality, are an essential component of population regulation through export of emigrants. (Ecologists too often seem to operate on the assumption that mortality is more or less uniform in all components of a population and in all habitats in a local area.)

It is convenient to designate the inhabitants of survival and colonization sites as distinct components of a population. These 2 components, as a result of their different environments and origins, would seem likely to differ genetically. Semeonoff and Robertson (1968) have recently published a study of esterase polymorphism frequencies in *Microtus agrestis* in which they unwittingly provide dramatic support for this idea. The study area in which they worked was triangular, each side being approximately one-half mile long, and was crossed by a small stream. They observed that one small subregion of the area showed a consistently divergent frequency of the esterase phenotype in question (Fig. 1). The comment was made in passing that this area was only seasonally inhabitable due to flooding. It is thus clearly a colonization habitat and the fact that it contained a genetically differentiated group of individuals is of great interest for the present discussion.

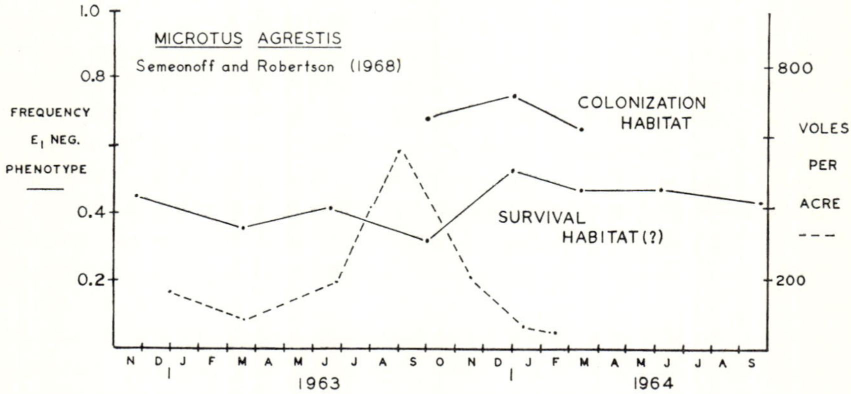

FIG. 1. Frequency of the E_1 negative esterase phenotype in a population of *Microtus agrestis*, modified from Semeonoff and Robertson (1968) to show the distinctiveness of the population in the colonization habitat ("Area X" of Semeonoff and Robertson). The remainder of the study area is referred to as survival habitat since populations appear to have been continuously present, although some individuals belonging to the colonization component of the population are undoubtedly included in their samples.

Summarizing some of the ecological and genetical implications of the ideas thus far advanced, the survival component of the population, in the survival habitat, is likely to display a demic structure. Under conditions of maximum environmental and demographic stability, it functions to maintain the ongoing population in both the ecological and the genetical senses. In contrast, the colonization component, in the colonization habitat, has exploitive, exploratory and experimental functions. It is variable and unstable in density and social structure. The conditions may provide for dissipation of the excess animals, but they also provide a test of the full range of ecological and genetical variation under all of the accessible variations of the fundamental niche of the species. While the survival of the colonization component may be low, the constant search for new genetical and ecological combinations is of crucial importance to the species population as a whole.

SEASONAL GENERATIONS AND ENVIRONMENTAL HETEROGENEITY IN TIME

The obvious criticism of this concept of habitat dichotomy is that a given species occupies a range of habitats along a continuum, rather than 2 distinctive habitat types. However, an effective dichotomy could still exist if populations produced 2 distinctive phenotypes or genotypes. There is, in fact, strong evidence for such bimodality in the qualities of individuals, exploiting and responding to the predictable seasonal periodicity in the geographical extent, quality, and carrying capacity of rodent habitats. Rodents, like daphnids, rotifers, and similar colonizing forms, can be cyclomorphic.

While investigators writing in English have occasionally noted seasonal differences in growth rates and in size at sexual maturity (Blair, 1948; Getz, 1960; Greenwald, 1956, 1957; Krebs, 1964; Pearson, 1967), the concept of seasonal generation in rodents of temperate and boreal zones is largely Russian in origin and application. The one contribution in English by Russian authors, that of S. S. Schwarz and his co-workers, was published in *Acta theriologica* in 1964. (Schwarz *et al.*, 1964). The same phenomena have been reported in German by Reichstein (1964) and in French by Martinet (1967).

Animals born in the spring and summer months can be shown to differ in many respects from those born later in the breeding season. The patterns of growth for spring and autumn generations of 3 species are shown in Fig. 2. The upper graph, modified from Schwarz *et al.* (1964) compares spring- and autumn-born *Microtus oeconomus*, and spring- and autumn-born *Lagurus lagurus*. The spring-born individuals grow in a single, rapid spurt and become sexually mature and begin reproduction within a very short time. Autumn-born individuals show a biphasic growth pattern. The lower part of Fig. 2, taken from Martinet (1967), shows seasonal fluctuation in the mean body length attained at an age of 2 to 4 months by *Microtus arvalis*. The mean body length of animals born in the spring is much greater than that of those born later in the breeding season.

The long delay in sexual maturity in autumn-born individuals is particularly striking and significant. In northern Russia, for example,

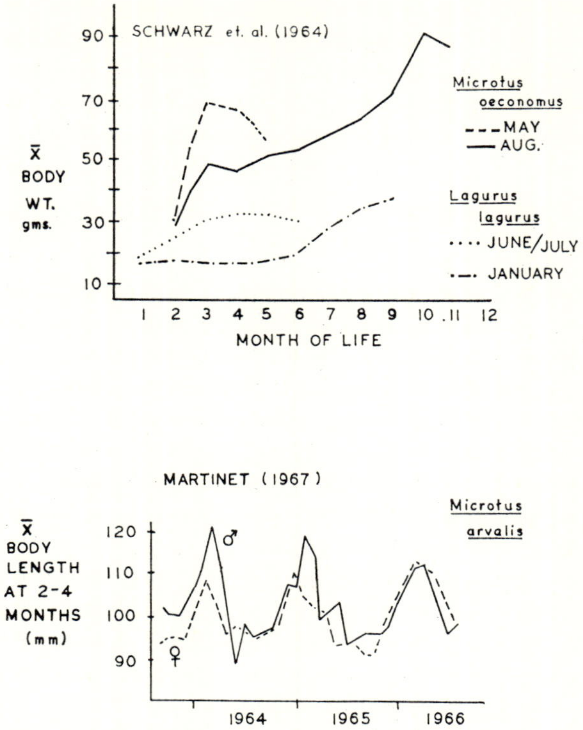

FIG. 2. Differential patterns of growth in seasonal generations of 3 rodent species. Graphs modified from Schwarz *et al.* (1964) and Martinet (1967) to illustrate the contrast between rapid growth to maturity of animals born in spring and early summer, and delayed growth of young born late in the breeding season.

Microtus gregalis matures at an age of one to one and a half months when born in April or May, but reaches sexual maturity only at an age of 8–9 months when born in July (Schwarz *et al.*, 1964). The same authors report similar delays in a number of other species. Similarly, Pearson (1967) described a seasonally correlated variation in age at maturity of from 2–7 months in *Akodon azare*, a field mouse of the southern hemisphere.

In addition to growth rate and age at maturity, seasonal generations may be distinctive in respect to a long list of other qualities. These include activity of various endocrine glands, life span, mitotic rate, rate of ageing, weight of certain internal organs (heart, liver, kidney, adrenal, thymus), eye lens weight, haematological indices, metabolic rate, thermal optima and response to climatic conditions, and vitamin content of various organs. These differences appear to be brought about primarily by photoperiod, and to be influenced only minimally, if at all, by temperature and diet (Schwarz *et al.*, 1964; Reichstein, 1964; Martinet, 1967).

The spring and summer generations, I submit, constitute the colonizing component of the population. Early sexual maturity invokes both internal and social pressures favouring dispersal (Howard, 1960). Early sexual maturity also maximizes their ability to exploit seasonally available habitats and seasonally abundant resources. Advancing the age at onset of reproduction is the most effective of conventional devices for increasing the intrinsic rate of natural increase (Lewontin, 1965), and in the spring generations this is further enhanced by greater size and frequency of litters (Martinet, 1967). While the life expectancy (Schwarz *et al.*, 1964) and survival rates of spring-born animals are low (Howard, 1949; Hoffmann, 1958) and their *per capita* contribution to the ongoing gene pool may be minor, their physiological and behavioural commitment to ecological and evolutionary exploration is crucial to the occupancy of niches in which nature appears particularly capricious (Lewontin, 1966). My own observations on Gull Island House mice (unpublished), and those of Dahl (1967) on *Microtus pennsylvanicus* in western Canada lead me to suspect that up to 80% of the annual reproductive effort of rodent populations may be devoted to the colonizing function.

The members of the autumn generation, by their delayed sexual maturity, are relieved of both internal and social pressures to disperse and do not do so (Naumov, 1940; Kalela, 1957). Their function is to perpetuate the survival component in the survival habitat. They do not undergo the rigours, the hazards, or the intense selection involved in the dispersal process, and after spending the winter in the immature state they reproduce towards the end of their long life spans. They seem, on a *per capita* basis, likely to contribute far more genetically to the ongoing species population than their spring and summer predecessors.

Particular emphasis should probably be placed on the high probability of assortative mating involved in these relationships. Members of the autumn and spring generations are very unlikely to mate with each other owing to the short physiological and ecological life spans of spring-born individuals, and the long delayed sexual maturity in autumn-born individuals. Further, the probability of sib mating is likely to be relatively high for the non-dispersing autumn generation, and relatively low (despite any tendencies toward inbreeding resulting from habitat selection or preferential mating) in the colonizing generations of spring and summer. Contributions by the colonizing component to the future species population will be derived from 3 groups:

those colonizers discovering unoccupied survival habitats; those which did not disperse and succeeded in establishing themselves in the over-wintering colonies where they originated; or those who (through survival of the selective ordeal, and innovation of new genetic combinations by outbreeding) have had the rare good fortune to stumble upon new ecological or genetical modes.

Figure 3 attempts to integrate a few of the complex ecological and genetical implications so far discussed. As indicated, although their numbers are great, most of the members of the spring and summer colonizing component come to bad ends. In most years there will be only a few descendants, but in exceptional years large numbers may survive and breed, so that they and their offspring make large contributions to

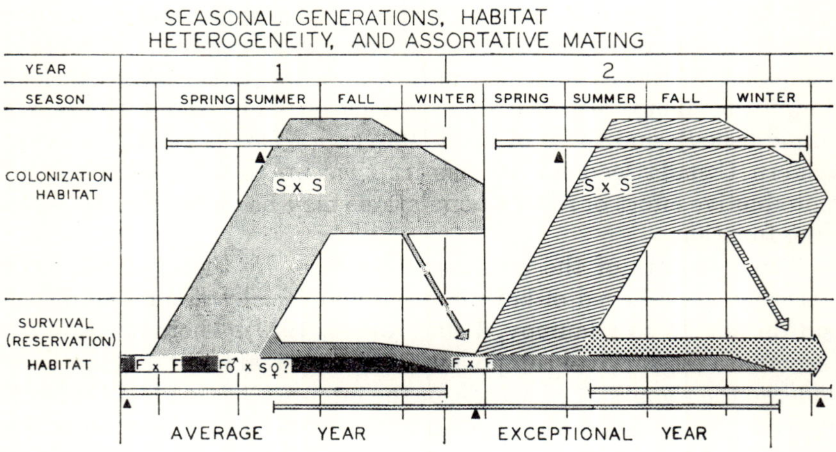

FIG. 3. Implications of seasonal generations and habitat heterogeneity. Horizontal bars indicate the survival times of seasonal generations and small triangles indicate the start of reproductive activity. Matings are indicated as being generally between members of the same generation with the exception that father-daughter matings occasionally involve autumn born males and spring born females. Widths of the hatched areas indicate relative numbers in survival and colonization components. Broken arrows indicate colonization of vacant survival habitat by members of the spring generation. In most years elimination of the remainder of the colonization component occurs in autumn or winter, but in exceptional years spring-born animals and their offspring may survive and continue to reproduce, contributing to an "outbreak" population.

"population highs" and to the continuing gene pool. It is through them that such factors as weather, in both density-dependent and density-independent aspects, have major effects on population fluctuations. In different years and in different habitats the undiscriminating student of populations will include varying proportions of the 2 types in his samples. During population peaks, the distinction between colonization and survival components may be submerged (Koshkina, 1967). Even in colonization habitats, failure to discriminate between established residents and transients, or failure to take the season in which the collection was made into account, will lead to anomalous results. Such mixing has undoubtedly helped to delay recognition of some of the phenomena described. It may also account for some of the differences in weight, body proportions, and other attributes reported for so-called "peak" and "crash" populations (e.g. Krebs, 1964) and, finally, may be involved in the apparent seasonal and density-related changes in gene frequency which Semeonoff and Robertson (1968), and Tamarin and Krebs (1969) have attempted to demonstrate.

It should be noted in passing that not all populations can be divided into survival and colonizing components. Long-lived hibernators, such as *Zapus*, which may produce only a single litter annually in northern parts of the range (Brown, 1967), furnish obvious examples among the smaller rodents.

HOME RANGE, SOCIAL ORGANIZATION AND THE DEME

While members of the colonizing component of rodent populations may spend an unknown (but possibly considerable) proportion of their life spans as transients, the establishment of home ranges appears to be of critical importance in the life of the individual, and in the biology of the population. The home range, once established, is permanent except in special cases (e.g. *Microtus oeconomus*, studied by Tast (1966) in northern Finland). It is an area in which routine activities are carried out (Burt, 1943) and which is probably systematically and regularly explored (Shillito, 1963). It is an area to which an individual will attempt to return if displaced. True home range is apparently an adult phenomenon and a *sine qua non* for reproductive contribution. The only apparent exceptions to this are found in the population structures of some insectivores, for example, *Sorex araneus*, where breeding females occupy territories, but males abandon the home range habit with sexual maturity in the spring and become transient for the reproductive season, remaining so until their deaths (Buckner, 1969). Such cases, unknown among rodents, may be the only instances where small mammals approximate to a simple isolation-by-distance breeding structure.

As has long been recognized (e.g. Blair, 1951, 1953), restriction of reproduction to adults occupying home ranges limits the number of potential mates, and has a major depressing effect on gene flow. The homing behaviour of accidentally displaced individuals similarly reduces the potential rate of spread of new alleles. For members of the colonizing component which survive to reproduce, the distance by which they have dispersed intervenes between the point of birth of the parent and the mean point of birth of its immediate descendants. The majority of dispersal distances reported in the literature are trivial, even in cases where mere physical movement has been equated with dispersal (e.g. French, Tagami and Hayden, 1968). Fenyuk (1937) cites recorded movements of up to 4 or 5 km in *Mus musculus*, and Kalela (1961) suggests that circumstantial evidence indicates effective dispersal of up to 100 km by *Lemmus lemmus* during the colonizing ("outbreak") phase. The implication of views expressed in this paper, however, is that even such movements are of primarily ecological significance and have little or nothing to do with "gene flow" since local survival populations are only rarely penetrated by new arrivals. Under the concepts developed here, gene flow would take place through competition between established demes in contributing to the establishment of new groups in survival habitats. The group process is probably a slow one and its role in maintaining species integrity very slight. From this vantage point it appears that measurement of gene flow in terms of individual "dispersal movements" is of little or no significance while the size, composition, and persistence of groups is of major evolutionary interest.

Groups, among rodents, appear largely of social origin. Our understanding of rodent social structure is still exceedingly unsatisfactory, despite heroic efforts at classification from primarily ethological viewpoints (Eisenberg, 1966, 1967; Fisler, 1969). Our views on social organization, like our understanding of population regulation, have suffered severely from uncritical reliance on observations made under conditions of confinement. Probably our understanding of social systems has suffered still more, however, from the fact that the human sense of smell is degenerate, while smell and chemical communication probably are the major bases of integration of rodent societies. Considering the reviews of Whitten (1966), Bruce (1967), and Ropartz (1968), it is clear that our understanding of the physiological and social effects of smell is rudimentary, yet we know that the information content of odours used by rodents includes individual identification, sex, a good deal about physiological and psychological state (including recent experiences such as defeat or fright), and indication of whether or not the individual releasing the odour is solitary or a group member, as well as whether or not the group in question is alien or familiar. This information travels with the individual in its movements, but it can also be posted at critical points along the way, where it will be available for a considerable period of time. During that interval it can be reviewed by neighbours and passers by, inducing them to leave messages of their own, or to take aggressive or evasive action. It appears that even a (small) group dispersed in such a way that home ranges were contiguous, but without overlap, could become an integrated social unit in the absence of bodily or visual contact. The evidence is considerable, therefore, that complex social organizations can be erected on the basis of rodent capabilities in chemical communication. Further, the fact that our ability to detect and evaluate chemical communication is so limited suggests that we are likely to underestimate the complexity and effectiveness of rodent social systems.

Two concepts underly the subsequent discussion. The first is that territory is best defined in the sense of Pitelka (1959) as an area occupied to the exclusion of some category of non-specific individuals. The second is that the distinctions between eco-geographically and socially defined groups, and between "kin selection" and "group selection" (Brown, 1966) are trivial insofar as the evolutionary consequences may be the same. In other words, the essential qualities of a Mendelian population or deme are internal panmixis, and restriction of genetical interchange with other units, not the mechanisms by which these results are achieved.

In the granaries, where sharp habitat discontinuities provide reference points for both the mice and the investigator, and in continuous habitats where reference points are recognizable only by the mice, the House mouse group functions as a deme in a survival habitat. The mechanisms conferring demic status appear to be social in this and many other rodent species.

The House mouse deme consists of so few individuals that drift may influence loci subject to even relatively strong selection (Lewontin, 1962; Anderson, 1965). What generalizations can be made about the size and composition of groups in other rodents?

In all such discussions it is necessary to remember that the total number of individuals in an aggregation is not necessarily a direct indication of the number actually participating in reproduction. The distinction between absolute group size and effective group size in the evolutionary sense is not always easily made.

In *Mus musculus*, demes may be family groups of 2–3 males and 4–5 females sharing a group territory. Equivalent structure is suggested for *Microtus californicus* by the observation of Pearson (1960) that runway systems were the shared and exclusive property of family groups of similar size, and by Batzli's (1968) analysis of dispersion patterns for this species in continuous grassland. Family groups have frequently been recognized as persistent units in populations of the genus *Rattus*. Even under conditions of confinement, Calhoun (1963) found that 120 adult *Rattus norvegicus* had formed 11 clear-cut colonies in his half-acre enclosure. Subordinate, non-reproductive individuals (which I would define as a colonization component frustrated in its function by confinement) were present, but the groups of individuals which Calhoun designated as "compact colonies" conform in size and composition to the *Mus* demes.

A slightly different pattern of organization, but with demes of similar size and composition, can be deduced from information available on a number of microtine and geomyid species. In *Clethrionomys rufocanus* Kalela (1957) concluded that populations were organized into inbreeding "clans". Breeding females are territorial, but the territories of up to as many as 30 females apparently can be integrated by the much larger overlapping home ranges of member males. Significantly, Kalela found that these breeding colonies were founded in autumn by immature animals. Similarly Tast (1966) found that in *Microtus oeconomus* small aggregations (about 10 immature individuals) formed in the autumn and overwintered on high ground. During the breeding season females occupied shifting home ranges, following the line of retreating flood waters, but groups appear to remain integrated by large male home ranges.

The geomyids have long been viewed by ethologists as solitary and are commonly cited as the extreme of dispersed social organization (Eisenberg, 1966), but Russell (1968), in reviewing the natural history of pocket gophers of the genus *Pappogeomys*, says that groups (colonies) are composed consistently of 1–3 adult males and a small number (apparently from 5–10) of mature females. Genelly (1965) has reported that in the similarly adapted bathyergid *Cryptomys hottentotus*, colonies consist of 10–12 individuals which form "probably socially exclusive units".

Significantly, geomyids have proved particularly "variable" as seen by systematists who tend to equate any difference in sample means, whether statistically significant or otherwise, with subspecific status (e.g. Russell, 1968). Whether they are truly more variable than other groups can now be evaluated with such techniques as electrophoresis.

In the much studied genus *Peromyscus*, Blair (1953) wrote that "freely interbreeding units may consist of only a few individuals" but stated that "in other cases there may be no limitation but distance on the interchange of individuals in a population of millions". Careful analysis is more likely to confirm, and extend to related groups, Rasmussen's (1964) estimate that Mendelian populations of *Peromyscus maniculatus* contain between 10 and 80 mice.

The examples which have been reviewed come from groups generally assumed to represent the lower end of the scale of complexity in rodent social organization (Eisenberg, 1966, 1967). In species such as *Microtus brandti*, *Rhombomys opimus* and *Psammomys obesus*, regarded by Eisenberg as colonial, the more elaborate social organization can reasonably be expected to emphasize demic structuring still further. Numerous specialized variants can also be expected, as indicated by the example of *Microtus pinetorum* for which Gentry (1968) has presented data suggesting that the population is organized into nomadic groups of up to 45 individuals.

Naumov (1963) has developed a concept of hierarchical organization in populations in which the basic unit is the "elementary population". This he defines as the smallest group capable of self-perpetuation. In my opinion this is equivalent to the deme as defined in the present review, although N. P. Naumov (personal communication) does not entirely agree.

To sum up, it appears likely that careful study will reveal demic organization in a number of species, and that the effective size of the Mendelian units may frequently be under 20 individuals. In demes of such small size, some loci will be subject to individual selection while others will be strongly influenced by drift and interdeme selection. Genetical evidence is the most compelling in this regard, and genetic data compatible with this hypothesis are now available for a number of species, including *Peromyscus maniculatus* (Rasmussen, 1964), *Rattus rattus* (Van Valen and Weiss, 1966; Tomich, 1968), and *Clethrionomys glareolus* (Corbet, 1963) as well as *Mus musculus* (Lewontin and Dunn, 1960; Anderson, 1964, 1965; Petras, 1967*a*, 1967*b*; Berry, 1968). Electrophoretic techniques will permit a rapid accumulation of relevant genetical information in the next few years.

HABITAT AND MATE SELECTION, AND THE ECOTYPE

Harris (1952) compared the behaviour of *Peromyscus maniculatus gracilis* (a woodland form) and *P.m. bairdi* (a grassland form) in a laboratory room subdivided into areas with simulated forest and simulated grassland. Each form tended to select the artificial environment most closely resembling the habitat characteristic of the subspecies. Wecker (1963) carried this analysis further and found that both field-caught animals and their laboratory-born-and-reared offspring showed preference for the appropriate habitat in a large outdoor enclosure. Animals from stocks of *P.m. bairdi* reared for 12 to 20 generations in the laboratory did not discriminate between grassland and forest habitats. When reared in the field, however, animals from these laboratory stocks showed preference for the habitat where their early experience had taken place. Wecker has therefore shown that both inherited and learned components are involved in habitat selection. The forms with which Harris (1952) and Wecker (1963) worked, and other forest and grassland forms of *Peromyscus maniculatus,* are able to maintain separate populations in areas of sympatry (Dice, 1968), even though laboratory crosses reveal no major genetic incompatibility.

Godfrey (1958) demonstrated that bank voles (*Clethrionomys glareolus*) were capable of discriminating in mate selection in favour of mates from the same island, and that mainland animals discriminated in favour of potential mates from relatively close, as opposed to more distant, mainland areas. Mainardi (1963*a*) has shown that *Mus musculus* are able to make precise distinctions in the selection of mates and are inclined to do so. Unfortunately his major summary of the role of sexual odours in evolution (Mainardi, 1968) has not been made available or abstracted in English.

With reference to the concept of population structure presented in this review, mate and habitat selection are of particular interest as they may be expressed by the colonizing component of populations. The material reviewed suggests that the members of this component will have the capacity and inclination to mate chiefly with individuals with (i) similar genomes, (ii) similar previous experience of the environment, and (iii) in habitats similar to those in which they were born.

Smith (1966) has shown on theoretical grounds that a stable polymorphism may be maintained if the environment is divided into two or more habitat types, if 3 conditions are satisfied: (i) if each of 2 alleles has a selective advantage for one habitat as compared with the other; (ii) if population sizes are separately regulated in the two habitats; and (iii) if the respective selective advantages are relatively large. Demic structure, in conjunction with mate and habitat selection, and the division of labour between seasonal generations, appears to fulfill these conditions and to provide a favourable opportunity for the development of alternative ecological forms (ecotypes) separated in both time and space.

Except in such relatively rare instances as the recognition of grassland and forest groups of subspecies in *Peromyscus* (Baker, 1968), western workers on mammals have taken little cognizance of the possible existence of ecotypes. In contrast, Russian investigators, for example Panteleyev and Terjechina (1968) and Koshkina (1967), have paid considerable attention to ecotypic variation, and Naumov (1963) recognizes the "ecological population" as a major category in his analysis of population structure.

The Russian work suggests that ecotypes may have important roles in the biology of some rodent populations. Investigation of ecotypic variation by western workers is badly needed. Seasonal generations are essentially temporal ecotypes, and their relation to spatial ecotypes is unclear. Modern techniques, particularly electrophoresis, must be applied to the identification of such units and analysis of their ecological and genetic basis.

SUBSPECIES, OR GEOGRAPHICAL POPULATIONS

Infra-specific groupings, such as those discussed in the present paper, must be expected to be part of a continuum: demes, ecotypes, and subspecies should be visualized as possible modal patterns of organization along this continuum. This does not, however, excuse the tendency to consider them generally equivalent, nor does it excuse the tendency to perpetuate unsophisticated taxonomy that takes slight deviations in sample means, the occurrence of a recognizable karyotype, or the presence or absence of a single electrophoretically recognizable protein, as adequate ground for a subspecific appellation (e.g. Youngman, 1967; Russell, 1968; Patton and Dingman, 1968; Waters, 1969). Lidicker (1962) has published a thoughtful consideration of the subspecies in mammalian practice. He concluded "the ability to prove that two populations are statistically different is only a measure of the persistence and patience of the systematist" and stated that since "categories based on a few arbitrary characters are themselves arbitrary," the use of the subspecies category for description of geographical variation in a few characters, or as a way of cataloguing geographical variants, means that nothing more than "artificial classifications of convenience" will be produced.

If demes have the qualities envisaged in the present paper, and if gene flow is as restricted as here implied, the major geographical variants to which trinomials can be reasonably applied owe little if anything to "geographical isolation" in the usual sense. Nevertheless, sharp character gradients are frequently correlated with geographical discontinuities, and this means that alternative explanations are required. One possibility is raised by the study of Selander, Hunt and Yang (1969) referred to earlier. If the integrity of coadapted genomes can maintain a sharp character gradient at the junction of independent dispersal routes, the association of geographical barriers with character gradients may be merely coincidental in some cases. In other cases geographical barriers may have an effect by preventing interdeme competition, analogous but not equivalent to restriction of gene flow in the traditional sense. In still other cases, barriers may have been involved in the disruption and reorganization of coadapted genomes through selection and founder effects. Strong selection and founder effects during colonization, as well as major reorientation of selective pressures, appear to characterize island situations (Anderson, 1960), and colonization across geographical barriers in non-island situations must involve many of the same phenomena. Finally, colonization components, with their tendencies towards outbreeding and specialization, are most likely to be involved in movements across barriers. This increases the possibility for reorganization of coadapted genomes. Where barriers do not exist, on the other hand, the ecological and evolutionary conservatism of the survival populations contributes to evolutionary stability.

These speculations indicate, at least, that the concept of population structure based on demes and ecotypes is not incompatible with those observed patterns of large-scale geographical variation below the species level which are properly recognized by application of the trinomial. From this viewpoint, the subspecies level of structure can be characterized by the possession of a widespread coadapted genome, by differentiation from other coadapted systems, and by sharp character gradients. An alternative definition could be based on evidence of substantially decreased viability in nature for the offspring of crosses between sympatric ecotypes which are reaching a high level of genetical differentiation (i.e. where sympatric speciation is in progress). Unfortunately, a considerable majority of the currently recognized rodent subspecies will not fit this or any other biologically reasonable definition (e.g. Hall and Kelson, 1959).

The category of "subspecies cluster" recently suggested by Russell (1969) for geomyids, is probably equivalent to the subspecies as conceived in the present paper.

THE SPECIES

The intention of this paper is to place emphasis on the question of how species populations maintain internal consistency, rather than on how they are isolated from other species. While these are, in one sense, part of the same question, only the former aspect will be touched upon here. If demes are as small and discrete, and gene flow so trivial, as the

Mus situation seems to indicate, how is species identity maintained? One alternative, suggested by Ehrlich and Raven (1969) is convergent or stabilizing (individual) selection. This can be regarded as an extension of the concept of the fundamental niche formulated by Hutchinson (1957) in that it would represent selection against departure from the niche parameters. The operation of selection in this way would be enhanced by the conservatism of the survival populations, and the fidelity of colonizers in habitat and mate selection. On the other hand, the example of *Mus musculus* provides a strong argument against an unqualified "selection hypothesis". It seems that species identity in *Mus musculus* is maintained in environments so diverse that it is difficult to conceive of much common ground for selection (for example, the Alberta granaries and the tropical islands of Guam or the Galapagos). In such cases one can invoke the Selander hypothesis of coadaptation of genomes (which, as Ehrlich and Raven would probably point out, is still a kind of selection). Indications are, however, that individual selection is not likely to be effective at all loci in demes of such small size. Therefore, genetic drift and interdeme selection, as well as the ecological and genetical experimentation in the dispersing and outbreeding colonizing component, must also be included in an equation of the factors maintaining identity in space and time, and bringing about change in species populations. This gives a complex picture of the species, and one which does violence to the definition of a species as "a group of actually or potentially interbreeding populations". Since it assigns some importance to group selection, however, the view of species structure presented in the present paper is also heretical to the "selectionist" approach of Ehrlich and Raven (1969). While is is not yet fully matured, and will certainly not prove applicable to all species, or even all small mammal species, I believe the present analysis has the potential for a considerable advance in our understanding of the nature and dynamics of natural populations.

CONCLUSIONS

1. Current knowledge of the structure of *Mus musculus* populations suggests a model applicable to analysis of the biology of populations of other small mammals.

2. In the House mouse the unit of population structure is a closed breeding group of small size which maintains a stable density through exclusion of immigrants and export of reproductive surpluses. Emigrants serve an important function in exploring new habitats and founding new groups.

3. The socially defined breeding groups are Mendelian populations or demes. As a consequence of their small size, genetic drift and inter-deme selection have influential roles in determining gene frequencies at some loci. Isolation of the demes is sufficient to eliminate "gene flow" as a major factor in the maintenance of species integrity.

4. Many rodent species are cyclomorphic (i.e. they have seasonal phenotypes with distinctive behavioural, physiological, morphological and genetical characteristics which have different ecological and evolutionary roles and occupy different habitats). Studies of the demography, genetics, and systematics of rodent populations must take account of the seasonal generation phenomenon.

5. The ecotype should be recognized to have potential ecological and evolutionary importance. The existence and nature of ecotypes should be investigated by western workers, who should also familiarize themselves with the Russian literature on the subject.

6. The structural system described provides for the ecological and evolutionary opportunism demanded by capricious environments, and it is inherent in the nature of the generalized "rodent niche".

7. Taxonomists must discard the notion that slight differences in sample means, or the presence or absence of one or two alleles, provide grounds for application of a trinomial. Since differences of the same order of magnitude exist between most demes, the practice can contribute nothing to our understanding of evolutionary processes or our knowledge of variation.

8. Organization of species populations into small demes, and division of labour among survival and colonizing components, in conjunction with the phenomena of home range and territoriality, reduce the potential role of "gene flow" in the maintenance of species integrity. The time has passed when facile assumptions regarding the existence and extent of gene flow are adequate for interpretation of observed patterns of variation.

Acknowledgements

In the summer of 1967 I participated in the scientific exchange sponsored by the National Research Council of Canada and Akademia Nauk, USSR. The opportunity, through personal contact, partially to overcome the barriers of language and inadequate literature exchange was influential in the genesis of ideas expressed in this paper.

References

Anderson, P. K. (1960). Ecology and evolution in island populations of salamanders in the San Francisco Bay region. *Ecol. Monogr.* **30**, 359–385.

Anderson, P. K. (1961). Density, social structure, and non-social environment in house-mouse populations and the implications for regulation of numbers. *Trans. N.Y. Acad. Sci. II*, **23**, 447–451.

Anderson, P. K. (1964). Lethal alleles in *Mus musculus*: Local distribution and evidence for isolation of demes. *Science, N.Y.* **145**, 177–178.

Anderson, P. K. (1965). The role of breeding structure in evolutionary processes of *Mus musculus* populations. In *Mutation in population*, Proceedings of the Symposium on the Mutational Process, Prague, August 9–11, 1965. 17–21. Academia.

Anderson, P. K., Dunn, L. C. and Beasley, A. B. (1964). Introduction of a lethal allele into a feral house mouse population. *Am. Nat.* **98**, 57–64.

Anderson, P. K. and Hill, J. L. (1965). *Mus musculus:* Experimental induction of territory formation. *Science, N.Y.* **148**, 1753–1755.

Andrzejewski, R., Petrusewicz, K. and Walkowa, W. (1963). Absorption of newcomers by a population of white mice. *Ekol. polska* **11A**, 223–240.

Baker, R. H. (1968). Habitats and distribution [of *Peromyscus*]. In *Biology of Peromyscus*. King, J. A. (ed.), 98–126. *Spec. Publ. Am. Soc. Mammal.* No. 2.

Batzli, G. O. (1968). Dispersion patterns of mice in California annual grassland. *J. Mammal.* **49**, 239–250.

Berry, R. J. (1963). Epigenetic polymorphism in wild populations of *Mus musculus*. *Genet. Res.* **4**, 193–220.

Berry, R. J. (1968). The ecology of an island population of the house mouse. *J. Anim. Ecol.* **37**, 445–470.

Berry, R. J. and Searle, A. G. (1963). Epigenetic polymorphism of the rodent skeleton. *Proc. zool. Soc. Lond.* **140**, 577–615.

Blair, W. F. (1948). Population density, life span, and mortality rates of small mammals in the blue-grass associations of southern Michigan. *Am. Midl. Nat.* **40**, 395–419.

Blair, W. F. (1951). Population structure, social behavior, and environmental relations in a natural population of the beach mouse. *Contrib. Lab. Vert. Biol. Univ. Mich.* **48**, 1–47.

Blair, W. F. (1953). Population dynamics of rodents and other small mammals. *Adv. Genet.* **5**, 1–41.

Brown, J. L. (1966). Types of group selection. *Nature, Lond.* **211**, 870.

Brown, L. N. (1967). Seasonal activity patterns and breeding of the western jumping mouse (*Zapus princeps*) in Wyoming. *Am. Midl. Nat.* **78**, 460–470.

Bruce, H. M. (1967). Effects of olfactory stimuli on reproduction in mammals. *Ciba Fdn. Study Gr.* No. 26: 29–42.

Buckner, C. H. (1969). Some aspects of the population ecology of the common shrew, *Sorex araneus*, near Oxford, England. *J. Mammal.* **50**, 326–332.

Burt, W. H. (1943). Territoriality and home range concepts as applied to mammals. *J. Mammal.* **24**, 346–352.

Calhoun, J. B. (1962). A "behavioral sink." In *Roots of behavior—Animal behavior*. Bliss, E. L. (ed)., 295–315. New York City: Harper and Bros.

Calhoun, J. B. (1963). The ecology and sociology of the Norway rat. *Publ. Hlth Serv. Publs. Wash.* No. 1008, 1–288.

Christian, J. J. and Davis, D. E. (1964). Endocrines, behavior and population. *Science, N.Y.* **146**, 1550–1560.

Corbet, G. B. (1963). An isolated population of the bank vole (*Clethrionomys glareolus*) with aberrant dental pattern. *Proc. zool. Soc. Lond.* **140**, 316–319.

Crowcroft, P. and Rowe, F. P. (1958). The growth of confined colonies of the wild house-mouse (*Mus musculus* L.): the effect of dispersal on female fecundity. *Proc. zool. Soc. Lond.* **131**, 357–365.

Dahl, A. E. (1967). *Responses of free-living vole populations to experimental modifications of density*. Unpubl. M.Sc. thesis, University of Calgary.

Dice, L. R. (1968). Speciation [in *Peromyscus*]. In *Biology of Peromyscus*. King, J. A. (ed.), 75–97. *Spec. Publ. Am. Soc. Mammal.* No. 2.

Dunn, L. C. and Levene, H. (1961). Population dynamics of a variant *t*-allele in a confined population of wild house mice. *Evolution, Lancaster, Pa.*, **15**, 385–393.

Ehrlich, P. R. and Raven, P. H. (1969). Differentiation of populations. *Science, N.Y.* **165**, 1228–1232.

Eibl-Eibesfeldt, I. (1950). Beiträge zur Biologie der Haus und der Ahrenmaus nebst einigen Beobachtungen an anderen Nagern. *Z. Tierpsychol.* **7**, 558–587.

Eisenberg, J. F. (1966). The social organizations of mammals. *Handb. Zool.* **8** (10), sect. 7, 1–92.

Eisenberg, J. F. (1967). A comparative study in rodent ethology with emphasis on evolution of social behavior, I. *Proc. U.S. nat. Mus.* **122**, 1–51.

Fenyuk, B. K. (1937). The influence of agriculture on numbers of mouse-like rodents and the biological foundations of rodent control. *Vest. Mikrobiol. Èpidem. Parazit.* **16**, 478–492.

Fisler, G. F. (1969). Mammalian organizational systems. *Contr. Sci.* No. 167, 1–32.

Frank, F. (1957). The causality of microtine cycles in Germany. *J. Wildl. Mgmt* **21**, 113–121.

French, N. R., Tagami, T. Y. and Hayden, P. (1968). Dispersal in a population of desert rodents. *J. Mammal.* **49**, 272–280.

Fuller, W. A. (1969). Changes in numbers of three species of small rodent near Great Slave Lake, N.W.T., Canada, 1964–1967, and their significance for general population theory. *Ann. Zool. fenn.* **6**, 113–144.

Genelly, R. E. (1965). Ecology of the common mole-rat *Cryptomys hottentotus* in Rhodesia. *J. Mammal.* **46**, 647–665.

Gentry, J. B. (1968). Dynamics of an enclosed population of pine mice *Microtus pinetorum*. *Researches Popul. Ecol., Kyoto Univ.* **10**, 21–30.

Getz, L. L. (1960). A population study of the vole, *Microtus pennsylvanicus*. *Am. Midl. Nat.* **64**, 392–405.

Godfrey, J. (1958). The origin of sexual isolation between bank voles. *Proc. R. phys. Soc. Edinb.* **27**, 47–55.

Greenwald, G. S. (1956). The reproductive cycle of the field mouse, *Microtus californicus*. *J. Mammal.* **37**, 213–222.

Greenwald, G. S. (1957). Reproduction in a coastal California population of the field mouse *Microtus californicus*. *Univ. Calif. Publs Zool.* **54**, 421–446.

Hall, E. R. and Kelson, K. R. (1959). *The mammals of North America*, 2. New York: Ronald Press.

Harris, V. T. (1952). An experimental study of habitat selection by prairie and forest races of the deer mouse *Peromyscus maniculatus*. *Contr. Lab. vertebr. Biol., Univ. Mich.* **56**, 1–53.

Hill, J. L. (1966). *Aspects of territorialism in Mus musculus*. Unpubl. M.Sc. thesis University of Calgary.

Hoffmann, R. S. (1958). The role of reproduction and mortality in population fluctuations of voles (*Microtus*). *Ecol. Monogr.* **28**, 79–109.

Howard, W. E. (1949). Dispersal, amount of inbreeding, and longevity in a local population of prairie deermice on the George Reserve, southern Michigan. *Contr. Lab. vertebr. Biol. Univ. Mich.*, **43**, 1–50.

Howard, W. E. (1960). Innate and environmental dispersal of individual vertebrates. *Am. Midl. Nat.* **65**, 257–259.

Hutchinson, G. E. (1957). Concluding remarks. *Cold Spring Harb. Symp. quant. Biol.* **22**, 415–427.

Justice, K. E. (1962). Ecological and genetical studies of evolutionary forces acting on desert populations of *Mus musculus*. *Report*, Environmental Sciences Branch Division of Biology and Medicine, United States Atomic Energy Commission. Arizona-Sonora Desert Museum, 1–68.

Kalela, O. (1957). Regulation of reproduction rate in subarctic populations of the vole *Clethrionomys rufocanus* (Sund.). *Ann. Acad. Sci. fenn.* Ser. **4A** (34/40), 1–60.

Kalela, O. (1961). Seasonal change of habitat in the Norwegian lemming, *Lemmus lemmus* (L.). *Ann. Acad. Sci. fenn.* **4A** (55), 1–72.

King, J. A. (1968). Psychology [of *Peromyscus*]. In *Biology of Peromyscus*. King, J. A. (ed.), *Spec. Publ. Am. Soc. Mammal.* No. 2, 496–542.

Koshkina, T. V. (1967). Ecological differentiation of species; the vole as an example. *Acta theriol.* **12**, 135–163.

Krebs, C. J. (1964). Cyclic variation in skull–body regressions of lemmings. *Can. J. Zool.* **42**, 631–644.

Levine, L. and Lascher, B. (1965). Studies on sexual selection in mice. II. Reproductive competition between black and brown males. *Am. Nat.* **99**, 67–72.

Lewontin, R. C. (1962). Interdeme selection controlling a polymorphism in the house mouse. *Am. Nat.* **96**, 65–78.

Lewontin, R. C. (1965). Selection for colonizing ability. In *The genetics of colonizing species*. Stebbins, G. L. (ed.). New York: Academic Press, 77—94.

Lewontin, R. C. (1966). Is nature probable or capricious? *BioScience* **16**, 25–27.

Lewontin, R. C. and Dunn, L. C. (1960). The evolutionary dynamics of a polymorphism in the house mouse. *Genetics* **45**, 705–722.

Lidicker, W. Z. (1962). The nature of subspecies boundaries in a desert rodent and its implications for subspecies taxonomy. *Syst. Zool.* **11**, 160–171.

Mainardi, D. (1963*a*). Experiment on the active part played by the female in sexual selection in *Mus musculus*. *Arch. Sci. biol.* **47**, 227–237.

Mainardi, D. (1963*b*). Speciazione nel topo. Fattori etologici determinonti barriere riproduttive tra *Mus musculus domesticus* e *M.m. bactrianus*. *Rc. Ist. lomb. Sci. Lett.* B **97**, 135–142.

Mainardi, D. (1968). *La scelta sessuale nell'evoluzione della specie*. Paolo Boringhieri.

Martinet, L. (1967). Cycle saisonnier de reproduction du campagnol des champs *Microtus arvalis*. *Annls Biol. anim. Biochim. Biophys.* **7**, 245–259.

Naumov, N. P. (1940). The ecology of the hillock mouse, *Mus musculus hortulanus*. (In Russian). *J. Inst. Evol. Morph.* **3**, 33–77.

Naumov, N. P. (1963). [*The ecology of animals*. A textbook for universities of the USSR.] (In Russain). Vysshaya Shkola, Moscow.

Newsome, A. E. (1969*a*). A population study of house-mice temporarily inhabiting a South Australian wheatfield. *J. Anim. Ecol.* **38**, 341–359.

Newsome, A. E. (1969*b*). A population study of house-mice permanently inhabiting a reed-bed in South Australia. *J. Anim. Ecol.* **38**, 361–377.

Panteleyev, P. A. and Terjechina, A. N. (1968). [The landscape variability of the water vole (*Arvicola terrestris* L.) as a new form of animal spatial variability.] (In Russian). *Zool. Zh.* **47**, 610–617.

Patton, J. L. and Dingman, R. E. (1968). Chromosome study of pocket gophers, genus *Thomomys*. I. The specific status of *Thomomys umbrinus* (Richardson) in Arizona. *J. Mammal.* **49**, 1–13.

Pearson, O. P. (1960). Habits of *Microtus californicus* revealed by automatic photographic recorders. *Ecol. Monogr.* **30**, 231–249.

Pearson, O. P. (1967). Age structure and reproductive dynamics in a population of field mice, *Akodon azarae*. *Physis* **27**, 53–58.

Petras, M. L. (1967*a*). Studies of natural populations of *Mus*. I. Biochemical polymorphisms and their bearing on breeding structure. *Evolution, Lancaster, Pa*, **21**, 259–274.

Petras, M. L. (1967*b*). Studies of natural populations of *Mus*. II. Polymorphism at the *T* locus. *Evolution, Lancaster, Pa*, **21**, 466–478.

Pitelka, F. A. (1959). Numbers, breeding schedule, and territoriality in pectoral sandpipers of northern Alaska. *Condor*, **61**, 233–264.

Polyakov, I. Ya. (1958). [Biological basis for the control of rodents.] (In Russian) *Trudy vses. Inst. Zashch. Rast.* **12**, 3–17.

Rasmussen, D. I. (1964). Blood group polymorphism and inbreeding in natural populations of the deer mouse *Peromyscus maniculatus*. *Evolution, Lancaster, Pa.* **18**, 219–229.

Reichstein, H. (1964). Untersuchungen zum Korperwachstum und zum Reproduktionspotenzial der Feldmaus, *Microtus arvalis* (Pallas 1779) *Z. wiss. Zool.* **170**, 112–222.

Reimer, J. D. and Petras, M. L. (1967). Breeding structure of the house mouse, *Mus musculus*, in a population cage. *J. Mammal.* **48**, 88–99.

Reimer, J. D. and Petras, M. L. (1968). Some aspects of commensal populations of *Mus musculus* in southwestern Ontario. *Can. Fld Nat.* **82**, 32–42.

Ropartz, P. (1968). Olfaction et comportement social chez les rongeurs. *Mammalia* **32**, 550–569.

Rowe, F. P., Taylor, E. J. and Chudley, A. H. J. (1963). The numbers and movements of house-mice (*Mus musculus* L.) in the vicinity of four cornricks. *J. Anim. Ecol.* **32**, 87–97.

Russell, R. J. (1968). Revision of pocket gophers of the genus *Pappogeomys*. *Univ. Kans. Publs Mus. nat. Hist.* **16**, 581–776.

Russell, R. J. (1969). Intraspecific population structure of the species *Pappogeomys castanops*. *Univ. Kans. Mus. nat. Hist. Misc. Publs* No. 51, 337–371.

Schwarz, E. and Schwarz, H. K. (1943). The wild and commensal stocks of the house mouse, *Mus musculus* Linnaeus. *J. Mammal.* **24**, 59–72.

Schwarz, S. S., Pokrovski, A. V., Istchenko, V. G., Olenjev, V. G., Ovtschinnikova, N. A. and Pjastolova, O. A. (1964). Biological peculiarities of seasonal generations of rodents, with special reference to the problem of senescence in mammals. *Acta theriol.* **8**, 11–43.

Selander, R. K., Hunt, W. G. and Yang, S. Y. (1969). Protein polymorphism and genic heterozygosity in two European subspecies of the house mouse. *Evolution, Lancaster, Pa*, **23**, 279–390.

Semeonoff, R. and Robertson, F. W. (1968). A biochemical and ecological study of plasma esterase polymorphism in natural populations of the field vole, *Microtus agrestis* L. *Biochem. Genet.* **1**, 205–227.

Shillito, E. E. (1963). Exploratory behaviour in the short-tailed vole *Microtus agrestis*. *Behaviour* **21**, 145–154.

Smith, J. M. (1966). Sympatric speciation. *Am. Nat.* **100**, 637–650.

Southwick, C. H. (1955*a*). The population dynamics of confined house mice supplied with unlimited food. *Ecology*, **36**, 212–225.

Southwick, C. H. (1955*b*). Regulatory mechanisms of house mouse populations: social behavior affecting litter survival. *Ecology*, **36**, 627–634.

Strecker, R. L. and Emlen, J. T. (1953). Regulatory mechanisms in house mouse populations: the effect of limited food supply on a confined population. *Ecology*, **34**, 375–385.

Tamarin, R. H. and Krebs, C. J. (1969). *Microtus* population biology. II. Genetic changes at the transferrin locus in fluctuating populations of two vole species. *Evolution, Lancaster, Pa*, **23**, 183–211.

Tast, J. (1966). The root vole, *Microtus oeconomus* (Pallas), as an inhabitant of seasonally flooded land. *Annl Zool. fenn.* **3**, 127–171.

Tast, J. (1968). Influence of the root vole, *Microtus oeconomus* (Pallas), upon the habitat selection of the field vole, *Microtus agrestis* (L.), in northern Finland *Suomal. Tiedeakat. Toim.* (4) No. 136, 1–23.

Tomich, P. Q. (1968). Coat color in wild populations of the roof rat in Hawaii. *J. Mammal.* **49**, 74–82.

Van Valen, L. and Weiss, R. A. (1966). Selection in natural populations. V. Indian rats (*Rattus rattus*). *Genet. Res.* **8**, 261–267.

Waters, J. H. (1969). The systematic position of white-footed mice, genus *Peromyscus*, of Nantucket, Massachusetts. *J. Mammal.* **50**, 129–132.

Wecker, S. C. (1963). The role of early experience in habitat selection by the prairie deer mouse, *Peromyscus maniculatus bairdi*. *Ecol. Monogr.* **33**, 307–325.

Whitten, W. K. (1966). Pheromones and mammalian reproduction. In *Adv. Reprod. Physiol.* **1**, 155–177.

Wright, S. (1943). Isolation by distance. *Genetics, Princeton*, **28**, 114–138.

Youngman, P. M. (1967). Insular populations of the meadow vole, *Microtus pennsylvanicus*, from northeastern North America, with descriptions of two new subspecies. *J. Mammal.* **48**, 579–588.

Symp. zool. Soc. Lond. (1970) No. 26, 327–333.

VARIATION AND POPULATION DENSITY

D. CHITTY

Department of Zoology, University of British Columbia, Vancouver, Canada.

SYNOPSIS

If a habitat is particularly favourable for reproduction and survival it will produce a surplus of animals that must die or emigrate if the food supply is to be preserved. Microtine rodents living in favourable habitats achieve this regulation by going through alternating phases of increase and decrease. These changes may be due to some genotypes being the fittest when there is plenty of space and others being the fittest when there is not. The greater fitness of the latter is thought to be due to their greater aggressiveness, which has the incidental effect of reducing population density. Being intolerant of their neighbours and less highly selected to survive local hazards, such animals will decline in numbers until other genotypes can increase once again, or return from marginal habitats. There is now evidence that selection is indeed associated with changes in the numbers of small mammals, but it may be that selection is just an incidental effect of regulation instead of *vice versa* as suggested here.

INTRODUCTION

The development of ideas in population ecology illustrates too well the truth of Popper's comment (1963) that: "It is easy to obtain confirmations, or verifications, for nearly every theory—if we look for confirmations". I shall therefore produce little evidence in favour of the following views, but try to show that they are worth testing.

Three groups of workers have found evidence of selection being associated with changes in population density, the frequency of certain alleles having changed during declines in numbers of 5 different species of rodents (Table I). The functional relationship between the gene and population changes is unknown, nor were these particular alleles suspected of having an influence on the regulation of numbers. Hence, as suggested by Tamarin and Krebs (1969), finding any association at all implies that selection goes on at many other loci as well.

It is too early to assess these new data; also there are other data, not yet published, that it would be wrong to refer to in print. None of them, as far as I can tell, contradicts the ideas put forward in a recent review (Chitty, 1967), which should be consulted by those wishing to check the story below. Thus the only excuse for the present paper is to expose some old ideas to the criticisms of a new audience.

TABLE I

Polymorphisms associated with population changes in wild rodents

Species	Locus	Reference
Microtus agrestis	Esterase	Semeonoff and Robertson (1968)
M. ochrogaster *M. pennsylvanicus*	Transferrin	Tamarin and Krebs (1969)
Clethrionomys gapperi *Peromyscus maniculatus*	Transferrin and Albumin	Canham (1969)

THE PROBLEM

Animals of most species presumably live in a variety of habitats, some so harsh that populations cannot maintain themselves without immigration, others so favourable that numbers would continue to rise if there were no induced processes to keep them down. Where numbers do in fact remain about the same in successive years, the relevant processes may be abstracted into 4 components: birth rates and death rates that vary according to local conditions, and the additional gains and losses that bring populations back to their starting points. (For present purposes we shall neglect the fact that habitats are unlikely to remain sufficiently constant for there to be a predictable level for populations to return to.)

Since the effects of all 4 of these components are likely to be confounded during the breeding season, and 3 of them are likely to be confounded during the rest of the year, we cannot rely on sorting them out by purely descriptive methods. Such descriptions, being difficult or impossible for anyone else to check, are unlikely to be of general interest. Also, knowing what kills an animal is often irrelevant; the processes to study may be those affecting its chances of dying from whatever happens to be around. What we need, therefore, is a population subjected only to the first 2 components of change. This would give us an unregulated population as a control and enable us to recognize the effects of the regulatory components in stationary and declining populations. Oddly enough the need for such controls is not widely recognized in population ecology.

One of the advantages of studying microtine rodents is that their populations sometimes fluctuate out of phase; some are then expanding in the same year that others are not; at worst they supply the appropriate contrasts within a year or two of one another. A second advantage is that, in spite of being made up of different species living in vastly different habitats, these populations are in many respects alike. They are alike in the length of their cycles, in their tendency to fluctuate in phase, and in associated peculiarities of body weight and survival. Since the same causes of death are unlikely to affect both Arctic lemmings and Continental voles, for example, our problem is to sort out, from a welter of local variables, the common antecedents to a common class of effects.

In pursuing this aim we consciously direct our attention to habitats in which numbers are capable of increasing rapidly, for this is where to expect the clearest evidence that they are prevented from doing so. By making this choice we neglect those instances in which the peculiarities of survival and reproduction are largely or entirely accounted for by accidents of the local scene, though, as suggested below, marginal habitats may play a part in determining the regional course of events.

The main point of this discussion is to dispel the idea that one should look for a single explanation for the way numbers are determined within a species: the most one can hope for is to generalize about populations of specified kinds, and to do this one must be able, by means of properly controlled observations, to sort out testable processes from the particulars with which they are confounded.

A POLYMORPHIC BEHAVIOUR HYPOTHESIS

In microtine populations the class of events that most obviously needs explaining is the recurrence of high losses in spite of seemingly favourable conditions (though with ingenuity after the event one can always find something abnormal). Recovery from the resulting scarcity also needs explaining, for although it is no faster than one might expect from the reproductive powers of small mammals it does not always begin as soon as numbers get low; yet when it does begin it may continue at times of year when stationary or declining populations are not breeding at all (Krebs, 1966).

When numbers are changing rapidly we cannot assume that all genotypes are equally fertile, or that losses are distributed at random, though, with the exception of Ford (1964), most authors doubt that natural selection has much effect on the regulation of numbers. Indeed, with the present lack of evidence it is still a matter of faith to suppose that the environment of an expanding population is so vastly different from that of a stationary or declining population that selection pressures are correspondingly different. It can be said, however, that this assumption leads to an explanation that is at least consistent with some puzzling features of the approximately four-year cycle in numbers. The argument runs as follows.

1. During the phase of rapid increase those genotypes are at an advantage that survive and multiply at the highest rates. Early maturity, a long breeding season, avoidance of enemies, tolerance of heat, cold, drought, disease, and poor quality food are among the characteristics that would increase an individual's chances of contributing to the gene pool.

2. As an area becomes crowded, animals spend more time running into each other or trying not to. Some animals will be at a selective disadvantage through being unduly susceptible to disturbance, or through being unduly precipitate in avoiding their neighbours and moving off the area. However high their intrinsic rate of natural increase, such animals are at a disadvantage if they can find nowhere to settle down and breed.

3. From observations that losses occur suddenly, that they afflict some cohorts and not others, that they are not consistently associated with changes in food supply, weather, predation, incidence of disease, or viability (as observed in the laboratory), that animals capable of reaching high body weights are eliminated or prevented from growing—from this and other evidence one can make a plausible guess that sudden changes in survival are due to changes in the behaviour of the animals.

4. If selection favours the more aggressive genotypes numbers will now fall (a) because intolerant animals are more widely spaced, and (b) because other genotypes, those previously selected for maximum resistance to local mortality factors, have been eliminated.

5. The consequent regulation of numbers is thus an incidental by-product of natural selection, in which success in driving others out of the best habitats increases the fitness of those that remain.

TESTING

The studies listed in Table I provide the first tests of these ideas, by at least seeming to dispose of the null hypothesis that there are no selection pressures associated with changes in the numbers of these populations. Next we have to discover the properties being selected, for the changes can be interpreted in several ways. They may be mere symptoms of non-random action of weather; or, at the other extreme, they may be both an effect of changing density as well as a cause of still further changes. Krebs (1970) has indeed found statistically significant differences between the behaviour of animals from expanding, peak, and declining populations; but he does not know whether these differences are heritable, how they are related to differences at the transferrin locus (Table I), or whether they will turn up in future replicates. He does know that the whole process looks exceedingly complex.

Since it will be some years before we understand the genetics of fluctuating populations we must consider other ways in which the present ideas may be tested. By removing animals from expanding and from declining populations we should be able to find differences between offspring born in the laboratory. A pilot test, under poorly controlled conditions, suggested that differences in weaning weight might be associated with differences in the phase of the cycle (Chitty and Phipps, 1966). We should also be able to introduce the 2 kinds of animals into trapped-out areas and get predictable differences in population trend. Similar differences in the wild should be predictable from differences in frequency of the supposedly relevant alleles.

In developing this type of work we should be aware of some recent ideas of Wallace (1968). According to one model (his Fig. 9) a population monomorphic at a certain locus increases in size until its Darwinian fitness falls to 1, the point at which, on average, each mother leaves only one surviving daughter. A population monomorphic for the other allele is shown as increasing until it reaches a higher density; and in a mixed population the heterozygotes can stand even more crowding. At this level of abundance neither homozygote can survive at all, which gives rise to a system of balanced lethals.

In another model, in which the heterozygotes are intermediate in fitness, Wallace (1968, Fig. 11) expresses the idea, so obvious from work on voles, that measurements of population size have no predictive value unless the kinds of animals are known as well. Fitness, in fact, depends both on population size and gene-frequency.

The general features of Wallace's models have something in common with ideas in the present paper. The two homozygotes might correspond to the extreme behavioural types supposed to occur in small mammals. Thus if monomorphic populations were to occur in nature the more docile animals might increase to plague proportions: but if aggressive animals alone survived a decline, they might keep numbers down indefinitely.

There are nevertheless substantial differences between these simplified models and the realities of natural populations. The model populations, for example, do not fluctuate, but reach an equilibrium both in population size and gene-frequency. Also, Wallace's monomorphs have less effect on each other than aggressive voles are likely to have on docile voles. Many of the latter presumably emigrate. (Tamarin and Krebs (1969), by preventing emigration, produced overcrowded populations that destroyed their food supply.) Furthermore, in any natural system the survival value of aggressive behaviour is unlikely to be the same at all times and places. Because of seasonal changes, selection will not go in the same direction all year long; and because of spatial diversity, selection will not go in the same direction in all habitats. Thus, where good and bad habitats are interspersed, gene flow, by preventing selection from going to extremes, may perhaps cause populations to fluctuate irregularly or not at all. Only at times of particularly high numbers may all areas become so crowded that docile animals are eliminated as a future source of immigrants.

Many other species are able to keep their breeding populations at fairly steady levels, perhaps through getting recruits from non-breeding surpluses in marginal habitats (see Watson and Moss (1970) for a review of spacing behaviour in vertebrates). It may also turn out that our own non-random selection of study areas has deceived us into thinking that cycles among microtines are commoner than they really are.

It would speed up testing the present ideas if fluctuating populations of microtines and the more stationary populations of other species were merely different instances of the same general phenomenon. We could then use a variety of different species in testing the one hypothesis. The unifying view presented here is that differences between microtine and other populations are due to differences in the success of emigrants at surviving in marginal habitats until they can return and fill up vacancies. If so, the problem becomes the same in both cases, i.e. to discover the kinds of behaviour responsible for selective elimination from favourable habitats.

Acknowledgements

Dr. R. P. Canham and Dr. C. J. Krebs were good enough to let me know about their unpublished work; I am also grateful to them and the following people for help with this paper: Mr. E. A. Carl, Mrs. Helen Chitty, and Dr. C. F. Wehrhahn.

References

Canham, R. P. (1969). *Serum protein variation and selection in fluctuating populations of cricetid rodents.* Unpublished. Ph.D. Thesis, University of Alberta.

Chitty, D. (1967). The natural selection of self-regulatory behaviour in animal populations. *Proc. Ecol. Soc. Aust.* **2**, 51–78.

Chitty, D. and Phipps, E. (1966). Seasonal changes in survival in mixed populations of two species of vole. *J. Anim. Ecol.* **35**, 313–331.

Ford, E. B. (1964). *Ecological genetics*. London: Methuen.

Krebs, C. J. (1966). Demographic changes in fluctuating populations of *Microtus californicus*. *Ecol. Monog.* **36**, 239–273.

Krebs, C. J. (1970). *Microtus* population biology. IV. Behavioral changes associated with the population cycle in *M. ochrogaster* and *M. pennsylvanicus*. *Ecology* **51**, 34–52.

Popper, K. R. (1963). *Conjectures and refutations: the growth of scientific knowledge*. London: Routledge & Kegan Paul.

Semeonoff, R. and Robertson, F. W. (1968). A biochemical and ecological study of plasma esterase polymorphism in natural populations of the field vole, *Microtus agrestis* L. *Biochem. Genet.* **1**, 205–222.

Tamarin, R. H. and Krebs, C. J. (1969). *Microtus* population biology. II. Genetic changes at the transferrin locus in fluctuating populations of two vole species. *Evolution, Lancaster, Pa*, **23**, 183–211.

Watson, A. and Moss, R. (1970). Dominance, spacing behaviour and aggression in relation to population limitation in vertebrates. In *Animal populations in relation to their food resources*. Watson, A. (ed.). London and Edinburgh: Blackwell Sci. Publ.

Wallace, B. (1968). Polymorphism, population size and genetic load. In *population biology and evolution*. Lewontin, R. C. (ed.). Pp. 87–108. Syracuse: University Press.

Symp. zool. Soc. Lond. (1970) No. 26, 335–349.

BIOCHEMICAL POLYMORPHISMS AND GENETIC STRUCTURE IN POPULATIONS OF *PEROMYSCUS*

DAVID I. RASMUSSEN

Department of Zoology, Arizona State University
Tempe, Arizona, U.S.A.

SYNOPSIS

The reality of local populations which are genetically differentiated in species of small mammals became apparent as systematists developed concepts of subspecies and geographical races. Ecologists became aware of the organization of individual small mammals within local populations as behaviourally defined spatial units such as home ranges and territories were investigated.

Currently, numerous cryptic biochemical polymorphisms are being defined within populations of small mammals. Investigations of biochemical polymorphisms in natural populations of the Deer mouse, *Peromyscus maniculatus*, present an opportunity to use Mendelian markers further to assess spatial organization and genetical structure between and within local populations. While the use of data on biochemical polymorphisms for the assessment of the movement of mammals is at best inferential, information suggests that gene flow is restricted between and within local populations of the Deer mouse. Restriction of gene flow and the existence of demic structure are indicated by: macrogeographical and microgeographical patterns of polymorphisms, frequency distributions of the occurrence of alleles, a marked reduction of polymorphisms at functionally apparently unrelated loci within certain populations, and zygotic frequencies in local populations. Such data add to the evidence that the genetical structure of certain mammalian species is a mosaic of small behaviourally defined demic units sufficiently restricted to determine some aspects of geographical variation.

INTRODUCTION

Within species of small rodents, the concept of subspecific populations, although often bothersome to biologists, resulted from the recognition of genetically distinctive local populations within the gene pool of a species. The nomenclatural status and standards for definition of such units within species can give cause for argument but the reality of genotypic differences and spatial structure at levels below the species is indisputable (Durrant, 1955). An appreciation of the complexity of local intra-specific variation in mammalian species was originally provided by the work of Francis B. Sumner (i.e. Sumner, 1915, 1930, 1932) with laboratory colonies and natural populations of mice of the genus *Peromyscus*. These investigations, which resulted in genetical and Darwinian explanations for geographical variation, were followed by the numerous studies of Lee R. Dice and his students (i.e. Dice, 1940, 1949; Blair, 1953) which emphasized the reality of variation in mammalian gene pools and the value of quantitative approaches to geographical variation in *Peromyscus*. In fact, the pioneering investigations of Sumner, Dice and their students form the conceptual foundation upon which much of this symposium rests.

As geneticists and taxonomists increased their awareness of the genetical individuality of local mammalian populations, the ecologist and behaviourist developed concepts related to spatial organization within local populations of small mammals. Concepts such as home range and territoriality developed from data indicating that mammalian movement within populations is not random and that behavioural patterns limit individual dispersal to distances less than those defined by environmental discontinuities. As the disciplines of ecology and genetics grasp for some synthetic interchanges (e.g. Lerner, 1965), questions regarding the effects of dispersal behaviour and environmental patchiness on the structure of gene pools become more pointed.

ESTIMATION OF GENETICAL STRUCTURE

The possible genetical and evolutionary effects of the genetical structure of natural populations have been explored and discussed by Sewell Wright for the last half century (Wright 1921, 1931, 1951). This wealth of theoretical background offers the mouse trapper a foundation for estimating and inferring the extent of genetical structure in mammalian species from several types of empirical data. The words estimating and inferring must be emphasized. Theoretical models have been developed which relate genetical neighbourhood or deme size, inbreeding coefficients, migrational and mutational introduction of genetical material and selection but our limited knowledge of the population dynamics and genetical variation of small mammals precludes precise evaluation of the genetical structure of mouse populations. For example, as Chitty (1970) has suggested, changes in population density may result in changes of individual dispersal patterns as well as changes in selection at specific loci. Changes in density may therefore affect gene dispersal and gene survival. The structure and content of genetical neighbourhoods cannot be considered static entities, but entities which vary within short periods of time. Thus any estimate of demic size must be considered only as an approximation. In some species of mammals, the genetical neighbourhood may be largely defined by environmental discontinuities, as might be suspected in populations of the pika or coney, *Ochotona princeps*, on talus slides in montane western North America or the gopher, *Thomomys bottae*, in mesic islands in the arid regions of western United States. The genetical differentiation of such patchy distributions within these species is reflected in a proliferation of designated subspecific names. On the other hand genetical neighbourhoods may be primarily defined by the behaviour limits of individual dispersal within large areas of favourable environment. In such situations the genetical neighbourhood becomes a statistical unit rather than a gene pool with absolute boundaries. Since no mammalogist can accurately interpret environmental patchiness as recognized by the species he is studying, the relative roles of environmentally and behaviourally defined limits of dispersal are impossible to assess precisely. The use of mathematical models for exact inferences regarding genetical structure in natural populations can be questioned on the basis of various assumptions used in the interpretation of empirical data, but before straining to reach precision, the accumulation of bits of evidence from various types of observations may lead to more realistic concepts of genetical structure in populations of small mammals. Some general questions can be explored. Is demic size sufficiently small to subdivide local mouse populations within isolated woods or continuous forests into a mosaic of panmictic units? If such limitations of gene flow exist, what relationship does such structure have on the patterns of allelic frequencies within a single subspecies in a homogeneous environment? Do behaviourally defined patterns of individual movement such as territoriality, homing, and home range lead to genetical evolutionary consequences?

I shall concentrate on the Deer mouse of North America, *Peromyscus maniculatus*, which is a species widely distributed both geographically and ecologically throughout most of North America with some discontinuity of distribution in the arid south-western United States and Mexico. The subspecific limits of recognized races are usually poorly defined and arbitrary. What types of information are available for the evaluation of genetical structure, and what general inferences can be made from such data?

Behavioural data

Gene flow depends upon individual dispersal and perhaps the most precise way to assess gene movement is to analyse individual movement. However, the movement of an individual need not be related to gene flow. An Arctic tern may travel 20 000 miles a year but if it returns to mate on the breeding ground on which it was conceived the movement of genes between generations may be a matter of a few feet. In the genus *Peromyscus*, a wealth of data related to individual movement has been gathered (e.g. Stickel, 1968), but the relationship of such movement to gene dispersal is usually unknown. One notable exception is Howard's study of Deer mice in southern Michigan (Howard, 1949; Dice and Howard, 1951). Observing marked individuals and nest sites, Howard gathered data related to dispersal distances from birth site to breeding site. Using these dispersal data (Dice and Howard, 1951) and Wright's method of relating the variance of intergeneration dispersal to genetical neighbourhood size (Wright, 1943, 1946), Rasmussen (1964) estimated the panmictic unit in this population to vary from 12 to 99 individuals, assuming various densities of mature mating individuals. This analysis indicated that dispersal of genes is restricted within local populations. Gene flow is limited by behaviourally defined limits of movement as well as by environmental discontinuity. The difficulties of gathering behavioural data related to dispersal of genes in natural populations are, however, formidable and survey methods for assessing promiscuity in *Peromyscus* have not come to my attention.

Genetical data

The development of techniques such as electrophoresis for identification of the molecular components of the fluids, extracts and tissues of the body present an array of phenotypes to the mammalogist interested in variation and genetic structure of populations. These molecular phenotypes are often under rather simple genetical control, thereby allowing gene pools to be characterized and analysed. Unfortunately the rapid application of data on molecular phenotypes to taxonomic questions has lead to a proliferation of typological evaluations of various taxa (Rasmussen, 1968, 1969). However, the increasing literature on molecular individuality provides the means of describing spatial patterns of allelic frequencies, zygotic frequencies and frequency distributions of allelic frequencies with a precision not available previously. During the past few years I have been investigating serum protein phenotypes in populations of *Peromyscus maniculatus* from various locations in Arizona. Protein phenotypes have been used to elucidate molecular structure (Shaw, 1964), and some data related to their use in elucidating the genetical structure of populations will be presented here. The data are derived from samples of *P. maniculatus rufinus* (as assigned by Cockrum, 1960) trapped from 14 montane localities. The sample size ranges from 18–85 mice (average 41) per site. The trapping areas were all ponderosa pine forests in the mountains. Ecological identity is impossible to prove, but all the localities showed marked ecological similarities. They were separated from one another by various distances and degrees of ecological discontinuity. The genetical markers used consisted of 4 groups of blood proteins with no obvious functional relationship: haemoglobins (one codominant locus with 2 known alleles: Rasmussen, Jensen and Koehn, 1968); serum transferrins (one codominant locus with 3 alleles: Rasmussen and Koehn, 1966); serum albumins (one codominant locus with 5 alleles: Brown and Welser, 1968; J. N. Jensen and D. I. Rasmussen (unpublished)); and serum esterases (one locus with dominance, one additive locus, and 2 alleles at each: D. I. Rasmussen and J. N. Jensen (unpublished)).

Spatial patterns of allelic frequencies

The frequencies of the 14 alleles showed a marked variation among the 14 forest sites (Fig. 1). Although some alleles were rare or absent,

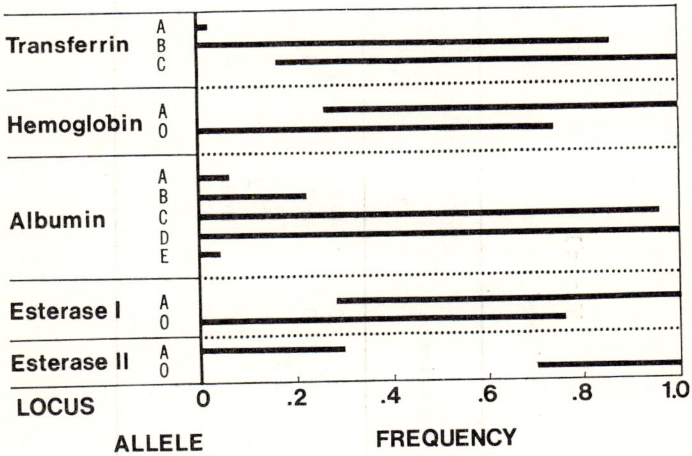

FIG. 1. Range of allelic frequencies in the Deer mouse, *Peromyscus maniculatus*, estimated from 14 ponderosa pine forest locations in Arizona. Average number of mice per location equals 41.

more than half the alleles identified had a range of frequency of over 50%. Alleles at 4 of the 5 loci varied in frequency in different samples by over 70%, and only 4 of the 14 alleles occurred in all samples. Quite obviously these montane populations do not represent a single gene pool, nor do these populations exhibit similar genetic contents despite their apparently similar environments.

What kind of spatial isolation exists between the populations and how is this reflected in allelic frequencies? No predictable pattern occurs. One of the most obvious features of geographical isolation in the state of Arizona is the Grand Canyon of the Colorado river. This massive barrier has been present since at least the Pliocene and this

TABLE I

Allelic frequencies and homogeneity probabilities estimated from montane populations of the Deer mouse in Arizona

		Protein Alleles									
		Albumins					Transferrins			Haemoglobins	
	N	A	B	C	D	E	A	B	C	A	O
Kaibab Plateau	85	0·01	0·04	0·95	0·01	0·00	0·01	0·81	0·18	0·61	0·39
Flagstaff	47	0·01	0·03	0·95	0·01	0·00	0·00	0·76	0·24	0·62	0·38
Homogeneity χ^2 probability				$P = 0·99$				$P = 0·15$		$P = 0·99$	
Combined homogeneity probability						$P = 0·69$					
Mingus Mtn. I	24	0·02	0·17	0·81	0·00	0·00	0·00	0·83	0·17	0·69	0·31
Mingus Mtn. II	34	0·00	0·04	0·96	0·00	0·00	0·00	0·63	0·37	0·54	0·46
Homogeneity χ^2 probability				$P = 0·01$				$P = 0·02$		$P = 0·14$	
Combined homogeneity probability						$P = 0·01$					

TABLE II

Estimated allelic frequencies in montane populations of the Deer mouse from Arizona

		Protein Alleles											
		Albumins					Transferrins			Haemo-globins		Front Esterase	
	N	A	B	C	D	E	A	B	C	A	O	A	O
White Mtns. 1966	37	0·03	0·11	0·86	0·00	0·00	0·00	0·85	0·15	0·36	0·64	0·58	0·42
White Mtns. 1968	18	0·00	0·14	0·83	0·00	0·03	0·00	0·72	0·28	0·42	0·58	0·56	0·44
White Mtns. 1969	18	0·06	0·17	0·78	0·00	0·00	0·03	0·69	0·28	0·61	0·39	—	—
Pinaleno Mtns.	56	0·00	0·00	0·00	1·0	0·00	0·00	0·00	1·0	1·0	0·00	0·65	0·35
Chiricahua Mtns.	62	0·00	0·00	0·00	1·0	0·00	0·00	0·00	1·0	0·94	0·06	0·59	0·41
Santa Catalina Mtns.	38	0·00	0·00	0·00	1·0	0·00	0·00	0·00	1·0	1·0	0·00	1·0	0·00

means that the trapping areas of the Kaibab Plateau and of Flagstaff have been separated by the canyon and 50 miles of scrub desert for at least 10 000 years. However, comparisons of the allelic frequencies at the 3 codominant loci indicate a marked homogeneity in the samples from these 2 sites (Table I). This similarity is not due to gene flow, and suggests evolutionary parallelism in these 2 isolated but similar ecological areas. However, an argument of strict selective maintenance of similar allelic frequencies in similar habitats falters when other montane samples are compared (Tables I and II). For example, 2 samples trapped on Mingus Mountain, from sites in a continuous montane forest habitat approximately 2 miles apart, show significant heterogeneity in spite of the lack of ecological isolation, Data such as these from Mingus Mountain indicate that behavioural restrictions on gene dispersal are sufficient to be reflected in allelic differences within continuous habitats. Furthermore, a comparison of the Kaibab-Flagstaff pair of estimates and the Mingus Mountain area pair implies that data showing differences in allelic frequencies may indicate restricted gene flow, but lack of difference need not be a result of gene exchange. Data on microgeographical differences in allelic frequencies in the genus *Peromyscus* are limited. However, fragmentary data on *Peromyscus maniculatus* in Arizona, on albumin types of *P. boylii* in Arizona (Jensen, 1969), on albumin types of *P. leucopus* in southern Michigan (Brown and Welser, 1968), and various proteins in *P. polionotus* from the south-eastern United States (C. J. Biggers and W. D. Dawson, unpublished; R. K. Selander, unpublished) indicate that genetical differences may occur in these species over very short geographical distances: distances not subdivided by rigorous environmental discontinuities but apparently by behaviourally defined restrictions of individual movement. Further investigations of microgeographical variation in populations of mice of the genus *Peromyscus* are certainly desirable. However, studies of allelic frequencies cannot lead to estimates of deme size without density estimates or recapture data.

Analysis of variation with time in allelic frequencies at a given locality offers possibilities of assessing changes in the composition of gene pools. Unfortunately, few such data exist (although see Table II).

Distribution of allelic frequencies

Most of my survey of genetic variation in Arizona populations has concentrated on macrogeographical comparisons, that is, populations in similar montane environments separated by relatively large distances and often by marked ecological discontinuities. The survey revealed a remarkable degree of homozygosity at independent loci in some populations. The White Mountains, Santa Catalina Mountains and Chiricahua Mountains are all separated from the Pinaleno Mountains by 50 to 75 miles of non-montane desert. Samples of Deer mice from the Pinalenos, Santa Catalinas and Chiricahuas all exhibit a marked reduction in polymorphism at different loci (Table II). The origin and maintenance of such monomorphism is perhaps the most unexpected observation of the survey. None of these mountains are small montane islands containing a limited total population of Deer mice. Each "island" supports more than 30 000 acres of ponderosa forests. The data from other mountainous areas (Tables I and II) indicate that other alleles are present in ponderosa pine habitats of the American south-west, yet these alternative alleles have not been incorporated into the gene pools of all populations from ponderosa forests. For example, the D albumin is rare or absent in some populations yet is the only albumin present in others, the C transferrin is rare in some forest populations yet omnipresent in others, and various other alleles at independent loci show similar patterns of frequencies. Sampling errors in the genetical composition of the founding populations may have caused the initial monomorphisms, but why have not recurrent mutations during the last 10 000 years restored polymorphism at such localities as the Santa Catalina Mountains?

The lack of alleles at some loci suggest that stochastic processes leading to allelic losses are operating in certain montane populations. The persistance in this reduction in the number of alleles at these places may result from continuing stochastic loss due to an existing restriction in deme size, or restriction of population size may have resulted in the origin of a set of unique coadapted alleles leading to the "predominance of combinations with selective advantages under prevailing conditions . . . [which] supersedes control by stochastic processes . . ." (Wright, 1951, p. 338). It is difficult to justify the selective control of the monomorphisms on the basis of the marked ecological similarity of the environments where the monomorphic and polymorphic populations were trapped. Nevertheless, the genetical structure of forest populations must play an important role in the pattern of allelic variation observed. How Wright's discussion of local occurrences of advantageous or favourable combinations due to pleiotrophic gene effects (i.e. Wright, 1951) differs from discussions of area effects due to genetical environment (i.e. Cain, 1963) I am not able nor willing to judge, especially in the light of the futile history of geneticists trying to distinguish between nonexclusive processes.

The marked variation in allelic frequencies (Fig. 1), and the fixation of one allele at certain localities (Table II) do not suggest the patterns of allelic frequency distributions expected from deterministic forces of selection and recurrent mutational introduction into large gene pools in similar environments.

If the frequencies of the 14 alleles from the 5 loci sampled in all montane localities are plotted as a histogram (Fig. 2), they bear an obvious similarity to the frequency distribution expected in populations with restricted gene flow due to limited deme size. (The right hand side of Fig. 2 represents the probable distribution of allelic frequencies

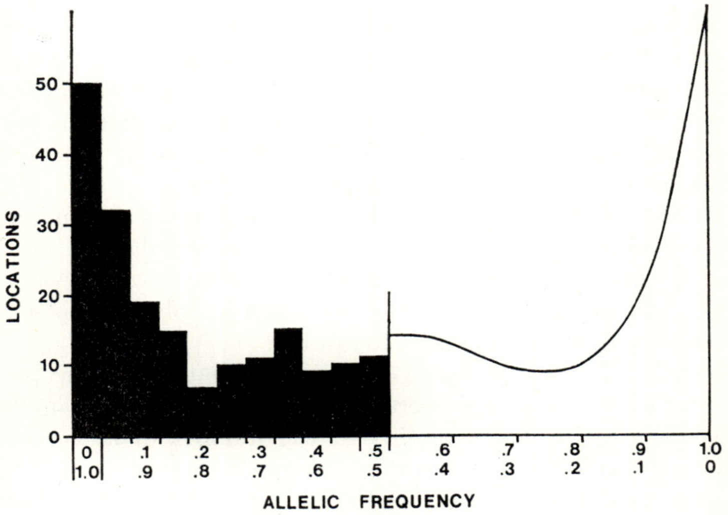

FIG. 2. Histogram distribution of frequencies of 14 alleles in Deer mouse populations from 14 montane locations in Arizona. Function on right from Wright (1951, Fig. 5) with variation in allelic frequencies resulting from interaction of drift due to small deme size, and selection favouring heterozygotes to varying extents.

resulting from conditions where local populations of restricted size increase the role of random drift and selection favours heterozygotes to varying extents: Wright, Fig. 5, 1951).

Zygotic frequencies

If a variance of allelic frequencies results from the genetical structure of different populations, the mouse trapper when collecting from a given area may be handling animals from adjacent demes. This results in the pooling of panmictic zygotes. If the populations in question have slightly different allelic frequencies, this will result in an increase of the proportion of homozygotes in the sample of animals collected

(Wright, 1943; Li, 1955). The excess of homozygosity can be measured by the coefficient of inbreeding (F) over the entire area surveyed. The value of F for the total population is a function of the population structure (i.e. the dispersal pattern, the genetically effective population size (N), and the rate of addition of "new" genetical material (m) into the area). Using the maximum likelihood estimate of F derived from zygotic frequencies at a single antigenic locus ($f = 0{\cdot}246$), Rasmussen (1964) estimated the size of the panmictic unit in populations of Deer mice from hardwood forests of northern Michigan to be of the order of 10–75 individuals. This estimate of N made use of families of curves (Fig. 3) derived from the relationships of F, N and m expected

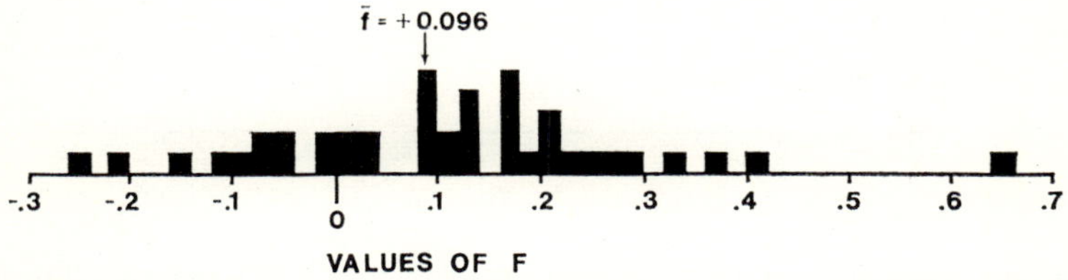

FIG. 3. Distribution of estimates of gametic correlation (F) in montane populations of Deer mice from Arizona. The 42 estimates are based on zygotic frequencies of co-dominant loci from 13 locations. The value $\bar{f} = 0{\cdot}096$, is the weighted average of the 42 locus-location estimates.

in 2 models of population structure: the isolation by distance model and the island model (Wright, 1943, 1946, 1951). The possible errors of estimating precisely genetical structure on the basis of a single locus are recognized. In my survey of Arizona populations, few loci in few samples exhibited a statistically significant excess of homozygosity. But the estimates of F derived from various forest areas where codominant polymorphic loci allowed zygotic types to be counted (42 locations-loci), obviously do not vary around a mean of $F = 0$ (Fig. 4). In fact the distribution of sample estimates of F exhibits a weighted mean of $\bar{f} = 0{\cdot}096$ (with 0·95 confidence interval, 0·093–0·099). Assuming the populations sampled exhibit a structure somewhere between Wright's island model and isolation by distance model (Fig. 3) and that the mean of these estimates represents the average level of gametic correlation (due to interdemic variance of allelic frequencies) in the populations sampled, the genetical neighbourhood size of *Peromyscus maniculatus* in ponderosa pine forest of Arizona is estimated to be of the order of $N = 30$ to 130 individuals. My reference to Wright's models does not apply to the relationship between the sampled

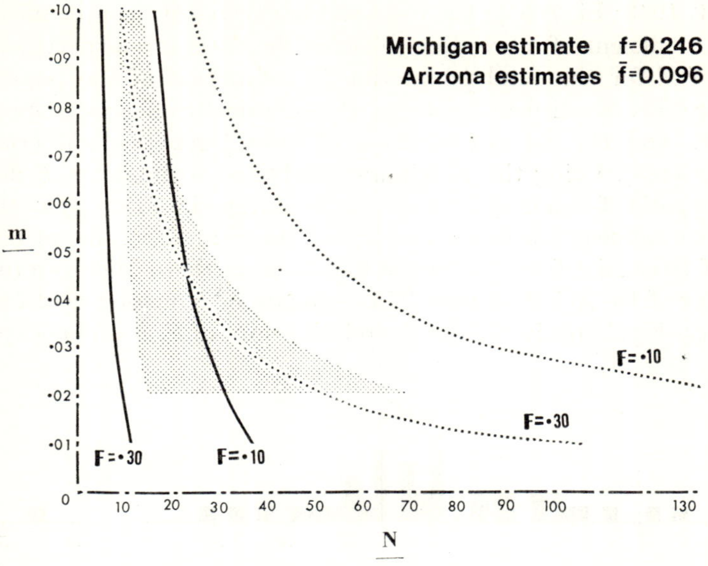

FIG. 4. Asymptotic values of gametic correlation (F). Vertical axis (m) = input rate of non-neighbourhood alleles; horizontal axis (N) = size of panmictic unit or neighbourhood. Solid lines, Wright's isolation by distance model; dotted lines, Wright's island model; shaded area lies between F = 0·20 for both models.

populations but to genetical structure within each forest area, since the zygotic frequencies were not pooled to estimate F. Instead the separate estimates of F for each location-locus were pooled to calculate an average value of gametic correlation for a Deer mouse population in ponderosa forests. I realise that the variance of allelic frequencies at different loci may not be equal, but my enquiry asked if the average variance of frequencies of a population of alleles from single forest areas was sufficient to suggest demic structure.

The genetical data used to infer population structure are all biased by selection. For phenotypes under rigorously similar selective control, intrapopulation variation of demic contents will be minimized even if interdemic gene flow is greatly restricted. For polymorphic loci exhibiting marked selective disadvantages against homozygotes, estimates of F derived from zygotic frequencies will be minimized. The covert polymorphisms investigated exhibit ranges of variation suggesting less stringent selective control than the limited overt variation leading to all these populations being included in a single subspecies. The unknown extent to which selection has reduced the genetical evidence of demic structure at any locus studied is the most serious difficulty of inferring demic structure from genetical data. This lack of precision in such estimates of N is recognized, but I also recognize that relative importance of evolutionary events and processes within a species differ if the neighbourhood size is 10, 100, 10 000 or 10 000 000 individuals. Data from Deer mouse populations indicate that the neighbourhood size within natural populations without obvious ecological fractionation is of the order of 20–120 mice. The importance of such structure in our interpretation of variation in these mammalian populations should not be confused by the question as to whether the exact neighbourhood size is actually 20 or 120 mice.

SUMMARY

Within the Deer mouse, *Peromyscus maniculatus*, behavioural data on individual dispersal, and genetical data related to differences between gene pools, the frequency distributions of allelic frequencies, and zygotic frequencies together indicate that the spatial organization of this species (related to ecological concepts such as home range and territoriality) plays a significant role in the genetical structure of the species. While such investigations are at best inferential, the data suggest that the behavioural patterns of individual mice produce a genetical structure in the species sufficient to present a mosaic pattern for the phenotypes examined. The cell was once considered a bowl of protoplasmic soup and is now thought of as an intricately combined complex of micro-structural units. With the concept of subspecies, some of the pieces in the gene pool of species became apparent. Our crude attempts to analyse the genetical microstructure of populations suggest that subspecies are not the smallest units involved in the cumulative change resulting in adaptive solutions to environmental challenges. My survey of a few genetical markers in populations of Deer mice indicates that within these populations sufficient genetical subdivision exists to allow stochastic processes to enter into the unpredictable process of evolutionary change and survival of this very successful species of North American rodent. The evolutionary advantages of such fine-scaled structure of partial isolation for evolutionary transformation have been extensively discussed by Wright (1931, 1951, 1960).

ACKNOWLEDGEMENTS

The investigations of Arizona mouse populations have been supported by the U.S. Department of Health, Education and Welfare, Public Health Research Grant GM 12190. Drs J. N. Jensen and

R. K. Koehn have given generously of their assistance in the laboratory and in the field. The laboratory assistance of Mrs. Martha Keller is gratefully acknowledged.

References

Blair, W. F. (1953). Population dynamics of rodents and other small mammals. *Adv. Genet.* **5**, 1–41.

Brown, J. H. and Welser, C. F. (1968). Serum albumin polymorphisms in natural and laboratory populations of *Peromyscus*. *J. Mammal.* **49**, 420–426.

Cain, A. J. (1963). The causes of area effects. *Heredity* **18**, 467–71.

Chitty, D. (1970). Variation and population density. *Symp. zool. Soc. Lond.* No. 26, 327–333.

Cockrum, E. L. (1960). *The recent mammals of Arizona: their taxonomy and distribution*. Tucson: University of Arizona Press.

Dice, L. R. (1940). Speciation in *Peromyscus*. *Am. Nat.* **74**, 289–298.

Dice, L. R. (1949). Variation of *Peromyscus maniculatus* in parts of western Washington and adjacent Oregon. *Contr. Lab. vertebr. Biol. Univ. Mich.* **44**, 1–34.

Dice, L. R. and Howard, W. E. (1951). Distance of dispersal by prairie deer mice from birthplace to breeding sites. *Contr. Lab. vertebr. Biol. Univ. Mich.* **50**, 1–15.

Durrant, S. D. (1955). In defense of the subspecies. *Syst. Zool.* **4**, 186–190.

Howard, W. E. (1949). Dispersal, amount of inbreeding, and longevity in a local population of prairie deer mice on the George Reserve, southern Michigan. *Contr. Lab. vertebr. Biol. Univ. Mich.* **43**, 1–50.

Jensen, J. N. (1969). *Blood protein polymorphisms in natural populations of the brush mouse* (*Peromyscus boylei*). Unpubl. D.Phil. thesis, Arizona State University.

Lerner, I. M. (1965). Ecological genetics: Synthesis. In *Genetics today*. Gerts, S. J. (ed.). **2**, 484–494. Oxford: Pergamon Press Ltd.

Li, C. C. (1955). *Population genetics*. Chicago: Univ. Chicago Press.

Rasmussen, D. I. (1964). Blood group polymorphism and inbreeding in natural populations of the deer mouse, *Peromyscus maniculatus*. *Evolution, Lancaster, Pa,* **18**, 219–229.

Rasmussen, D. I. (1968). Genetics. In *Biology of* Peromyscus (*Rodentia*). King, J. A. (ed.). pp. 340–372. Spec. Publ. 2, American Society of Mammalogists, Lawrence, Kansas.

Rasmussen, D. I. (1969). Molecular taxonomy and typology. *BioScience* **19**, 418–420.

Rasmussen, D. I., Jensen, J. N. and Koehn, R. K. (1968). Hemoglobin polymorphism in the deer mouse, *Peromyscus maniculatus*. *Biochem. Genet.* **2**, 87–92.

Rasmussen, D. I. and Koehn, R. K. (1966). Serum transferrin polymorphism in the deer mouse. *Genetics* **54**, 1353–1357.

Shaw, C. R. (1964). The use of genetic variation in the analysis of isozyme structure. *Brookhaven Symp. Biol.* **17**, 117–130.

Stickel, L. F. (1968). Home range and travels. In *Biology of* Peromyscus (*Rodentia*). King, J. A. (ed.). pp. 373–411. Spec. Publ. 2, American Society of Mammalogists, Lawrence, Kansas.

Sumner, F. B. (1915). Genetic studies of several geographical races of California deer-mice. *Am. Nat.* **49**, 688–701.

Sumner, F. B. (1930). Genetic and distributional studies of three subspecies of *Peromyscus*. *J. Genet.* **23**, 275–376.

Sumner, F. B. (1932). Genetic, distributional, and evolutionary studies of the subspecies of deer mice (*Peromyscus*). *Bibliogr. Genet.* **9**, 1–106.

Wright, S. (1921). Systems of mating. *Genetics* **6**, 111–178.

Wright, S. (1931). Evolution in Mendelian populations. *Genetics* **16**, 97–159.

Wright, S. (1943). Isolation by distance. *Genetics* **28**, 114–138.

Wright, S. (1946). Isolation by distance under diverse systems of mating. *Genetics* **31**, 39–59.

Wright, S. (1951). The genetical structure of populations. *Ann. Eugen.* **15**, 323–354.

Wright, S. (1960). Physiological genetics, ecology of populations and natural selection. In *Evolution after Darwin*. Tax, S. (ed.). **1**, 429–475. Chicago: Univ. Chicago Press.

Symp. zool. Soc. Lond. (1970) No. 26, 351–367.

VARIATION IN RODENT POPULATIONS IN RESPONSE TO CONTROL MEASURES

D. C. DRUMMOND

Infestation Control Laboratory, Ministry of Agriculture Fisheries and Food, Tolworth, Surrey, England

SYNOPSIS

The ability of populations of organisms to survive various control measures has become more noticeable in the last few decades, particularly amongst such groups as bacteria and insects. The potential for the same phenomenon to occur also among rodent populations seems to be fairly widespread. Its apparent absence to any important extent in practice where many control methods are concerned can probably be largely attributed to the lack of sustained effort with any one of these methods for a long period. On the other hand, where such effort has been made, as it has been in some countries with warfarin against rats and mice for almost 20 years, some populations of animals resistant to the poison have developed. In some instances a change in frequency of a single gene has been responsible and in others more than one gene seems to be involved. At least some of the genes concerned apparently affect the relationship between warfarin and vitamin K. The genetical and physiological nature of warfarin resistance is discussed with particular reference to its distribution and ability to spread.

INTRODUCTION

Man in his efforts to house and feed himself better has created a whole range of new environments and in so doing has often inadvertently favoured the increase of some species of animals rather than others. In many cases a favoured species has increased only at the expense of man's health or food and has thereby earned for itself the title of "pest" and subsequent retribution in the form of control measures. Few of these measures however have been aimed at or been capable of completely eradicating a pest population and repeated application of a particular method has not infrequently resulted in the eventual return of the population to pest proportions in a form capable of withstanding the method of control. In some cases alternative control methods then employed have met with a similar fate. Also, often associated with the failure of a control method, is a tendency for it to be applied, at any rate initially, with a high frequency or in more concentrated form than is necessary. Particularly when a poison is involved, this may increase the hazard to non-target organisms either immediately or as a result of long-term contamination of the environment. Today, an increasing effort is being made to find techniques that can be coordinated in such a way as to minimize all these various undesirable effects of control, whilst continuing to keep the pest population more or less permanently at the desired level—an ideal approach currently known as "integrated pest control". Meanwhile, however, man in his role as pest controller remains one of the most potent forces in exposing variation amongst populations of many organisms.

Examples of these man-invoked variations are to be found most frequently amongst such groups as bacteria and insects, no doubt largely because of their rapid reproductive rates, which enhance their ability both to become pests and to evolve strains that will withstand control pressures. Similarly the relatively few cases of variation of a comparable origin that are known in mammals occur amongst rapidly-breeding rodent populations.

The development of a population resistant to a control method depends in the first place on there being variation within the population which allows certain types of individuals to survive rather than others. Of course, as soon as control is applied to such variation, the residual population will tend to be different from any other population in which the same control method has not been used. Moreover, such inter-population differences will remain so long as control continues. In what follows however, discussion will centre around the extent to which survivors can pass on the ability to survive to their offspring and the extent to which such a process has led to increased difficulties in rodent control in the field, by allowing an increasing proportion of survivors in succeeding generations.

The control methods to be discussed will be confined to the use of poisons and disease-causing organisms. Other methods such as trapping, which generally have played only a minor role, and those still under development, such as chemosterilants, will receive no further mention.

DISEASE-CAUSING ORGANISMS

One of the earliest attempts to control rodents by spreading a lethal disease was apparently the use of *Salmonella typhimurium* against Field mice in Greece (Loeffler, 1892). This was followed in a few years by *Salmonella enteritidis* and particularly its var *danysz* for the control of the Norway rat (*Rattus norvegicus*) and the Roof rat (*R. rattus*). These various salmonellas subsequently had rather a chequered career in rodent control, and largely due to the danger of them giving rise to human food poisoning, their use was more or less discontinued in many countries by the early 1930's. However, *S. enteritidis* var *danysz* was used in Poland for a year or two on a large scale immediately following the second world war (Brodniewicz, 1968) and the use of various other *Salmonella* strains is now apparently becoming commonplace in the USSR against a wide range of rodent species (Vashkov and Polezhaev, 1965).

Right from the start of their use it was clear that many individuals in rodent populations did not succumb to *Salmonella* infections. Bahr (1947), for example, estimates the average kill of *R. norvegicus* with even the most virulent cultures of *S.e.* var *danysz* to have been about 80%. Laboratory investigations with domestic strains of the House mouse (*Mus musculus*) have demonstrated at least 2 genetically controlled mechanisms for resistance against *Salmonella* infection (Webster, 1937; Hill, Hatswell and Topley, 1940), neither of which is directly concerned with the acquisition of resistance through the production of antibodies. Other work, largely with other animals and other disease-causing organisms, has indicated the importance of such inherited mechanisms either in preventing disease at an early stage or in delaying its onset until antibodies have been produced. It also appears that the ability to produce antibodies and thereby acquire resistance to a disease is itself subject to considerable genetically-based variation (see for example, Wilson and Miles, 1964).

It seems likely therefore that many rodent populations are capable of increasing their level of resistance to *Salmonella*. That they are not recorded as having done so in practice may well be due to the often rather poor and variable results in initial treatments, which even led in some areas to following up the use of *Salmonella* cultures with an acute poison such as red squill as a routine (Bahr, 1947). It is perhaps understandable in such circumstances that, if any increase in resistance had occurred, it would have been overlooked. Somewhat in contrast to the *Salmonella* situation in rodents is the result of using myxoma virus since 1950 to control the European rabbit in Australia. The disease has apparently been largely self-perpetuating and less virulent strains of the virus have appeared as well as more resistant strains of rabbits (Marshall and Fenner, 1958, 1960).

NON-ANTICOAGULANT POISONS

Poisons used for killing rodents are, for the purposes of the present discussion, considered as belonging to one of two groups. One group, the anticoagulant poisons, which all seem to have a similar mode of action, are considered in the next section. The remainder form a miscellaneous assemblage with a variety of physiological effects, none of which is concerned with the clotting mechanism of blood. Of these nonanticoagulant poisons that are still to some extent in use today, arsenious oxide, strychnine, and red squill have been available for at least 200 years, whilst phosphorus and barium carbonate had already put in an appearance before the middle of the nineteenth century (Anon, 1837; Freeman, 1954). The remainder are comparatively recent introductions—as will be seen from the following references describing some of the earliest investigations of their properties and use—zinc phosphide (Grandi, 1912), thallium sulphate (Munch and Silver, 1931), sodium fluoroacetate (Kalmbach, 1945), antu (Richter, 1945), castrix (DuBois, Cochran and Thomson, 1948), DDT (Sergent and Sergent, 1948), fluoroacetamide (Chapman and Phillips, 1955), toxaphene and endrin (Frank, 1956), norbormide (Roszkowski, Poos and Mohrbacher, 1964), alphachloralose (Cornwell, 1966), and gophacide (DuBois, Kinoshita and Jackson, 1967).

The majority of these poisons are used by first mixing them with a bait and then siting the poison bait in places where the rodents are most likely to find and eat it. Most of the poisons used in this way are very quick-acting and may often affect an animal sufficiently within about $\frac{1}{2}$ to 1 hour to prevent it feeding further. Thus the chance of an animal surviving will depend not only on its physiological resistance to the particular poison, but also on its feeding behaviour during its first contact with it. If, for example, the animal is a slow or sporadic feeder, or if it is put off by the taste of the poison, it will tend to survive. Naturally the operator does his utmost to get the rodents to eat the poison bait as rapidly as possible—for example by laying unpoisoned bait for a few days first, using a bait that he knows the rodents will tend to eat quickly, selecting the right concentration of poison and making sure that the environment remains undisturbed during the critical period. Nevertheless, the variability within most rodent populations seems to be adequate to circumvent to some extent the first efforts of even the most experienced of operators and for most quick-acting poisons 100% success must be regarded as the exception rather than the rule. In addition there are 2 possible mechanisms which give survivors of a first treatment with a particular poison an even greater chance of surviving subsequent ones. Firstly, physiological resistance to a poison may be increased as a result of one or more sublethal doses. This phenomenon has been recorded, for example, in the Norway rat in response to antu (Richter, 1945) and norbormide (Roszkowski, 1965), and in the Pine mouse (*Pitymys pinetorum*) to endrin (Webb and Horsfall 1967). Some other examples of acquired physiological resistance are mentioned by Chitty (1954). Secondly, a rodent may learn to associate the effect of the poison with the poison itself or with the bait in which it is contained and these may then be subsequently avoided. Such acquired behavioural resistance is documented for the Norway rat in response to the effects of red squill, arsenious oxide and barium carbonate (Rzóska, 1954), antu (Gaines and Hayes, 1952) and norbormide (Greaves, 1966). The same phenomenon has been used to protect artificial sowings of pine seeds from the depredations of the 2 species of White-footed mice (*Peromyscus maniculatus* and *P. truei*), by first laying seed poisoned with an appropriate amount of sodium fluoroacetate (Tevis, 1956).

At present apparently the only definite case of increased resistance to any non-anticoagulant poison as a result of the selection of inherited characters has been the artificial production in the laboratory of a strain of House mouse more resistant to DDT than the stock from which it was derived (Ozburn and Morrison, 1964). There is, however, no special reason to believe that the various types of physiological and behavioural characteristics mentioned above which allow rodents to survive poison treatments are not generally genetically based and capable of being selected for by treatment of rodents with poisons in the field. On the other hand, apparently in only one case in the literature has such selection been offered as an explanation for treatment failure. An investigation of the lack of effectiveness of endrin after 11 years use against the Pine mouse in apple orchards showed that the mice were more resistant than an untreated control population, but the investigators did not rule out the possibility of acquired resistance, although no treatments had apparently been done just prior to the investigation (Webb and Horsfall, 1967). This example is particularly instructive because the poison was applied as a spray, thereby apparently eliminating many of the difficulties mentioned earlier that are generally associated with the use of non-anticoagulant poisons, and the kill that could generally be expected from such treatment was 100% (Horsfall, 1956).

The resistance situation amongst non-anticoagulant poisons thus seems to be more or less what might be expected. Whenever resistance to a normally very effective control method appears, the selection pressure to increased resistance will tend to be maintained at a relatively high level and the operator is likely to notice the appearance of resistance because he can readily appreciate a drop in effectiveness from say 100% to 80%. On the other hand where the expected success is variable and relatively low, the already lower selection pressure tends to be made lower still by the use of alternative control methods and the chances of an operator noticing and remarking upon a drop in effectiveness of the order of 20% becomes small.

ANTICOAGULANT RODENTICIDES

Although O'Connor (1948), after some preliminary work with dicoumarin, was apparently the first to publicize the potential of blood anticoagulants as rodenticides, this potential was not fully realized until the more effective anticoagulant, warfarin, became generally available in the early 1950's following its development as a rodenticide by the Wisconsin Alumni Research Foundation (Link, 1959). Initially warfarin was used mainly against *Rattus norvegicus, R. rattus* and *Mus musculus*, but its use and that of other subsequently-developed anticoagulants (e.g. diphacinone, chlorophacinone, fumarin and coumatetralyl) have now been extended to a variety of rodent pests in many parts of the world (Bentley, 1966). In many countries anticoagulants have relegated other rodenticides to a relatively minor role. For example, enquiries in 1961 from a random sample of local authorities in England and Wales indicated that about half the Public Health Departments of these authorities were using nothing but warfarin to control rats whilst a further 41% were using it for 80% of their treatments against these rodents.

This general preference for using anticoagulants is due partly to the fact that they are less likely to harm other animals, but more particularly to their greater effectiveness than other poisons. Their success seems to be mainly due to their cumulative and slow acting effect which makes it possible to use them at concentrations in a bait low enough to be acceptable by most, if not all rodents, and ensures that almost invariably a lethal dose has been ingested before an animal becomes too ill to feed. Thus the behavioural features associated with failures of some of the other poisons already mentioned have been more or less eliminated and, provided a satisfactory bait can be found, a 100% kill can be expected as the normal outcome of a treatment. This statement still remains largely true for many rodent populations today. However, treatments are not always successful and over the last decade it has become clear that some failures of warfarin against populations of *Rattus norvegicus* and *Mus musculus* can be attributed to the appearance within them of heritable physiological resistance to the poison.

Warfarin resistance in Rattus norvegicus

The inability to control populations of *R. norvegicus* with warfarin was first reported in Scotland in 1958 (Boyle, 1960). Since then, a similar phenomenon has been recorded in a number of other localities. All these localities, apart from one mentioned by Jackson (1969) for

Hungary which may refer to *R. norvegicus* but for which in any case no confirmatory evidence is apparently available, are sited in western Europe and are shown in Fig. 1 together with the years in which they were first recognized and code numbers to facilitate reference to them in the text.

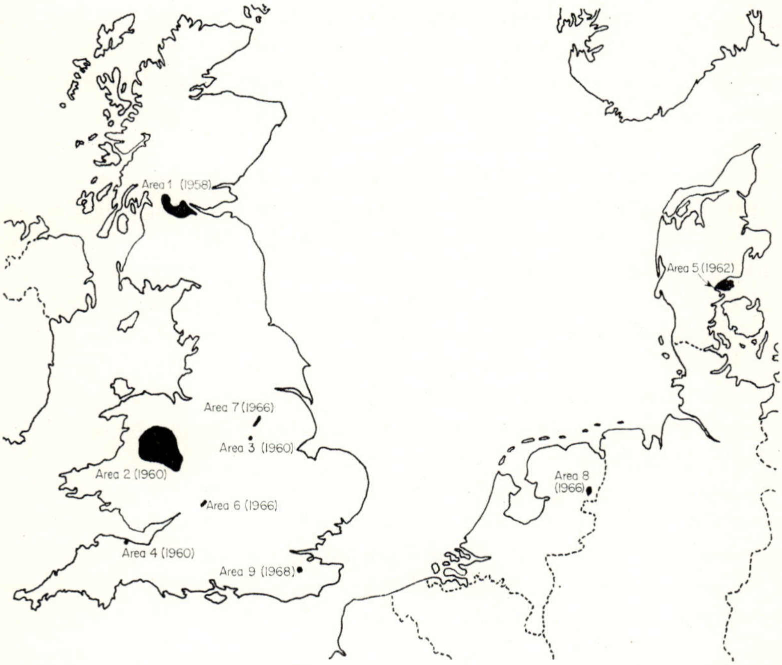

FIG. 1. World distribution of areas in which warfarin resistant populations of *Rattus norvegicus* have been reported.

The largest area involved (Area 2, Fig. 1) currently covers some 1200 square miles (3100 km^2) of largely agricultural land in the Welsh county of Montgomeryshire and the neighbouring English county of Shropshire. In this area, as in others in Britain, preliminary investigations were concerned with demonstrating as far as possible in the field that reported failures of warfarin against rats could not reasonably be attributed to causes other than resistance to the poison (Drummond, 1966). Subsequently rats were trapped and fed in the laboratory for 6 days on oatmeal containing 0·005% warfarin, a regimen under which it had already been shown that few, if any, normally susceptible rats could be expected to survive. This 6-day test was found to separate the survivors of field treatments into 2 more or less distinct types: those that were affected by the poison in a few days and succumbed apparently as readily as normally susceptible animals and those that continued to feed at a reasonable level throughout the test period and survived. In addition, the latter were for the most part able to survive much longer feeding periods on the same concentration of poison and some even survived feeding on concentrations as high as 1%, which led to them taking doses of poison over 100 times higher than that survived by animals from a non-resistant population (Drummond and Wilson, 1968). Clearly these surviving animals were physiologically resistant to warfarin.

The next main step in the investigation was to cross resistant wild rats with a domesticated albino strain, test the offspring with the 6-day test for resistance already referred to, and backcross the survivors to the albino strain. This procedure was followed for 5 generations and the ratio of resistant to susceptible rats in each generation indicated that the resistance was due to a single dominant autosomal gene, which has been denoted by the symbol Rw^2 (Greaves and Ayres, 1967, 1969*a*). This result has since been confirmed by the work of Pool *et al.* (1968) using a different technique to discriminate between susceptible and resistant animals.

As soon as the genetical nature of the warfarin resistance was appreciated it became evident (Drummond and Bentley, 1967) that the explanation for resistance in rats might be similar to, if not identical with, that given for warfarin resistance in man (O'Reilly *et al.*, 1964). In addition such an explanation readily fitted in with the suggestion of Olson (1964) that one of the genetical models of Jacob and Monod (1961) could explain the role of vitamin K in inducing the production of certain blood-clotting factors. In this a regulator gene produces a substance which represses the activity of an operon responsible for synthesizing blood-clotting proteins. Vitamin K combines with the repressor, prevents it from acting, and thereby stimulates the production of blood-clotting proteins. In the non-resistant animal, warfarin competes with vitamin K and allows the repressor to act. In the resistant animal however, the regulator gene (i.e. the gene for resistance) produces a different repressing substance for which warfarin fails to compete successfully with vitamin K. It now appears that this explanation is oversimplified in that vitamin K may only be concerned in the production of a protein which then competes with warfarin (Hill *et al.*, 1968) or perhaps with a metabolite of warfarin (Pool *et al.*, 1968). On the other hand, recent work on resistant rats, has eliminated a number of possible mechanisms for resistance suggested by the latter authors —such as accelerated excretion of warfarin (Pool *et al.*, 1968 and Hermodson, Suttie and Link, 1969)—and left intact the idea of a changed "repressor", or at any rate a site of eventual interaction between vitamin K and warfarin. Further evidence in favour of a change in this site is the fact that rats heterozygous and homozygous for resistance that were made vitamin K deficient required respectively 2 and 20 times as much vitamin K to restore their K-dependent blood clotting factors to comparable levels as did similarly-treated animals homozygous for susceptibility (Hermodson *et al.*, 1969).

So far in this paper warfarin resistance in rats has been described as though it were a single phenomenon with a single explanation. As will be shown presently in dealing with other sites this is clearly not the case. It is therefore important to recognize that all the work on warfarin resistant rats already referred to, including that done in the U.S.A. by Poole and Hermodson and their respective colleagues has been done with rats that came, or whose ancestors came, from Area 2.

Area 1 is in Scotland and covers a fairly extensive region mainly between Edinburgh and Glasgow. Resistant rats from this area are similar in many ways to those from Area 2. Some of them are able to withstand long periods of feeding on food containing warfarin (Cuthbert, 1963). The resistance is due to a single dominant autosomal gene which does not affect the rate of excretion of warfarin (Price-Evans and Sheppard, 1966). Rw^2 has recently been mapped (Greaves and Ayres, 1969*a*) and the gene conferring resistance upon rats from Area 1 appears to lie at the same locus. On the other hand, Area 1 rats heterozygous for resistance succumb much more readily to large doses of warfarin given by stomach tube than do similar Area 2 rats. It seems possible therefore that the genes for resistance in Areas 1 and 2 are not identical, but may be alleles (J. H. Greaves, personal communication).

There is relatively little that can be said about Areas 3 and 4. Both were reported before any very satisfactory methods had been developed for evaluating resistance. Nevertheless field and laboratory evidence did indicate the possibility of some degree of warfarin resistance. At each site only a single rat population was involved in a small area and each was apparently subsequently successfully controlled with acute poisons and no further evidence of resistance in the same or neighbouring places has come to light during the intervening 10 years. Perhaps in retrospect the main surviving interest in these areas is that Area 3 was a city block and Area 4 was a municipal refuse tip, whereas in all other areas suspected-resistant rat populations have initially been exclusively reported in and around farm buildings.

Areas 6 and 7 involved respectively 2 farms in Gloucestershire and 4 farms in Nottinghamshire. The 6-day test failed to make a clear distinction between susceptible and resistant rats from these 2 areas. The survivors of the tests, 59% and 23% from Areas 6 and 7 respectively, were fewer in number than would be expected if they had come from Areas 1 and 2, and had also, unlike rats from these latter areas, reduced their feeding to a low level by the 6th day, thereby suggesting that they would not have survived very much longer periods of feeding on warfarin. Thus it appears that not only may the resistance involved in Areas 6 and 7 be somewhat different from that in Areas 1 and 2 but the resistance in Area 6 may be different from that in Area 7. In contrast preliminary investigations of rats from Area 9 indicate a similarity between this site and Areas 1 and 2 (J. H. Greaves, personal commication).

Surveys of other farms around those affected in Areas 6 and 7 revealed no further evidence of resistance and, since the control of rats on the affected farms, no further cases of resistance in or near these sites have appeared. The situation however in the more recent Area 9 is still under investigation.

The first area reported outside Britain (Area 5, Fig. 1) was noticed in Jutland, Denmark, in 1962 (Bang and Lund, 1964). The survival of rats from this site after feeding on bait containing 0·05% warfarin for as long as 51 days indicated that the level of their resistance was comparable to that of rats from Areas 1 and 2. It was also soon confirmed that the resistance was genetically based (Bang and Lund, 1965) and later that, unlike Areas 1 and 2 more than one gene was involved (Lund, 1968). On the other hand, in their ability to withstand the vitamin K- depleting effect of mineral oil in their food the Danish rats seem to resemble rats of Area 1 rather than those of Area 2 (Drummond and Wilson, 1968). The information so far available from Area 8 in Holland (Ophof and Langeveld, 1968) indicates resistance, but for the present remains insufficient to make comparisons with other areas worthwhile. Recently, Pyörälä and Nevanlinna (1968) have shown the rate of excretion of warfarin by the Norway rat to be under polygenic control and to be susceptible to increase as a result of selection. It may be that some of the resistance which has already been shown to be unlike that found in Areas 1 and 2 is concerned with changed rates of excretion or metabolism of warfarin.

There is obviously still a great deal to be done to elucidate the nature of warfarin resistance in *Rattus norvegicus* in the various places in which it occurs. Nevertheless, what has been learnt so far accords very well with the situation in the field. Area 2, for example, started at a point about 5 km south-west of Welshpool and over the years 1962–65 spread radially at an average rate of about 4·6 km per year (Bentley and Rennison, 1966). The spread in this area is now readily explained by supposing that the gene for resistance was present in one original population where it was selected as a result of rat control treatments with warfarin. Resistant rats subsequently moved elsewhere and set up new populations or introduced the gene for resistance into existing populations where again it was selected for, and so on. Interestingly enough, the rate of spread of the resistance gene in this way seems to be comparable to the rate at which the Norway rat colonizes completely new territory. For example, advances of about 32 kms in 6 years in south-west Georgia and 64 kms in 10 years in Idaho can be inferred from the reports of Ecke (1954) and Harmston and Wright (1960) respectively.

Although the periods for which warfarin resistance has been present at the various sites is probably known with adequate accuracy, the lack of knowledge about the selection pressure in terms of the frequency and extensiveness of the use of warfarin in each site and about the success of other control methods to eliminate resistant rats, makes comparisons of the rates of spread of resistance rather difficult. Nevertheless it seems worth pointing out that in Area 1, where although a single gene is involved it confers a lower degree of resistance than that in Area 2, and in Area 5, where more than one gene is concerned, the rates of spread of resistance have apparently been much lower than in Area 2. In addition the apparent success in eliminating the resistance in Areas 6 and 7 may have been largely due to a much lower ability to spread the types of resistance present in these particular areas.

There are still other important questions concerned with warfarin resistance for which only partial or rather tentative answers are yet available. Why, for example, should such an apparently uniform type of selection pressure as warfarin poisoning, produce such a number of different responses in different populations? Presumably each population has a different combination and frequency of genes for warfarin to act upon, a situation which perhaps might be expected in an animal such as the rat where whole populations are continually being eliminated and restarted probably by only one or two individuals. Variation in frequency of 2 alleles responsible for the production of 2 forms of a polymorphic enzyme has recently been demonstrated in wild rat populations in Britain (Carter and Parr, 1969). Again, why should genes or combinations of genes which confer resistance on an animal be so rare? Presumably in the absence of warfarin poisoning, they are deleterious and selected against. Some support for this assumption comes from Lund (1968) who observed that an artificially enclosed population of rats from Area 5 failed to increase in numbers as readily as expected for non-resistant animals and that the proportion of resistant individuals dropped over several generations. In addition the finding already referred to that resistant rats from Area 2 are more susceptible to vitamin K deficiency than non-resistant animals indicates a possible mechanism for selecting against it. If resistant rats from other sites are also found to be particularly susceptible to vitamin K deficiency, this would go some way to explaining the apparent relationships between warfarin resistance and the farm environment, since rats in such places probably have more than average supplies of vitamin K available—particularly in the form of vitamin-fortified animal feeds.

Finally, why should warfarin resistance in the rat apparently be confined to the north-western part of western Europe? There are a number of possible and interconnected reasons for this situation which seems to depend primarily on climate. The generally mild climate in this area allows relatively large numbers of rats to live outdoors throughout the year and to reinfest areas such as farm buildings, villages and towns where most rat control work is done. Apart from providing a large gene pool on which selection can act, this situation has almost certainly in the past played a major role in prompting the Governments of the countries concerned to enforce rodent control and provide inspection, advisory, research and in some cases control services, thereby increasing the use of particularly effective control methods such as warfarin poisoning and at the same time creating an organization able to discover and investigate cases of suspected resistance. On the question of selection pressure, Jackson (1969) points out that in the U.S.A., pest control operators, at any rate in urban areas, continue to use acute poisons as well as several anticoagulants. On the effect of using more non-anticoagulant poisons there can be little doubt. However, whether the use of a variety of anticoagulants will markedly reduce the rate of selecting resistance to any one of them is less certain. A number of anticoagulants other than warfarin have been tested against warfarin resistant rats from a number of sites and in every case at least some degree of cross resistance has been detected (e.g. Drummond and Bentley, 1967; Greaves and Ayres, 1969*b*; Bang and Lund, 1966). Thus it seems likely that the present distribution of warfarin resistance in *R. norvegicus* is largely a matter of selection pressure and the ability to detect resistance and that the resistance of this rodent to warfarin and other anticoagulants is almost certain to appear elsewhere in the future.

Warfarin resistance in Mus musculus

Apparently the first cases of failure to control the House mouse with warfarin that could be attributed to physiological resistance to the poison, were in some food stores in Harrogate, England, in 1960 (Dodsworth, 1961). Laboratory investigation of mice from these premises, as well as from further ones in several other English towns where warfarin resistance was suspected, showed most of the mouse populations to have a greater degree of resistance to warfarin than was exhibited by 13 populations untreated by warfarin (Rowe and Redfern, 1964, 1965). On the other hand, 9·4% of mice from the untreated populations were able to survive 21 days feeding on food containing 0·025% warfarin, suggesting the widespread nature of the ability of some mice to withstand warfarin treatments in the field. This suggestion was subsequently supported by reports of resistant mice mainly from other urban areas in Britain. Since however the untreated mice already referred to came from rural areas, the apparent association between resistance in mice and urban habitats is almost certainly due to a greater use of warfarin in towns rather than any direct effect of the environment. Although few details are available, similar difficulties in controlling House mice with warfarin seem to have occurred in Denmark (Bang and Lund, 1964) and Germany (Roll, 1966).

Breeding experiments with wild resistant mice have indicated the resistance to be inherited, with more than one gene involved (Rowe and Redfern, 1965), and Roll (1966) has demonstrated that the level of resistance to warfarin can be increased by selection in 2 laboratory strains of the House mouse. The physiological nature of the resistance is not yet understood, but Rowe and Redfern (1968*a*) have pointed out 2 very interesting clues in their comparison of the effect of warfarin on the blood clotting factors of resistant and non-resistant wild mice. Firstly, the level of the vitamin K-dependent blood clotting factors II, VII and X were initially reduced but then tended to return to a more normal level while the animals continued to feed on oatmeal containing 0·025% warfarin over a period of 84 days, thus indicating that inherited resistance in mice may be at least partly due to an increased ability to acquire resistance. Secondly the blood of some wild mice fed food containing mineral oil and 0·025% warfarin, was found to have become unclottable and to have remained so over a period of 21 days without any sign of haemorrhage, suggesting that resistance in mice may also be partly due to an increased ability of their blood vessels to remain intact—in contrast to non-resistant animals in which break down of blood vessels and internal haemorrhages generally accompanies uncoagulability of the blood.

As with rats, experiments with warfarin resistant mice have shown them to be resistant to many other anticoagulants (Rowe and Redfern, 1968*b*). Also the explanation for the apparent confinement of resistant mice to western Europe almost certainly follows the same lines already given for the situation in rats and while anticoagulant poisons continue to be used against them in the field, resistant House mouse populations can be expected to be encountered elsewhere.

References

Anon. (1837). *The new family receipt book*. London: John Murray.

Bahr, L. (1947). The ratin bacillus and the "ratin system" through 40 years. *Maanedsskr. Dyrloeg.* **59**, 161–192.

Bang, P. and Lund, M. (1964). Testing of poisons for control of rodents. *Årsberetn. St. Skadedyrlab.* **1963**, 68–76.

Bang, P. and Lund, M. (1965). Testing of poisons for control of rodents. *Årsberetn. St. Skadedyrlab.* **1964**, 65–70.

Bang, P. and Lund, M. (1966). Testing of poisons for control of rodents and moles. *Årsberetn. St. Skadedyrlob.* **1965**, 65–69.

Bentley, E. W. (1966). Review of currently used anticoagulants. WHO/Vector Control/66.217, 89–96. Unpublished working document.

Bentley, E. W. and Rennison, B. D. (1966). The spread in the United Kingdom of resistance by *Rattus norvegicus* to anticoagulants. WHO/Vector Control/66. 217. Unpublished working document.

Boyle, C. M. (1960). Case of apparent resistance of *Rattus norvegicus* Berkenhout to anticoagulant poisons. *Nature, Lond.* **188**, 517.

Brodniewicz, A. (1968). Microbial control of pest rodents in Poland during 1926–1950. *Sudan med. J.* **6**, 90–95.

Carter, N. D. and Parr, C. W. (1969). Phosphogluconate dehydrogenase polymorphism in British wild rats. *Nature, Lond.* (in press.)

Chapman, C. and Phillips, M. A. (1955). Fluoroacetamide as a rodenticide. *J. Sci. Fd Agric.* **6**, 231–232.

Chitty, D. (1954). The study of the brown rat and its control by poison. In *Control of rats and mice* **1**, 160–305. Chitty, D. & Southern, H. N. (eds). Oxford: Clarendon Press.

Cornwell, P. B. (1966). Warfarin resistance with particular reference to mice and their control with "alphakil". *Trans. R. sanit. Ass. Scotl.* **1966**, 1–7.

Cuthbert, J. H. (1963). Further evidence of resistance to warfarin in the rat. *Nature, Lond.* **198**, 807–808.

Dodsworth, E. (1961). Mice are spreading despite such poisons as warfarin. *Munic. Engng, Lond.* No. 3746, 1668.

Drummond, D. C. (1966). The detection of rodent resistance to anticoagulants. WHO/Vector Control/66.217. Unpublished working document.

Drummond, D. C. and Bentley, E. W. (1967). The resistance of rodents to warfarin in England and Wales. In *Report of the international conference on rodents and rodenticides*. 57–67. European and Mediterranean Plant Protection Organisation.

Drummond, D. C. and Wilson, E. J. (1968). Laboratory investigations of resistance to warfarin of *Rattus norvegicus* Berk. in Montgomeryshire and Shropshire. *Ann. appl. Biol.* **61**, 303–312.

DuBois, K. P., Cochran, K. W. and Thomson, J. F. (1948). Rodenticidal action of 2-chloro-4-dimethylamino-6-methyl-pyrimidine (castrix). *Proc. Soc. exp. Biol. Med.* **67**, 169–171.

DuBois, K. P., Kinoshita, F. and Jackson, P. (1967). Acute toxicity and mechanism of action of a cholinergenic rodenticide. *Archs int. Pharmacodyn. Ther.* **169**, 108–116.

Ecke, D. E. (1954). An invasion of Norway rats in southwest Georgia. *J. Mammal.* **35**, 521–525.

Frank, F. (1956). Die neue Entwicklung der chemischen Bekämpfung von Mäuseplagen. *NachrBl. dt. PflSchutzdienst. Berl.* **8**, 105–109.

Freeman, R. B. (1954). Properties of the poisons used in rodent control. *In Control of rats and mice.* **1**, 25–146. Chitty, D. & Southern, H. N. (eds). Oxford: Clarendon Press.

Gaines, T. B. and Hayes, W. J. (1952). Bait shyness to antu in wild Norway rats. *Publ. Hlth Rep. Wash.* **67**, 306–311.

Grandi, G. (1912). Alcune notizie sui topi campagnoli che infestanole terre dell' Italia meridionale e sul modo di combatterli. *Boll. uff. Minist. Agric. Ind. Comm. Roma* **11**, 20–23.

Greaves, J. H. (1966). Some laboratory observations on the toxicity and acceptability of norbormide to wild *Rattus norvegicus* and on feeding behaviour associated with sublethal dosing. *J. Hyg. Camb.* **64**, 275–285.

Greaves, J. H. and Ayres, P. (1967). Heritable resistance to warfarin in rats. *Nature, Lond.* **215**, 877–878.

Greaves, J. H. and Ayres, P. (1969*a*). Linkages between genes for coat colour and resistance to warfarin in *Rattus norvegicus*. *Nature, Lond.* **224**, 284–285.

Greaves, J. H. and Ayres, P. (1969*b*). Some rodenticidal properties of coumatetralyl. *J. Hyg., Camb.* **67**, 311–315.

Harmston, F. C. and Wright, C. T. (1960). Distribution and control of rats in 5 Rocky Mountain states. *Publ. Hlth Rep. Wash.* **75**, 1077–1084.

Hermodson, M. A., Suttie, J. W. and Link, K. P. (1969). Warfarin resistance in the rat. *Fedn Proc. Fedn Am. Socs. exp. Biol.* **28**, 386.

Hill, A. B., Hatswell, J. M. and Topley, W. W. C. (1940). The inheritance of resistance, demonstrated by the development of a strain of mice resistant to experimental inoculation with a bacterial endotoxin. *J. Hyg., Camb.* **40**, 538–547.

Hill, R. B., Gaetani, S., Paolucci, A. M., Ramakao, P. B., Alden, R., Ranhotra, G. S., Shah, D. V., Shah, V. K. and Johnson, B. C. (1968). Vitamin K and biosynthesis of protein and prothrombin. *J. biol. Chem.* **243**, 3930–3939.

Horsfall, F. (1956). Rodenticidal effect on pine mice of endrin used as a ground spray. *Science, N.Y.* **123**, 61.

Jackson, W. B. (1969). Anticoagulant resistance in Europe. *Pest Control* **37**, (3) 51–55, (4) 40–43.

Jacob, F. and Monod, J. (1961). Genetic regulatory mechanisms in the synthesis of proteins. *J. molec. Biol.* **3**, 318–356.

Kalmbach, E. R. (1945). 1080, a war-produced rodenticide. *Science, N.Y.* **102**, 232–233.

Link, K. P. (1959). The discovery of dicumarol and its sequels. *Circulation* **19**, 97–107.

Loeffler, F. (1892). Die Feldmausplage in Thessalien und ihre erfolgreiche Bekämpfung mittels des *Bacillus typhi-murium*. *Zentbl. Bakt. Parasitkde* **12**, 1–17.

Lund, M. (1968). Testing of poisons for control of rodents. *Årsberetn. St. Skadedyrlab.* **1967**, 69–74.

Marshall, I. D. and Fenner, F. (1958). Studies in the epidemiology of infectious myxomatosis of rabbits. V. Changes in the innate resistance of Australian wild rabbits exposed to myxomatosis. *J. Hyg., Camb.* **56**, 288–302.

Marshall, I. D. and Fenner, F. (1960). Studies in the epidemiology of infectious myxomatosis of rabbits. VII. The virulence of strains of myxoma virus recovered from Australian wild rabbits between 1951 and 1959. *J. Hyg., Camb.* **58**, 485–488.

Munch, J. C. and Silver, J. (1931). The pharmacology of thallium and its use in rodent control. *Tech. Bull. U.S. Dept. Agric.* No. 238.

O'Connor, J. A. (1948). The use of blood anti-coagulants for rodent control. *Research, Lond.* **1**, 334.

Olson, R. E. (1964). Vitamin K induced prothrombin formation: antagonism by actinomycin D. *Science, N.Y.* **145**, 926–928.

Ophof, A. J. and Langeveld, D. W. (1968). Warfarin resistance in the Netherlands. WHO/VBC/69. Unpublished working document.

O'Reilly, R. A., Aggeler, P. M., Hoag, M. S., Leong, L. S. and Kropotkin, B. A. (1964). Hereditary transmission of exceptional resistance to coumarin anticoagulant drugs. *New Engl. J. Med.* **271**, 809–815.

Ozburn, G. W. and Morrison, F. O. (1964). The selection of a DDT-tolerant strain of mice and some characteristics of that strain. *Can. J. Zool.* **42**, 519–526.

Pool, J. G., O'Reilly, R. A., Schneiderman, L. J. and Alexander, M. (1968). Warfarin resistance in the rat. *Am. J. Physiol.* **215**, 627–631.

Price-Evans, D. A. and Sheppard, P. M. (1966). Some preliminary data on the genetics of resistance to anticoagulants in the Norway rat. WHO/Vector Control/66. 217. Unpublished working document.

Pyörälä K. and Nevanlinna, H. R. (1968). The effect of selective and non-selective inbreeding on the rate of warfarin metabolism in the rat. *Annls Med. exp. Biol. fenn.* **46**, 35–44.

Richter, C. P. (1945). The development and use of alpha-naphthylthiourea (antu) as a rat poison. *J. Am. med. Ass.* **129**, 927–931.

Roll, R. (1966). Über die Wirkung eines Cumarinpräparates (warfarin) auf Hausemäuse (*Mus musculus* L.). *Z. angew. Zool.* **53**, 277–349.

Roszkowski, A. P. (1965). The pharmacological properties of norbormide, a selective rat toxicant. *J. Pharmac. exp. Ther.* **149**, 288–299.

Roszkowski, A. P., Poos, G. I. and Mohrbacher, R. J. (1964). Selective rat toxicant. *Science, N.Y.* **144**, 412–413.

Rowe, F. P. and Redfern, R. (1964). The toxicity of 0.025% warfarin to wild house-mice (*Mus musculus* L.). *J. Hyg., Camb.* **62**, 389–393.

Rowe, F. P. and Redfern, R. (1965). Toxicity tests on suspected warfarin resistant house mice (*Mus musculus* L.). *J. Hyg., Camb.* **63**, 417.

Rowe, F. P. and Redfern, R. (1968*a*). The effect of warfarin on plasma clotting time in wild house mice (*Mus musculus* L.). *J. Hyg., Camb.* **66**, 159–174.

Rowe, F. P. and Redfern, R. (1968*b*). Laboratory studies on the toxicity of anticoagulant rodenticides to wild House mice. *Ann. appl. Biol.* **61**, 322–326.

Rzóska, J. (1954). The behaviour of white rats towards poison baits. In *Control of rats and mice.* **2**, 374–394. Chitty, D. & Southern, H. N. (eds). Oxford: Clarendon Press.

Sergent, Edmond and Sergent, Etienne (1948). Le DDT peut servir à la destruction de rongeurs nuisibles (souris, rats d'egout, merions). *C.r. hebd. Sêanc. Acad. Agric. Fr.* **34**, 954–960.

Tevis, L. (1956). Behavior of a population of forest-mice when subjected to poison. *J. Mammal.* **37**, 358–370.

Vashkov, V. I. and Polezhaev, V. G. (1965). Investigations in the USSR concerning the biological control of rodents. WHO/EBL/30.65. Unpublished working document.

Webb, R. E. and Horsfall, F. (1967). Endrin resistance in the pine mouse. *Science, N.Y.* **156**, 1762.

Webster, L. T. (1937). Inheritance of resistance of mice to enteric bacterial and neurotropic virus infections. *J. exp. Med.* **65**, 261–286.

Wilson, G. S. and Miles, A. A. (1964). *Topley and Wilson's principles of bacteriology and immunity*, 5th edition. London: Edward Arnold.

CONCLUSION

Symp. zool. Soc. Lond. (1970) No. 26, 371–383.

THE CAUSES OF POLYMORPHISM

J. MAYNARD SMITH

School of Biological Sciences
University of Sussex, Brighton, England

SYNOPSIS

Electrophoretic studies of proteins have shown that a large proportion of loci in natural populations are polymorphic. There are 2 possible explanations for these polymorphisms:

(i) they are maintained by a balance of selective forces. The possible types of selective balance are reviewed.

(ii) the majority of polymorphisms are selectively neutral. If this "neutral mutation" theory is true, then, at equilibrium between selection and mutation, there should be a higher proportion of polymorphic loci in large populations than small ones. Unfortunately, this prediction cannot easily be used to test the theory because the approach to equilibrium is very slow. An attempt is therefore made to derive the gene frequency distribution expected on the neutral mutation theory for a population which has recently increased in size. The method is applied to haemoglobin variants in man, and it is concluded that *either* few or no neutral mutations occur at the haemoglobin loci, *or* that human numbers passed through a bottleneck during the past million years.

INTRODUCTION

In this paper I shall try to give a general picture of the causes of genetic variability in natural populations from the point of view of a population geneticist. I will introduce algebra only when it is very simple and when it seems likely to be helpful to non-mathematicians. Mostly what I have to say is not new. However, the ideas on genetic variability in populations, like that of man, which have increased rapidly in numbers in the recent past, have not previously been published and indeed are not yet fully worked out.

From the beginning of population genetics there have been two schools of thought. On one hand, followers of R. A. Fisher have attempted to account for all changes and all variation in selective terms; on the other, followers of Sewall Wright have argued that chance effects arising because many actual breeding populations are relatively small are also important. Both views have been represented at this Symposium. Those who like myself are students of J. B. S. Haldane pride ourselves on our ability to see both sides of the question.

This classic argument has broken out with renewed vigour as a result of the discovery of the vast amount of variation which can be shown to exist in proteins, mainly by electrophoresis. This variability, which is greater than population geneticists had expected, can be interpreted in 2 ways. On the one hand, it can be taken to show that selective mechanisms capable of maintaining polymorphism are more widespread than anyone had supposed. On the other, it has been argued that most of the protein variation which is found is selectively neutral, and is able to persist precisely because there are no selective forces acting on it. Along with the view that most protein variation is selectively neutral goes the view that most amino acid substitutions which have occurred during evolution were likewise selectively neutral, and were established by genetic drift.

THE SELECTIVE INTERPRETATION OF POLYMORPHISM

Before pursuing the theory of selective neutrality further, it is worth considering briefly the selective mechanisms which have been suggested as capable of maintaining a genetic polymorphism.

Transient polymorphism

It may be that polymorphism observed in a natural population is not stable, and that the population has been caught in a moment of transition from one common allele at a locus to another. Although most species may be in a state of transient polymorphism at some loci, it is most unlikely that a majority of the protein polymorphisms recently described fall into this category.

Heterosis

If at any locus the heterozygote *Aa* is fitter than either homozygote *AA* or *aa*, it can easily be shown that a stable equilibrium exists, with both alleles common (Fisher, 1930). A classic case of heterosis at a single locus is that for sickle-cell haemoglobin (Allison, 1956), since the heterozygote dies neither of anaemia nor of malaria. Heterosis for chromosomal inversions in *Drosophila pseudo-obscura* has been fully investigated by Dobzhansky (1951). Whenever a polymorphism is long-lasting in a relatively small population in a stable and uniform environment, it is natural to suspect heterosis.

Spatially varied environment plus migration

If in one part of the range of a species the homozygote *AA* is fitter and in another part *aa* is fitter, and if a degree of migration occurs, then there is likely to be a cline of gene frequencies, and local populations will be found to be polymorphic.

Frequency-dependent selection

If for any reason each of 2 homozygotes *AA* and *aa* at a locus is the fitter when it is rare, then there will be a stable polymorphism at some intermediate frequency. A number of ecological situations may give rise to such equilibria:

Predation

Predators may have a frequency-dependent effect in either of 2 ways. Ford (1953) pointed out that in Batesian mimicry there may be a balance between the frequencies of the mimic and cryptic forms of the mimicking species, because the mimic must not be too common relative to its models if predators are to leave it alone. Clarke (1962) has emphasized that this formation of "search images" by predators could give an advantage to the rarer morph of a species, even if it were more conspicuous, because individual predators would be more likely to form a search image of the commoner morph.

Disease

Haldane (1949) discussed the possibility that some biochemical variants of a species might be favoured because parasites will tend to be better adapted to attack the common variants; by the same argument, rare variants of the parasite may be favoured because the host species is less likely to evolve resistance to them.

Selection in a varied environment

Maynard Smith (1962) showed that if 2 ecological niches are available to a species, and if 2 genotypes exist such that *AA* is fitter in one niche and *aa* in the other, then there can be a stable polymorphism even if one allele is fully dominant in both niches. The equilibrium requires that the population density be separately regulated in the 2 environments, and that the selection pressure be high. This is a case of frequency-dependent selection for the following reason. When a particular genotype is rare, it finds few efficient competitors in its favoured niche, and so has a higher chance of surviving.

Cyclical selection

Haldane and Jayakar (1963) investigated the possibility of a stable equilibrium between 2 alleles *A* and *a* if different homozygotes are favoured in different generations. They showed that such equilibria are possible, but are likely to exist only if selective pressures are large. Those interested in changes in gene frequency during cycles of population abundance should refer to this paper.

Other mechanisms maintaining stable polymorphisms are known; one is gametic selection favouring a gene which would otherwise be eliminated, as in the *t*-alleles in the House mouse (Bruck, 1957; Lewontin and Dunn, 1960; Anderson, 1964). A major difficulty in analysing particular cases of polymorphism arises from the phenomena of linkage: it can be difficult to decide whether the stability arises from selection acting on the locus whose phenotypic effects are being observed or on some other locus closely linked to it.

Most participants in this symposium would probably attempt to interpret any case of polymorphism in terms of one of the above mechanisms, rather than as a result of neutral mutation or non-Darwinian evolution—or whatever phrase it is now fashionable to call what Wright would have called genetic drift. This may merely reflect the fact that this symposium is being held in the country of Fisher and not that of Wright. Whatever the reason, it may help to redress the balance if I next present what seems to me the strongest argument in favour of the neutral mutation theory; this is that the neutral mutation theory predicts that the rate of evolution for any class of proteins should be constant, and that there is some evidence for such uniformity.

THE RATE OF NON-DARWINIAN EVOLUTION

First, what is the evidence in favour of a uniform rate of evolution of proteins? The evidence is strongest for haemoglobin (Kimura, 1970). Figure 1 shows a phylogenetic tree of haemoglobin molecules, with an approximate time scale. The lines in this diagram represent descent by DNA replication, the polypeptide chains whose sequences have been determined being direct translations of these DNA molecules. Most bifurcations in the diagram represent (as is customary in phylogenetic trees) a splitting of a single ancestral species into two; one of them represents the duplication, in an ancestor of the jawed vertebrates, of the gene which specified a single globin molecule into two initially identical genes whose descendants now specify the α and β chains of modern haemoglobins.

The amino acid sequences of existing polypeptide chains have been determined. By appropriate comparisons, it is possible to calculate the number of amino acid substitutions necessary to convert one into another. By dividing this number by twice the time to a common ancestor, one obtains an estimate of evolutionary rate in terms of amino acid substitutions per polypeptide chain per year. A further division by the number of amino acids per chain gives the rate in substitutions/

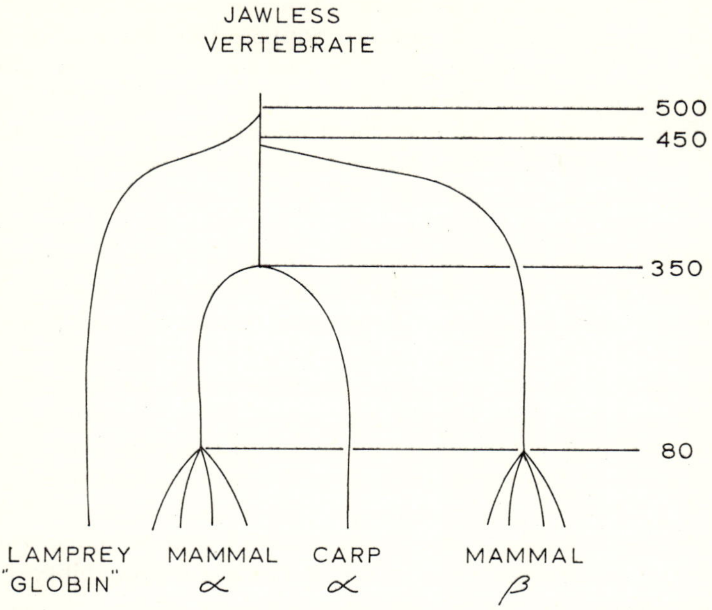

FIG. 1. The phylogeny of haemoglobins. The figures on the right are approximate times in millions of years.

site/year. A number of such estimates, based on different comparisons, are given in Table I. Some of these estimates are completely independent of one another, and others mainly so. The similarity of the estimates is striking.

TABLE I

Average rates of amino acid substitution in the evolution of haemoglobins

Comparison			Substitutions/site/year $\times 10^{10}$
Human β	vs.	Lamprey Globin	12·8
Human β	vs.	Human α	8·9
Human β	vs.	Other Mammal βs	11·9
Mouse β	vs.	Other Mammal βs	14·0
Human α	vs.	Carp α	8·9
Human α	vs.	Other Mammal αs	8·8

The argument from the neutral mutation theory predicting this uniformity is so simple that it is worth giving (Kimura, 1968*a*). The basis of the argument is the assumption that the majority of these substitutions are selectively neutral. Suppose that the rate of neutral mutation per generation per gene (i.e. cistron) is u, and is constant for a given class of proteins. The idea is that most mutations will be harmful and will be eliminated. An occasional mutation is selectively neutral, and the fraction of mutations which are neutral is constant for a given type of protein; it will differ for different proteins, depending on the variety and stringency of the selective requirements the protein must satisfy. Very rarely, a favourable mutation may occur and be established; these cases, though important in functional evolution, will be too rare to influence estimates of the rate of amino acid substitution.

Suppose that in any generation the population size is N. There are then $2N$ genes in this population, and $2Nu$ newly arising neutral mutations. If now we travel far enough into the future, we shall find that all genes in the population are ultimately derived from a single one of the $2N$ genes. Hence each gene has a probability $1/2N$ of ultimately being established; since a newly arising mutant makes the same contribution to fitness as any other, it too has a probability $1/2N$ of being established. Hence the number of new genes which arise and which ultimately are established per generation is $2Nu \times 1/2N = u$, a constant.

One complication is that the observational evidence suggests uniformity per year, not per generation as suggested by the preceding argument. This difficulty will disappear if it turns out, as may well be the case, that the mutation rate per generation in different mammals is proportional to generation time.

Evidence for uniformity of evolutionary rate for proteins other than haemoglobin has been reviewed by King and Jukes (1969). For some proteins the rule breaks down. For example, insulin has changed very little in most evolutionary lines, but is very variable among hystricomorph rodents. This latter fact could be explained on the neutral mutation theory only if it turns out that the selective restraints on insulin are less severe in hystricomorphs; I know of no reason why they should be.

Before leaving this topic, it will be useful later on to estimate for haemoglobin the value of u_e, the rate of neutral mutations per generation which are also electrophoretically recognizable. If changes at the nucleotide level were random, then about one third of all amino acid substitutions would involve a change in charge. An examination of the changes which have actually occurred in the evolution of haemoglobin shows that approximately the expected proportion of one third were charge changes. It is therefore justifiable, when testing the neutral mutation theory, to assume that electrophoretically recognizable changes are typical of all changes, at least for haemoglobin.

From Table I, the rate of substitution per site per year is approximately 10^{-9}. The number of sites per chain is 140, and a human generation lasts approximately 25 years. Hence the approximate neutral mutation rate is:

$$u_e = \tfrac{1}{3} \times 10^{-9} \times 25 \times 140 \simeq 10^{-6} \text{ per cistron per generation.}$$

EVIDENCE FROM FIELD STUDIES

It is natural to attempt to use field data on protein variability to settle the argument between the proponents of selection and of drift. Some of the ways in which this might be done are as follows:

Measurements of selection

A popular way of demonstrating heterosis as a mechanism maintaining polymorphism is to show an excess of heterozygotes compared to expectation on the Hardy-Weinberg ratio. This was done (Dobzhansky and Levene, 1948) for *Drosophila pseudo-obscura* males carrying inversions. The major snag to this method is that the ratio rests on the assumption of random mating. It is probably rare for there to be mating preferences based solely on morph differences (but see Sheppard (1952) for an example). Unfortunately, if a sample contains members of different small inbreeding groups, it will depart from the Hardy-Weinberg ratio, although in the opposite direction to that expected in the case of heterosis. Examples of this difficulty have been given at this symposium (Anderson, 1970; Selander, 1970).

An alternative method of measuring selection in natural populations is to compare gene or morph frequencies in young and old individuals. This was one of the first methods used in the measurement of natural selection (Weldon, 1901). It has been applied to mammalian populations by Van Valen (1965), and at this symposium by Lush (1970).

Association between environmental and gene frequency changes

If the frequency of a particular allele is found to vary consistently with the environment, this strongly supports a selectionist interpretation. An example, quoted by Lush at this symposium, concerns a locus in highland and lowland breeds of sheep. In general, Selander considers that the data he has presented to this symposium fit this interpretation, whereas Rasmussen (1970, this symposium) considers that the data on *Peromyscus* show gene frequency changes not associated with corresponding environmental change.

Environmental variation may be temporal rather than spatial. In *Drosophila*, which can manage several generations a year, Dobzhansky (1951) has found annual cycles in gene frequency. A mammal must evolve a genotype which will work the whole year round, but may nevertheless show cyclical changes in gene frequency in time with fluctuations in population numbers. Evidence has been presented at this symposium by Chitty (1970) to show that such cyclical changes in gene frequency occur; if it can be established that the changes are genuinely associated with the population cycle, then the differences cannot be selectively neutral. Of course, it does not follow that the changes in gene frequency are a necessary driving component of the population cycle itself. Before this aspect of Chitty's theory can seriously be investigated, it will be necessary to have a precise model of the theory, either simulated or analytical, which can be shown to lead to oscillations.

Variability and population size

It is a necessary consequence of the neutral mutation theory that small populations should be less variable genetically than large ones. This affords a powerful method of testing the theory. Unfortunately, there are difficulties, which are discussed in more detail in the next two sections of this paper.

POPULATION SIZE AND POLYMORPHISM

It is much easier to work out the consequences of the neutral mutation theory than of its selectionist alternatives. Consequently the strategy I shall adopt is to assume the truth of the neutral mutation theory and to work out the consequences of this assumption. When these consequences do not agree with observation, then some explanation in terms of selection is called for.

In a finite population, alleles will continuously be eliminated by chance, and new ones will arise by mutation. It was shown by Kimura and Crow (1964) that when equilibrium has been reached between mutation and elimination,

$$I = \frac{1}{1+4N_e u} \quad (1)$$

where I = probability that, at a particular locus, an individual chosen at random will be a homozygote,

u = rate of neutral mutation per generation at that locus, and
N_e = effective population size.

We usually cannot estimate I if all possible alleles at a locus are treated as different, but often have a fair idea if electrophoretically recognizable variants only are taken into account. Suppose then that $u \simeq 10^{-6}$, as estimated above for electrophoretic variants of haemoglobin.

Then if $N_e = 10^6$, $I = 0{\cdot}20$.
and if $N_e = 10^4$, $I = 0{\cdot}96$.

There is therefore a large difference between the degree of polymorphism to be expected in populations of large and small effective size. This difference can be used to test the theory, provided that populations can be found in which there is sufficient gene flow for the above formula to hold, and that the populations have been in existence for long enough for the equilibrium to be reached.

The time taken to reach the equilibrium may be the most serious qualification. It can be shown that it approximates, in generations, to the effective population size N_e. If the human population had kept at its present size since the Cambrian, we should still not have approached the equlibrium given in (1). A possible way out of this difficulty is discussed in the next section. The objection does not always prevent the use of (1). For example, suppose that a population of mice of effective number 10^4 or less has been reproductively isolated for 10 000 years from a much more abundant and effectively interbreeding population, one would, according to the neutral mutation theory, expect the isolate to show a markedly lower degree of polymorphism.

Before considering in more detail the problem of fluctuating population size, I must say more about what is meant by "effectively interbreeding population" in the previous paragraph. What is required is that there should be sufficient gene flow to ensure that equation (1) can be used with N_e equal to total population number, and not to the number in individual demes. The degree of gene flow to ensure this is much less than is required to ensure a good fit to the Hardy-Weinberg ratio.

I have considered this problem (Maynard Smith, 1970) for the "Island" model, in which a species is divided into a number of demes between which some migration takes place. If migration does occur, an individual is equally likely to migrate to any other deme whether close or distant. For mammals, a "stepping stone" model in which migration occurs only between neighbouring demes would be more plausible, and should if possible be analysed.

For the island model, if u is the neutral mutation rate, r is the number of local populations and m is the probability that an individual born in one deme will breed in another, then there are two possible states of affairs. If $m > ur$, there is sufficient gene flow to ensure that (1) can be used with N_e corresponding to the number of individuals in the species as a whole. If $m < ur$, there will be insufficient gene flow to ensure genetic similarity between demes and almost all hybrids between populations would be heterozygous. Hence if Anderson's suggestion (this symposium) that in the House mouse there is insufficient gene flow to ensure genetic similarity between demes is correct, then if at any locus neutral mutations are possible, almost all inter-deme hybrids would be heterozygotes for that locus. In a sense, this conclusion is obvious, since if species integrity were not dependent on gene flow, it would have to depend on selection.

Summarizing this section, the following conclusions can be drawn:

(i) Equation (1) can be used only if an effectively interbreeding population of size N_e has been isolated for at least N_e generations.

(ii) If, as is usually the case, hybrids between demes include an appreciable proportion of homozygotes at any locus, then one of three interpretations are possible. First, no neutral mutations occur at that locus. Secondly, there is sufficient migration to make it appropriate to use equation (1) with N_e corresponding to the population as a whole and not to individual demes; this interpretation is of course possible only if N_e is small enough for equation (1) to give a reasonable value of I. Thirdly, the population has arisen in the recent past from a smaller and effectively interbreeding population.

(iii) The conclusions under (ii) rest in part on an analysis of the "island" model of population structure. It would be desirable to analyse the "stepping stone" model in a similar way.

POLYMORPHISM IN A POPULATION WHOSE SIZE IS CHANGING

The most serious difficulty with equation (1) is the great length of time required for the equilibrium to be reached—i.e. a number of generations of the same order as N_e. The human population has increased by a factor of perhaps 10^4 in the past 500 or so generations. What type of variability could one expect from the neutral mutation theory in such a population? In view of the mathematical complexity of the equilibrium case (e.g. Kimura, 1968*b*), it may seem foolhardy to tackle the transient one. However, an approximate but adequate picture can be obtained in the following way.

If the present population consists of N_o individuals, there are $2N_o$

genes. Of these genes, what fraction are copies, without further mutational change, of genes arising exactly n generations ago?

If N_n was the population size n generations ago, the number of new mutant genes arising then was $2N_n u$, where as before u is the neutral mutation rate per generation.

For each gene (including the new mutants) present n generations ago the expected number of copies now is N_o/N_n. (This ignores mutations occurring in the last $(n-1)$ generations. This is permissible provided that $nu \ll 1$). Hence the number of copies now of genes arising n generations ago is $2N_n u. N_o/N_n = 2N_o u$.

Hence the fraction of genes now which arose in any past generation is u. If $F_{n,m}$ is the expected fraction of genes now which originated between m and n generations ago, we have

$$F_{n,m} = (n-m)u. \tag{2}$$

This equation holds only if $nu \ll 1$; in the case of man, the equation can safely be applied over periods of up to one million years.

This equation will now be applied to electrophoretically distinguishable variants of haemoglobin in man. For such variants, it was estimated on p. 376 that $u = 10^{-6}$ per generation. Hence approximately 4 genes in every 10 000 originated in the last 400 generations, or 10 000 years; i.e. since the invention of agriculture and the subsequent increase in human numbers. Each individual allele would be very rare. Such rare variants of haemoglobin are known to exist (Sick *et al.*, 1967) in about the right frequency. Their presence does not support the neutral mutation theory, since much the same result would be expected if there were a similar mutation rate to slightly deleterious alleles, since the probability of survival of a mutant for the first few hundred generations does not depend very critically on whether it is neutral or slightly harmful.

Consider now genes originating between 10^4 and 10^6 years ago, a period of about 50 000 generations. According to equation (2) approximately 5% of existing genes should have originated during that period. A further 5% should have originated between one and two million years ago, and so on. Now it seems clear that no such electrophoretically recognizable variants of haemoglobin exist. There are some common electrophoretic variants but they appear to be maintained in the population by heterosis, because of the resistance of the heterozygotes to malaria.

At first sight, the absence of such variants seems to be the death blow of the neutral mutation theory, at least as far as haemoglobin is concerned. Unfortunately, there is another possible explanation for their absence. $F_{n,m}$ is the expected fraction of genes arising between m and n generations ago. Hence there are two possibilities:

either (i) 5% of existing haemoglobin genes arose between 10^4 and 10^6 years ago,

or (ii) there is a 5% chance that *all* existing haemoglobin genes arose during that period.

The latter possibility arises if there has been, during the past million years, a bottleneck in human numbers sufficiently small to give a reasonable chance that the haemoglobin locus became homozygous.

It is possible to give some idea of how small the effective population size would have to be to account for the absence of variability. Using the approach given by Kimura (1968*b*), and taking $u = 10^{-6}$, it can be calculated that N_e must fall below 10^5 before there is any approach to homogeneity. For $N_e = 10^4$ or less the population would ultimately have a high probability of being genetically homogeneous at a given locus. Using the "diffusion" method developed by Kimura (1964), it can be shown that after a number of generations equal to approximately $2N_e$ an initially heterogeneous population would have an evens chance of becoming homogeneous.

Thus if the neutral mutation theory is true, the haemoglobin data requires us to suppose that during the past million years there has been a "bottleneck" in human numbers. This bottleneck could have been a single pair for one generation, or an effective population of 10 000 for half a million years, or something in between. The alternative is to conclude that there are no electrophoretically recognizable mutants of haemoglobin which are selectively neutral.

As on a previous occasion, we can choose between Darwin and the Garden of Eden.

References

Allison, A. C. (1956). The sickle-cell and haemoglobin C genes in some African populations. *Ann. hum. Genet.* **21**, 67–89.

Anderson, P. K. (1964). Lethal alleles in *Mus musculus:* local distribution and evidence for isolation of demes. *Science, N.Y.*, **145**, 177–178.

Anderson, P. K. (1970). Ecological structure and gene flow in small mammals. *Symp. zool. Soc. Lond.* No. 26, 299–325.

Bruck, D. (1957). Male segregation ratio advantage as a factor in maintaining lethal alleles in wild populations of house mice. *Proc. natn Acad. Sci. U.S.A.* **43**, 152–158.

Chitty, D. (1970). Variation and population density. *Symp. zool. Soc. Lond.* No. 26, 327–333.

Clarke, B. (1962). Balanced polymorphism and the diversity of sympatric species. In *Taxonomy and geography*. 47–70. Nichols, D. (ed.). London: Systematics Association.

Dobzhansky, T. (1951). *Genetics and the origin of species*. 3rd ed. New York: Columbia Univ. Press.

Dobzhansky, T. and Levene, H. (1948). Genetics of natural populations. XVII. Proof of operation of natural selection in wild populations of *Drosophila pseudoobscura*. *Genetics* **33**, 537–547.

Fisher, R. A. (1930). *The genetical theory of natural selection*. Oxford: Clarendon Press.

Ford, E. B. (1953). The genetics of polymorphism in the Lepidoptera. *Adv. Genet.* **5**, 43–87.

Haldane, J. B. S. (1949). Disease and evolution. *Ricerca scient.* Suppl. **19**, 68–76.

Haldane, J. B. S. and Jayakar, S. D. (1963). Polymorphism due to selection of varying direction. *J. Genet.* **58**, 237–242.

Kimura, M. (1964). *Diffusion models in population genetics*. London: Methuen.

Kimura, M. (1968*a*). Evolutionary rate at the molecular level. *Nature, Lond.* **217**, 624–626.

Kimura, M. (1968*b*). Genetic variability maintained in a finite population due to mutational production of neutral and nearly neutral isoalleles. *Genet. Res.* **11**, 247–269.

Kimura, M. (1970). The rate of molecular evolution considered from the standpoint of population genetics. *Proc. natn. Acad. Sci. U.S.A.* (in press).

Kimura, M. and Crow, J. F. (1964). The number of alleles that can be maintained in a finite population. *Genetics* **49**, 725–738.

King, J. L. and Jukes, T. H. (1969). Non-Darwinian evolution. *Science, N.Y.* **164**, 788–798.

Lewontin, R. C. and Dunn, L. C. (1960). The evolutionary dynamics of a polymorphism in the house mouse. *Genetics* **45**, 705–722.

Lush, I. E. (1970). The extent of biochemical variation in mammalian populations. *Symp. zool. Soc. Lond.* No. 26, 43–71.

Maynard Smith, J. (1962). Disruptive selection, polymorphism and sympatric speciation. *Nature, Lond.* **195**, 60–62.

Maynard Smith, J. (1970). Population size, polymorphism and the rate of non-Darwinian evolution. *Am. Nat.* (in press).

Rasmussen, D. I. (1970). Biochemical polymorphisms and genetic structure in populations of *Peromyscus*. *Symp. zool. Soc. Lond.* No. 26, 335–349.

Selander, R. K. (1970). Biochemical ploymorphism in populations of the House mouse and Old-field mouse. *Symp. zool. Soc. Lond.* No. 26, 73–91.

Sheppard, P. M. (1952). A note on non-random mating in the moth *Panaxia dominula* (L.). *Heredity* **6**, 239–241.

Sick, K., Beale, D., Irvine, D., Lehmann, H., Goodall, P. T. and MacDougall, S. (1967). Haemoglobin $G_{\text{Copenhagen}}$ and Haemoglobin $J_{\text{Cambridge}}$. Two new β-chain variants of Haemoglobin A. *Biochim. Biophys. Acta* **140**, 231–242.

Van Valen, L. (1965). Selection in natural populations. III. Measurement and estimation. *Evolution, Lancaster, Pa,* **19**, 514–528.

Weldon, W. F. R. (1901). A first study of natural selection in *Clausilia laminata* (Montagu). *Biometrika* **1**, 109–124.

AUTHOR INDEX

Numbers in italics refer to pages in the References at the end of each article.

SYSTEMATIC INDEX

SUBJECT INDEX